ARCHITECTURAL COMMON SENSE: SUN, SITE AND SELF

Rad Dike

VNR VAN NOSTRAND REINHOLD COMPANY
NEW YORK CINCINNATI TORONTO LONDON MELBOURNE

Library of Congress Catalog Card Number: 82-19982
ISBN: 0-442-21364-6
ISBN: 0-442-21805-2 pbk.

Manufactured in the United States of America

Published by Van Nostrand Reinhold Company Inc.
135 West 50th Street, New York, N.Y. 10020

Van Nostrand Reinhold Publishing
1410 Birchmount Road
Scarborough, Ontario MIP 2E7, Canada

Van Nostrand Reinhold
480 Latrobe Street
Melbourne, Victoria 3000, Australia

Van Nostrand Reinhold Company Limited
Molly Millars Lane
Wokingham, Berkshire, England

15 14 13 12 11 10 9 8 7 6 5 4 3 2 1

Library of Congress Cataloging in Publication Data

Dike, Rad.
Architectural common sense.

Includes index.
1. Architectural design. 2. Architecture – Composition, proportion, etc. 3. Architecture – Environmental aspects. I. Title.
NA2750.047 1983 729 82-19982
ISBN 0-442-21364-6
ISBN 0-442-21805-2 pbk.

DEDICATION

This volume is dedicated to my father who was always building and to my mother who was always explaining.

INTRODUCTION

These notes are collected from my notebooks, which I attempted from childhood, although I did not find them necessary until I began to teach architectural design at Pratt in 1970. They were drawn from what I hadn't been taught but wanted to learn.

Common sense is the art uniting our sensations in a general perception. Architectural common sense is the art and science of designing and building true accounts of the actual.

Einstein advised: "No popular book on science should be published before it is established that it can be understood and appreciated by an intelligent and judicious layman."

Various friends, students, and associates have used one or another of these notes, and have either requested copies or offered encouraging comments. I thank you all. Sue Ann Snyder first encouraged gathering the notes into notebooks. Stanley Salzman suggested making prints of them to hang on construction sites. Elizabeth Wright-Ingram said that Peter Van Dresser must see them. Peter said that Steve Baer and Keith Haggard should see them. Steve admonished me to self-publish, and sent me off to meet Paoulo Soleri, who to this day still doesn't have a copy of the philosophy set. Keith introduced me to David Wright, who seven years ago introduced me to Jean Kofoed, then of VNR, who offered to publish a volume. And six years and many sketches later, Gene Falken offered to publish two volumes, of which this is the first.

I have never met Lewis Mumford, Kevin Lynch, and Stewart Brand, but their books are my old friends. Thank you, Dan Sankey, Colin James, Stanley Birge, Niels Diffrient, and Sid Katz for "investing" in my work.

PREFACE

Plato advised: "First, the taking in of scattered particulars under one idea, so that everyone understands what is being talked about. Second, the separation of the idea into parts, by dividing it at the joints, as nature directs, not breaking any limb in half as a bad carver might." This book is about nonvisible architecture. It is divided into three parts: sun, site, and self.

The principles selected for this publication cannot be successfully applied impersonally. That is, if one uses them to help someone else, it is helpful to think about them as if for oneself. Your own measures must be compared with these principles, for there is no perfection in the principles alone. Your comparison must be shared with others, whose differences propel living order. Chaos is unshared principles; order is the sharing of principles.

In the moment that we approach the wind or wall or any part of architecture, we lose the ability to see the whole architecture as if for the first time. These sketches are more for staring at than for organizing architecture. In this way a perfection can be felt about an architecture you may want to use these for. As useful approximations they offer us a living order. Each new stare creates a new imbalance of order, more in either part or whole, inviting life to refind a balance of a newer perfection.

When an organism, tissue, or cell affects another, the effective result in the other affects itself. Multicellular organization provides the basics for studying the development of life from preexisting life.

To design and build architecture is to further develop life where life already exists. This collection about sun, site, and self observes some of the cells, tissues, and organs of design possibilities. These possibilities each speak to me of possible form-giving forces in organizing the potential of the site and the ideas of the inhabitants. You might say that architecture of this kind is determined by a sort of genetic code found in local climate, local situation, and local society.

Growth does not extend equally or simultaneously in all directions or from only one point. So it is with the usefulness of this book.

All but these few ideas have appeared in literature before:

The Shadowlabe, pp. 15-30.
The Calstice, Equical, etc., pp. 25-48.
A Golden Symbol of Woman, Man, and Child Unity, pp. 143-158.
A Golden Rectangle of a Tree pp. 160-162.
A Golden Visual Octave pp. 163-166.

These notes do not form a system. I took the notes while I was wondering if I would ever think originally. At best, they are an artist's fun of drawing the obvious.

CONTENTS

This typeset text at the top of each page has been written to be read continuously by turning page after page without having to explore the collection of notes on each page. In this way the thread of the "story" is introduced.

I. SUN

1. The three chapters, of this book are Sun, Site and Self.

2. To me, good design images the truth of a proposition.

3. A significant solution integrates climate, situation, and society.

4. Universal parameters do not result in architecture that is similar everywhere.

5. The forces and forms of each different self in each different situation will result in appropriately different architectures.

6. The study of birth is a lesson in architecture that we have all experienced, yet few of us understand.

7. Architectures we can see or measure are the parts of appearance. Parts which affect appearance but do not appear are the architecture of existence.

8. This collection of notes represents obvious connections of figure to ground.

9. Because of what does not exist (rooms, doors, windows, reservoirs), we fit into what does exist.

10. Thus architecture verifies the necessary complimentarity of being and nonbeing.

11. The wonder of the birth process, as in design, is that something larger than the aperture comes through to see the light. The sun is the lantern of architecture.

12. The earth goes around the sun.

13. The sun goes around the earth.

14. A shadow is cast by the sun into the earth whenever any visible thing is between them.

15. For all practical purposes, the sun's rays are considered parallel when they strike the earth.

16. Shadows can deter people like walls. Diagramming the possibilities of shadows is an infinitely tedious process. The journal of our yearly cycles can be drawn on our surroundings by tracing shadow edges at commemorable times.

17. Shadow path diagrams for the base of a homemade solar-sextant (shadowlabe) are given on the following pages for six latitudes, 30°, 32°, 40°, 48°, 50°, 56°. Any moment in the year can then be modeled or predicted.

18. Climate is the ratio of sun to slope. Both the earth's atmosphere and the surface of the earth cause the parallel light rays to be changed.

19. Architectural climates are created by various conversions of light, but first the designer must know where the sun is.

20. Any point in the celestial sphere (the sky) can be located by two numbers, the altitude and the azimuth.

21. These two numbers can be used to make a shadowlabe to predict shadows at any time and angle.

22. Shadow and sun area predicuons can be made with the shadowlabe on the site or on a model of the site.

23. These sun and shadow patterns offer form possibilities of interesting and useful configurations, especially in a complex landscape such as the city.

24. The greater the latitude, the longer the shadowcaster's shadows.

25. The lesser the latitude, the shorter the shadowcaster's shadows.

26. The calstices are the extreme heat points of the year, just as the solstices are the extreme angle points of the year. The equicals are the midpoints of the heat extremes, just as the equinoxes are the midpoints of the sun angle extremes.

27. Five lines on the shadowlabe show the shadowcaster's lengths through six of the most architecturally significant days of the year. (2 equinoxes = one line, 2 solstices, and 2 calstices)

28. The summer sun moves around the shadowcaster in an arc from the northeast to the northwest.

29. The winter sun's arc moves towards and away from the shadowcaster from the southeast to the southwest.

30. Sunlight and the angle at which it strikes the earth are measurable as heat potential to change temperature.

31. The area of sunlight and its angle define the maximum energy (in this case, heat as BTUs) available on a site from the sun at a certain time.

32. For every angle at which the sun strikes a slope, a maximum amount of BTU's has been calculated by ASHRAE, AIA, and others.

33. The diagrams at left show a variety of slopes (0°, 30°, 40°, 50°, 60°, 90°) per each hour of sunshine. Each square on the slope represents 10 BTU.

34. The plans of these slopes are shown square with true north, which is not the case, for instance, on New York's streets and avenues.

35. Imagine designs that accomodate the complexity of positions ideal for gathering sunshine.

36. Though these diagrams are shown for forty degrees latitude, they can be adapted to any latitude.

37. By filling in the uncompleted celestial hemispheres and veiws of the earth, the reader can familiarize himself with the sun's apparently complex relationship with earth.

38. The lowest sun angle is not exactly at the time of coldest temperatures (the winter calstice).

39. The calstices and equicals follow the solstices and equinoxes because of the earth's thermal lag.

40. The sun-arc angle changes quickly at the time of the equinoxes; the length of day changes noticeably from week to week.

41. Any one thing casts equinox shadow lengths whose sum of endpoints through the day forms a straight line east and west.

42. Sunlight is not uniform lighting.

43. Orientation proportions shadows, contrasts, images, borders, diffusions, and photosynthesis.

Most of the double page spreads are meant to be read as a painting, whole unto themselves. Turning the pages will illumine other pages further, but the book can be opened and read at any point.

44. The highest sun angle is not exactly at the time of hottest temperatures (the summer calstice).

45. The calstices change their attractive polarity like magnets.

46. The sun-arc angle changes slowly at the time of the solstices.

47. For a month before and after the solstice, the sun-arc does not seem to be much higher or lower; the days do not seem much longer or shorter.

48. The solstices and equinoxes are the extremes of the sun angles during the year.

49. The calstices, equicals, and calperis are the temperature extremes of the seasons.

50. Various climates are created at various sites because natures differ.

51. The darkened proportions show the monthly movement of a temperature as the sun changes on the slopes of America.

52. Sun and slope create large climates.

53. Sun and slope create small climates.

II. SITE

54. Sun and slope create wind. The air is moistened and headed by the sun's reaction with the slope and is made tangible as wind.

55. Wind wooes water, flower, and flame to bring about movement.

56. Wind not wound to hopes and expectations profits no one and is likely to be an annoyance if not a harm.

57. Sails propel ships around the earth, as vanes propel the shaft.

58. The primal year for water continues in the humidity ocean about us.

59. Sun powers wave motion, evaporates water to the clouds, and precipitates it to the ground.

60. All the water that flows is detained in some way by something.

61. The lowering of water can raise a wheel and the lowering of a wheel can raise water.

62. Your right and left hands hold enough soil to understand the slope you stand on.

63. Distinguish gravel from sand, visually; clay from silt, tensilely; plastic from organic, sensorially.

64. Dirt plus sun, water, wind, and life equals soil.

65. As soil is the obvious living material, so the material beneath the soil is the obvious building material.

66. A root is the seed's downward thrust from the sun.

67. A branch is the seed's upward thrust from the slope to the sun.

68. Speculations of electromagnetic and genetic control may someday architecture growth.

69. But fundamental facts of climate must be respected if we are the selectors of what grows where.

70. Both branches and leaves need protection and are protection.

71. Protected roots are contained, but must be protected from their containment.

72. Stem begets stem, a system of sunlight exploitation.

73. All trees follow the same branching laws to define different patterns.

74. Shrubs can be arranged to move color throughout a space as the seasons change.

75. Fruiting trees are ornamental with usefulness. All trees shown here are well known for their hardiness.

76. Small deciduous trees are reasonably pruned and offer transitional scale between shrubs and giants.

77. Large deciduous trees speckle light on the ground or densely shade, causing great visual commotion or color.

78. Small evergreen trees can form platforms, ledges, and walls.

79. Large evergreen trees can form columns, walls, windows, ceilings, and skylights.

80. A natural architectural order can be developed for tree ceilings and tree walls.

81. Weave and density can be arranged as living "wallpaper."

82. Unrelated species are related by similarity of canopy textures.

83. For example, Manna Ash, Black Locust, Olive, Rowan, China Berry, Ailanthus, and Chinese Cedar are a group of canopies of different species consistently similar in overall texture.

84. Weave and density tree classes will be detailed just as are masonry, wood, and steel.

85. Branches will become the mullions in our views and leaves the frames. To see is to move and movement is ordered by windows.

86. Finding one's way around a site is the beginning of a map.

87. A mental map can be verified with a compass and a length.

88. Circular timepieces and the circles of the sun verify the compass.

89. The lines and times of the shadowlabe verify the pendulum of day per season.

90. Surveying is the practical art of measuring changes which you consider significant in the horizon.

91. What you choose to survey becomes your map of the world; that is, what isn't considered can't be part of the intended design.

92. Surveying instruments are better used to allow for the unplannable, rather than to master plan.

93. Cutting and filling is not bulldozing, but accommodating incongruous ideas with the delights of the site.

94. Structure originates in sun/slope.

95. Each sun/slope suggests its own true comfortable architecture.

III. SELF

96. Truth-finding originated in practicality, and had faith in the unanswerable, which diverged religion and science.

97. I believe that what a being creates is shaped in part by his reason for living and creating, difficult as it may be to verbalize.

98. If there is a meaning to life, it seems most likely to be found in the ordinary basics of life.

99. This provocative list of statements may help to relate design intention to beliefs, or at least to stretch the imaginative use of hardware.

100. It does not satisfy me to add the parts of technology together to make architecture or metaphors for organisms.

101. The psyche and its processes may prove yet that the parts we don't see are the parts we use most.

102. However, truth is of the whole human, not only of the intellect.

103. Thermodynamics, the mechanical relations of heat, is certainly the basis of our comfortable existence.

104. Each living body is its own source of heat energy, emmeshed in the heat flow of earth and sun.

105. The sun is the source of greenhouse-effect heat energy, which architects the heat flow of earth and sun.

106. The ever changing sky, by clouds or seasons, is the primary stimulation of living organisms' receptor organs, photosynthesis, sight, and sensation.

107. The value of reflection can be seen on a cloudy day with unobstructed sun, for that sky is brighter than a clear sky. (How to get sunburned on a cloudy day.)

108. The sun's angle of incidence is the angle that the sun ray makes with the perpendicular to the slope.

109. When a slope is perpendicular to the sun ray, the area of the slope equals the area of the solar radiation.

110. The heat gain benefits of the greenhouse effect are understood by calculating the expected heat losses.

111. The atmosphere of the earth is its greenhouse glass and its clouds are its nighttime insulation.

112. Earthen materials buffer our heat capacity with their own.

113. Water materials have the greatest heat capacity and the most uniform distribution of that capacity.

114. Our comfort range is 65 to 70°F; fluctuations above and below this can be minimized or encouraged by proportions of glass to material mass.

115. Overheating and underheating are prevented by design based on the division of effective mass by sunshine area.

116. Generally, the greater the wall thickness, the less the air temperature fluctuation.

116. (alternate) The seven medians of winter calstice outdoor temperatures can be expressed in seven degree days to provide one indoor comfort range.

117. The difference in heights of the sums of % of wall area; 1 sq ft floor between a water wall and a masonry wall is the area of a man's chest, the human heat regulator.

118. Radiation exchange between us and our walls affects us more than the air temperature of the room.

119. The mean radiant temperature that we feel from ourselves to our space surfaces is the average impact of all their radiations.

120. The radiation we see is a different vision cone when we move or do not move.

121. The plan and section of our vision cones on the floor, ceiling, and walls of a space resemble the lines on a shadowlabe.

122. Space is not a single note; space is a span of time; space is the movement of sight.

123. Simultaneous verticals and horizontals counter each other in a vocabular akin to music.

124. Balance is an economy of motion; a musical octave is the harmonic combination of two tones an octave apart, apparently blending as one.

125. The visual octave is the harmonic combination of detail and visual field where the two blend as one.

126. When the distance to an object from the eye equals the height of the object, the whole object is not seen, but rather a detail one third of the cone of vision is seen clearly without being lost in or dominating the surrounding object.

127. When the sight distance is twice the height of the object, the object is comfortably seen. When distance is three times height, the object dominates the visual field. When distance is four times height, the object becomes nondominant/not-lost detail.

128. If the width of a small room becomes more than the greater side of a golden rectangle to its height, and certainly if its width is more then twice its height, it will "feel" too low when crowded.

129. A room for gathering is a gathering of alcoves.

130. When ascending and descending space, the step becomes the stair and the stair becomes the staircase, i.e., the human step dimensions the stair step, and the staircase is in the same proportions as its stair step.

131. Most of our life we are on stairs, or in spaces affected by stair spaces or near walls hiding stair spaces; one flight is two stories.

132. Our movement is structured by bones, and bones define the laws of structuring space.

133. Bones offer logic for detailing and joining.

134. Our spine challenges us to design a flexible column, a responsive post.

135. The interrelation of the skeleton and organs is an organizational message.

136. The posture of the spine is primary to the durability of the seating.

137. The length of the bones is primary to the convenience of the moving.

138. The flesh is primary to the surfaces and atmospheres, required for delight.

139. Architecture is built by numbers in the hopes of suiting our hopes and capacities, bound by our culture as it really is.

140. The sins of architecture are not caused by the natural laws of proportion, but by their unnatural institutionalization.

141. Though we are confined within an infinite variety of body form, we are apparently not limited to categories correlative with our possibilities.

142. Clothing styles not only suit our culture; clothing is the first layer of shelter.

143. When architecture requires bandaging, structure responsive to sun, site, and self is the only acceptable aesthetic retrofit.

144. Stand Da Vinci's man on his toes and upstretch his hands, then you will have begun to visualize the reasonable extremities of man.

145. The regular divisions of man are halves or doubles of each other.

146. The extremities of man can be shown as a cube and sphere about the navel in top, front, and side views.

147. Great horizons may be seen in the splatter of plaster, but there is no scale to them without human proportions.

148. Rules of human proportion are not necessary when we see and remember our basic behavior.

149. However, elegance is the dignity and grace which restrains the great wealth of our behavior.

150. Expression of the logic symbolized by woman is one of my basic intentions.

151. The irrational proportions of the pentagon's harmony (golden ratio, divine proportion, etc.) are expressed by a woman giving birth.

152. Le Corbusier's relaxed modular man was drawn to fit divine proportions rather than found in the proportions of actual man.

153. One day I found that a drawing of woman's extremities fit the divine proportions.

154. This discovery only proved that my ideal fit the divine proportions.

155. An anthropometric study of Americans was used to show that neither our male nor female extremities fit the divine proportion.

156. It then seemed that only pregnancy, which changes the position of the navel, might allow the divine proportion to be found in anthropometrics.

157. A lower than normal navel is bad posture and doesn't allow the divine proportion; good pregnant posture required new control of the natural postural reflex.

158. The healthy posture of the fully pregnant estate of woman does find the divine proportion.

159. This moment is the full term symbol. This full term moment is the symbol of the closest unity of man, woman, and child.

CONCLUSION

160. Significant relations of the minimums and maximums of different invisible architectural parts can be symbolized by finding their divine proportions.

161. For instance, a tree's height is in golden ratio with its roots' spread if the soil volume is arranged as the thickness of the tree's golden rectangle.

162. The average tree's height is four stories; in human extremity terms, four stories is four triangles high, each formed about the four golden rectangles about the human navel.

163. A medium sized tree's golden rectangle is based on the triangle base formed by the human's golden rectangles. A small sized tree's golden rectangle is based on the triangle base formed by the squares originating the human rectangles.

164. One eighth of the visual octave is determined by a golden rectangle the height of the viewer and the length of the intersection of the peripheral vision cone with the ground. One quarter of the visual octave is confirmed by the focal cone's intersection with the ground. One half is twice one fourth, etc.

165. One sixth of the visual octave is the bisection of the angle between the peripheral and the focal cones with the ground. One fifth and one seventh are the bisections of the previous bisection iwth the ground.

166. The Golden Visual Octave's fundamental tone is an octave itself, ranging from 1/8 of the closest focal distance to 1/8 of the distance where that detail blends with the whole focal field, whose diameter is four human triangles high, the height of a mature tree's golden rectangle and a fair approximation of the average townhouse.

167. Everything in the world fits. Design it all together at once, and living order goes; relate a little here and a little there, a little now and a little then, and living order comes.

168. We are shadowcasters. Our focal and peripheral cones intersect our shadowlabe and fix sixteen shadow lengths in divine proportion to our height, eight to our left and eight to our right.

169. These divine shadow fans are formed of different azimuth combinations for each latitude. We need only remember our latitude to know the sun on a site by our self.

170. The side view of the Golden Visual Octave teaches the relation of the focal and peripheral cones to the intervals of the octave and the ranges of branches and roots. The plan view teaches relations of our golden shadows to our radiant comfort to trees' canopy range.

171. The end view relates man and woman geometric harmonies to their Golden Shadow Sun Spots fixed in their Octave's focal area. Three-grace-like positions fix the Golden Visual Octave: The beloved's head fills the focal cone, stands at the first and second interval and blends with the whole visual field at the eighth interval.

172. Another collection of notes will be published which will demonstrate design results from the interactions of principles in this collection. To me, though these principles are old, they are symbols of something to come.

173. Symbols make visible what we've learned to be true and useful. Though the symbols in this book are not avant garde, I believe they should be the cutting edge of art & architecture.

1. The three chapters of this book are Sun, Site and Self.

2. To me, good design images the truth of a proposition.

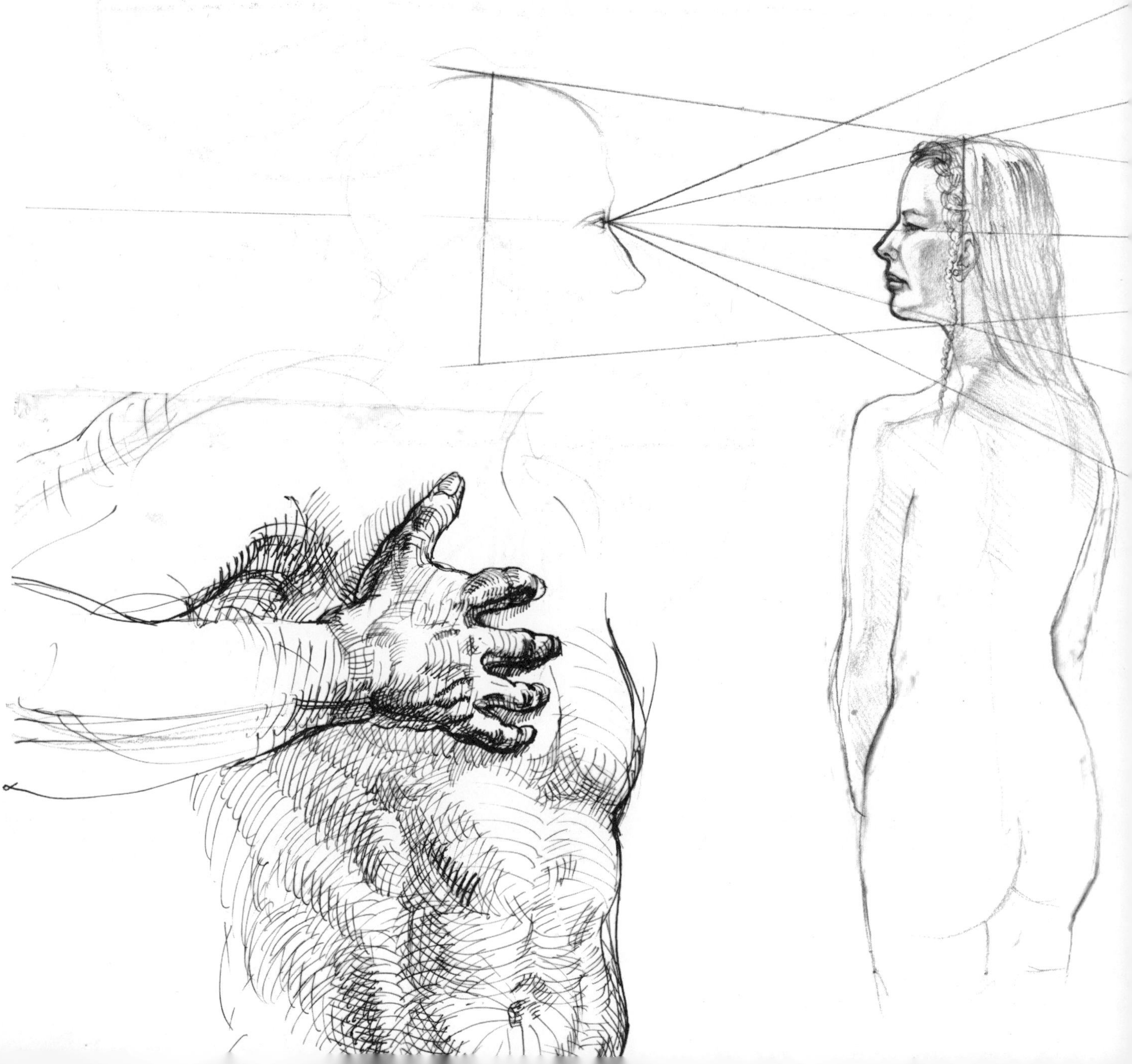

3. A significant solution integrates climate, situation, and society.

4. Universal parameters do not result in architecture that is similar everywhere.

SUN is the measure of CLIMATE

SITE is the measure of SITUATION

SELF is the measure of SOCIETY

architecture is the attempt to resolve their requirements

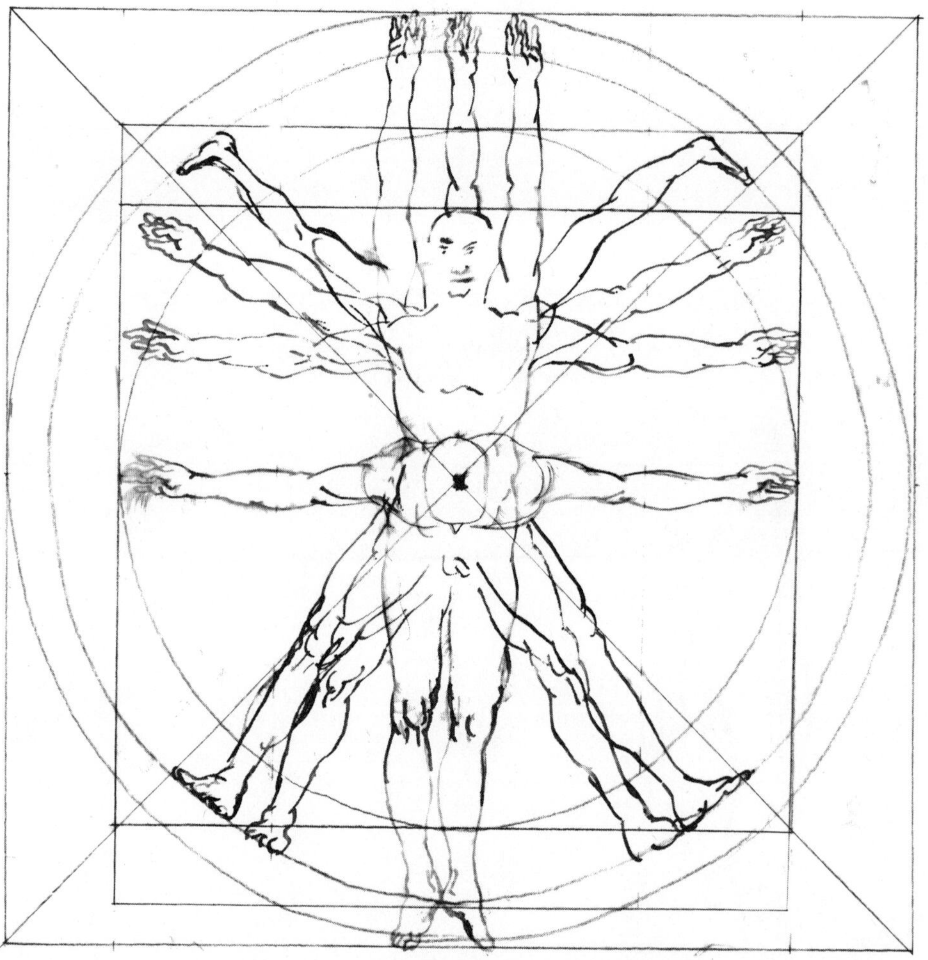

This drawing is a combined top, front & side view of man's basic proportions in relation to the circle and the square. By themselves, these proportions do not make architecture. With relation to the concerns on the opposite page, they suggest reasonable minimum and maximum limits within which humans can predict and make architecture work for livability.

Certainly there are more concerns than I have listed. I have selected those subjects which most concern me as universal and perpetual form-suggesters.

5. The forces and forms of each different self in each different situation will result in appropriately different architectures.

HOW SUN + EARTH MAKE SEASONS 3-4
SIMPLE WAY TO PREDICT SUN POSITIONS 5-6
SIMULATING SUN WITH MODEL + LAMP 1-2
IMAGINING SUN WHILE STANDING/SITE 7-8
DRAWING SUN SHINE & SHADOWS 9-10
SOLAR POSITION & INSOLATION DATA 11-12
SHADOWCASTERS PLAN & SECTION 13-14
SOLAR POSITION & INSOLATION ILLUSTRATED -28

SOLAR COLLECTION PRINCIPLES
SPACE AS SOLAR COLLECTOR
SPECIFIC HEAT & HEAT CAPACITY
HEAT STORAGE FOR SPACE
WALL CONDUCTIVITY/DEGREE DAYS
MASS/SURFACE/SPACE PROPORTIONS

HOW SUN + SLOPE MAKE CLIMATE 9-10
SIMPLE WAY TO EXPECT WEATHER 11-12
MACRO & MICRO CLIMATE FACTORS
SPATIAL PROPORTIONS / CLIMATE TYPES

THERMODYNAMICS AND
HOW OUR BODIES FEEL HOT OR COLD
HOW TO TAKE TEMPERATURE OF SPACE
CALCULATING HEAT GAIN
CALCULATING HEAT LOSS

HOW SUN + SLOPE MAKE WIND
HOW TO CALCULATE WIND'S ENERGY
AIR MOTIONS PER FORMS
AIR CHANGES & INFILTRATION
AIR CIRCULATION & STRATIFICATION
WIND CONVERSION PROCESSES
WIND CONTROL PROCESSES

HUMANS & ELECTROMAGNETIC SPECTRUM
HUMANS & STRESS PROPORTIONS
HUMAN HEALTH & LIFE SUPPORT FACTORS
HUMAN BREATHING & SMELLING
HUMAN LIFE CYCLE
HUMAN BIRTHING

HOW TREES GROW 49-50
BRANCHING STUCTURE 51-52
ROOTING STRUCTURE 53-54
TREE CEILINGS & WALLS 55-56
TREE TEXTURE CATEGORIES 57-68
TREE FORM CATEGORIES

VEGETATION TYPES
VEGETATION USAGES

ANIMALS & TERRITORY

ECOLOGICAL THEORY

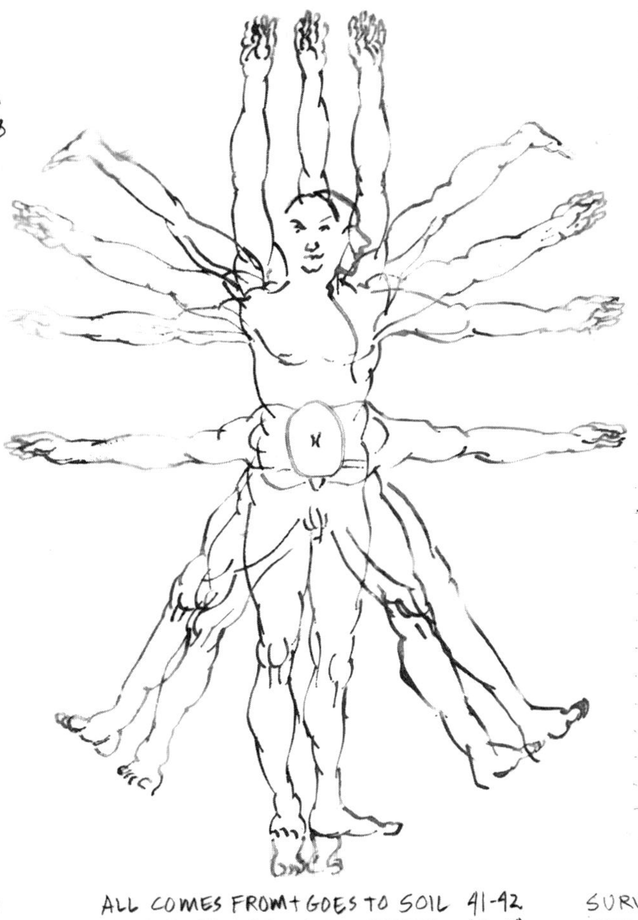

WOMAN'S PROPORTIONS
MAN'S PROPORTIONS
CHILD GROWTH PROPORTIONS
WHEEL CHAIR PERSON
M & F AVERAGE DIMENSIONS
M & F AVERAGE MOVEMENTS
M & F AVERAGE VISION SPACES
M & F GENERAL SENSES
M & F AVERAGE SEATING
M & F AVERAGE CLIMBING
M & F LIGHT REQUIREMENTS
FAMILY PROPORTIONS

WELCOMING & FAREWELLING
PRIVACY WITHIN GROUP SPACE
APPROACH & DEPARTURE
PASSING BY
MOVING THROUGH
TRAFFIC
REST
PLAY
STORING
WORK
EDUCATION
WAITING
EATING
GARDENING
COOKING
ENTERTAINING, BOTH SIDES
BUYING & SELLING
FROM PUBLIC TO PRIVATE
BATHING
REVERING

AESTHETIC POSTULATES
AXIS & ORIENTATION
VISUAL PROPORTION SYSTEMS
POINT LINE PLANE VOLUME
CENTER DENSITY ACTION
CIRCULATION
HEIGHT WIDTH & DEPTH
COMPLEXITY

HOW WATER IS FOUND
WATER COLLECTION PRINCIPLES
CALCULATING WATER'S ENERGY
WATER ENERGY CONVERSIONS
HUMIDITY & PRECIPITATION

ALL COMES FROM + GOES TO SOIL 41-42
EYE + HAND DEFINE SOIL TYPES 43-44
SOIL TYPES + APPROPRIATE USE 45-46
LAND TYPES & GEOLOGY
CRITICAL SECTIONS & IMPLICATIONS
TOPOGRAPHY
REGION
LOCALE
NEIGHBORHOOD
SITE

ANALYSIS TECHNIQUES
SYNTHESIS TECHNIQUES

SURVEYING INSTRUMENTS
ORIENTING ONESELF
MAPS & MAPMAKING
EXCAVATION & FILLING IN

Volume two presents material, structural & mechanical concerns as universal & perpetual form-suggesters.

6. The study of birth is a lesson in architecture that we have all experienced, yet few of us understand.

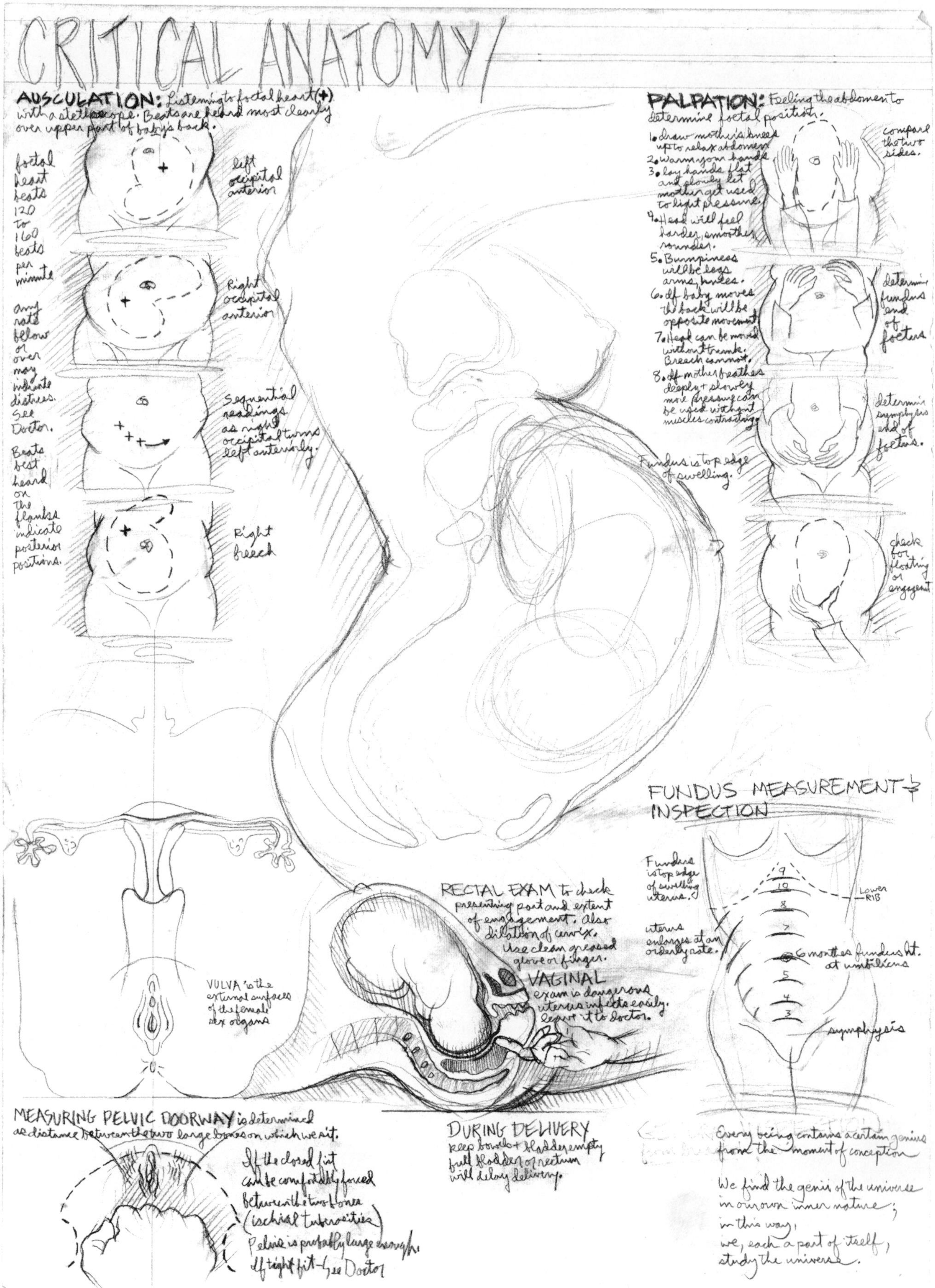
CRITICAL ANATOMY
AUSCULATION: Listening to foetal heart (+) with a stethoscope. Beats are heard most clearly over upper part of baby's back.
foetal heart beats 120 to 160 beats per minute
any rate below or over may indicate distress. See Doctor.
Beats best heard on the flanks indicate posterior positions.
left occipital anterior
Right occipital anterior
Sequential readings as right occipital turns left anteriorly.
Right breech
PALPATION: Feeling the abdomen to determine foetal position.
1. draw mother's knees up to relax abdomen
2. Warm your hands
3. lay hands flat and slowly let mother get used to light pressure.
4. Head will feel harder, smoother, rounder.
5. Bumpiness will be legs, arms, knees.
6. If baby moves the back will be opposite movement.
7. Head can be moved without trunk. Breech cannot.
8. If mother breathes deeply + slowly more pressure can be used without muscles contracting.
compare the two sides.
determine fundus end of foetus
determine symphysis end of foetus.
check for floating or engagement
Fundus is top edge of swelling.
FUNDUS MEASUREMENT & INSPECTION
Fundus is top edge of swelling uterus.
uterus enlarges at an orderly rate.
9
10
8
7
6 months fundus ht. at umbilicus
5
4
3
Lower RIB
symphysis
RECTAL EXAM to check presenting part and extent of engagement. Also dilation of cervix. Use clean greased glove or finger.
VAGINAL exam is dangerous uterus infects easily. Leave it to Doctor.
VULVA is the external surfaces of the female sex organs
MEASURING PELVIC DOORWAY is determined as distance between the two large bones on which we sit.
If the closed fist can be comfortably forced between the two bones (ischial tuberosities) Pelvis is probably large enough. If tight fit—See Doctor
DURING DELIVERY keep bowels + bladder empty full bladder or rectum will delay delivery.
Every being contains a certain genius from the moment of conception
We find the genius of the universe in our own inner nature; in this way, we, each a part of itself, study the universe.

7. Architectures we can see or measure are the parts of appearance. Parts which affect appearance but do not appear are the architecture of existence.

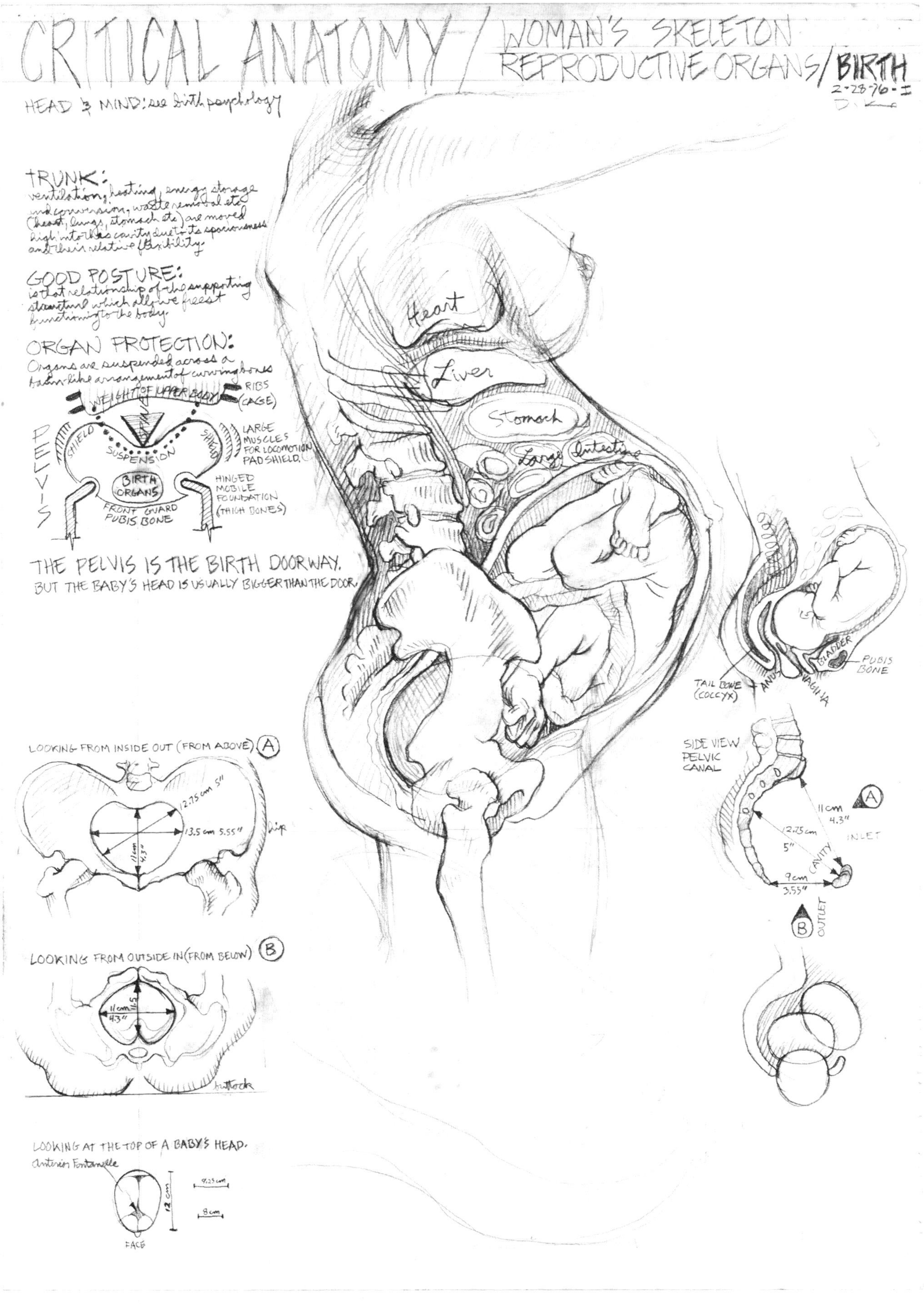

8. This collection of notes represents obvious connections of figure to ground.

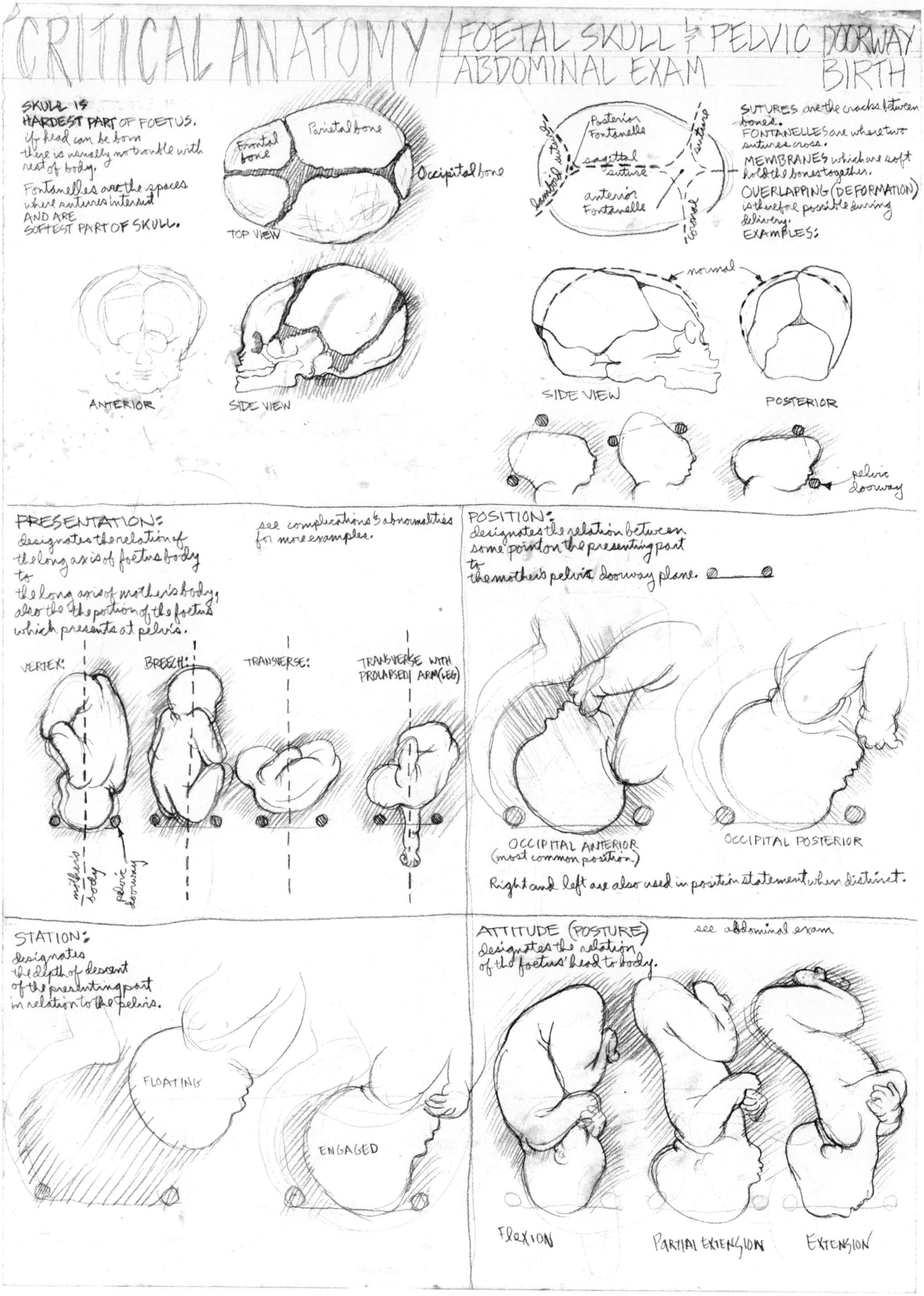
CRITICAL ANATOMY / FOETAL SKULL & PELVIC DOORWAY
ABDOMINAL EXAM
BIRTH
SKULL IS HARDEST PART OF FOETUS.
if head can be born there is usually no trouble with rest of body.
Fontanelles are the spaces where sutures intersect AND ARE SOFTEST PART OF SKULL.
Frontal bone
Parietal bone
Occipital bone
TOP VIEW
Posterior Fontanelle
sagittal suture
anterior Fontanelle
lambdoid suture
suture
coronal
SUTURES are the cracks between bones.
FONTANELLES are where two sutures cross.
MEMBRANES which are soft hold the bones together.
OVERLAPPING (DEFORMATION) is therefore possible during delivery.
EXAMPLES:
normal
ANTERIOR
SIDE VIEW
SIDE VIEW
POSTERIOR
pelvic doorway
PRESENTATION:
designates the relation of the long axis of foetus body to the long axis of mother's body, also the portion of the foetus which presents at pelvis.
see complications & abnormalities for more examples.
VERTEX:
BREECH:
TRANSVERSE:
TRANSVERSE WITH PROLAPSED ARM (LEG)
mother's body
pelvic doorway
POSITION:
designates the relation between some point on the presenting part to the mother's pelvis doorway plane.
OCCIPITAL ANTERIOR (most common position)
OCCIPITAL POSTERIOR
Right and left are also used in position statement when distinct.
STATION:
designates the depth of descent of the presenting part in relation to the pelvis.
FLOATING
ENGAGED
ATTITUDE (POSTURE)
designates the relation of the foetus' head to body.
see abdominal exam
FLEXION
PARTIAL EXTENSION
EXTENSION

9. Because of what does not exist (rooms, doors, windows, reservoirs), we fit into what does exist.

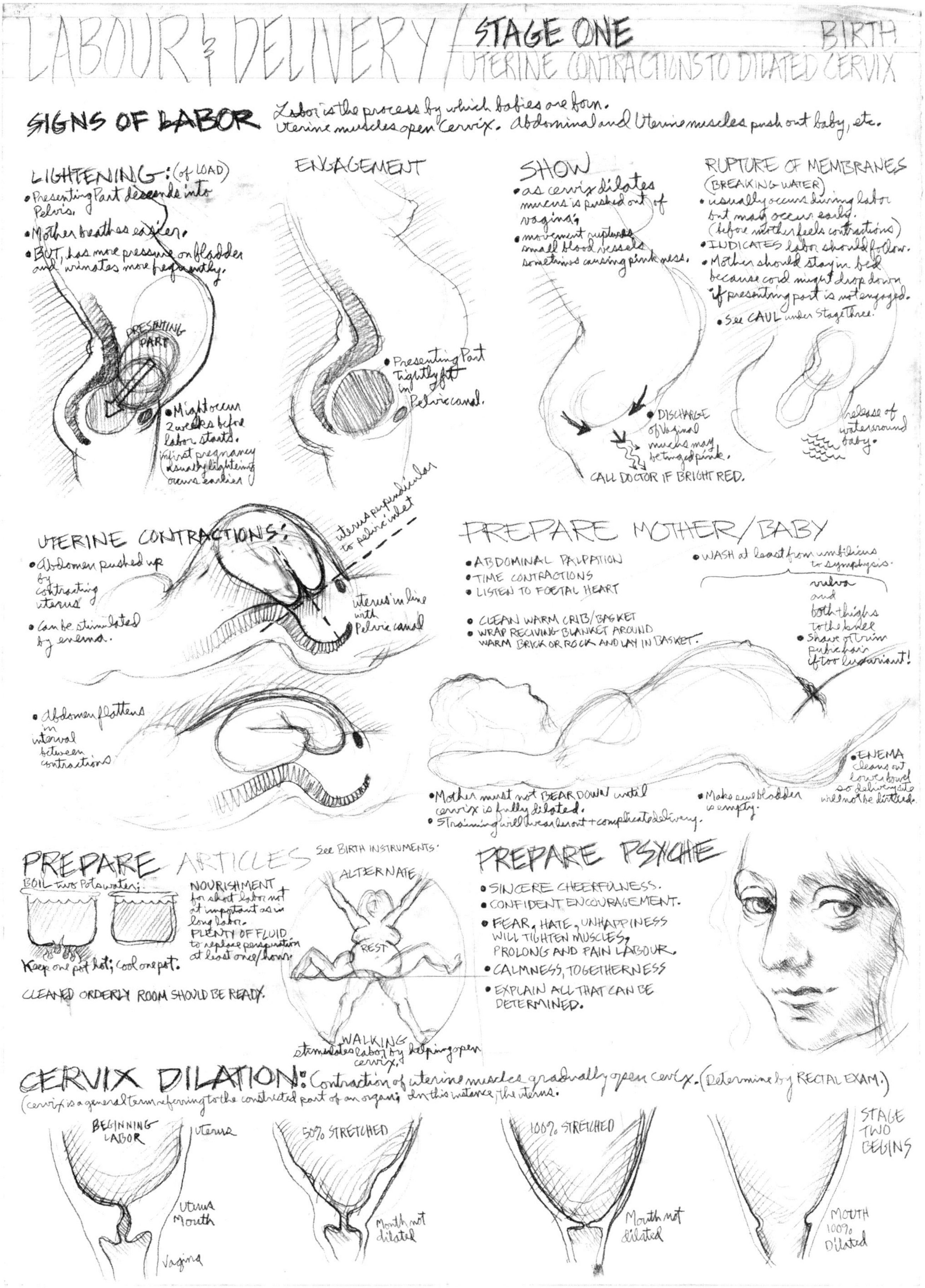

LABOUR & DELIVERY / STAGE ONE
BIRTH
UTERINE CONTRACTIONS TO DILATED CERVIX
SIGNS OF LABOR
Labor is the process by which babies are born. Uterine muscles open Cervix. Abdominal and Uterine muscles push out baby, etc.
LIGHTENING: (of LOAD)
• Presenting Part descends into Pelvis.
• Mother breathes easier.
• BUT, has more pressure on bladder and urinates more frequently.
PRESENTING PART
• Might occur 2 weeks before labor starts. In first pregnancy usually lightening occurs earlier
ENGAGEMENT
• Presenting Part tightly fit in Pelvic canal.
SHOW
• as cervix dilates mucus is pushed out of vagina;
• movement ruptures small blood vessels sometimes causing pinkness.
DISCHARGE of vaginal mucus may be tinged pink.
CALL DOCTOR IF BRIGHT RED.
RUPTURE OF MEMBRANES (BREAKING WATER)
• usually occurs during labor but may occur early. (before mother feels contractions)
• INDICATES labor should follow.
• Mother should stay in bed because cord might drop down if presenting part is not engaged.
• See CAUL under Stage Three.
release of water around baby.
UTERINE CONTRACTIONS:
• Abdomen pushed up by contracting uterus
• Can be stimulated by enema.
uterus perpendicular to pelvic inlet
uterus in line with Pelvic canal
• Abdomen flattens in interval between contractions
PREPARE MOTHER/BABY
• ABDOMINAL PALPATION
• TIME CONTRACTIONS
• LISTEN TO FOETAL HEART
• CLEAN WARM CRIB/BASKET
• WRAP RECEIVING BLANKET AROUND WARM BRICK OR ROCK AND LAY IN BASKET.
• WASH at least from umbilicus to symphysis. vulva and both thighs to the knee
• Shave or trim pubic hair if too luxuriant!
• ENEMA cleans out lower bowel so delivery site will not be dirtied.
• Mother must not BEAR DOWN until cervix is fully dilated.
• Straining will wear her out + complicate delivery.
• Make sure bladder is empty.
PREPARE ARTICLES
See BIRTH INSTRUMENTS.
BOIL two Pots water;
Keep one pot hot; cool one pot.
CLEANED ORDERLY ROOM SHOULD BE READY.
NOURISHMENT for short labor not as important as in long labor.
PLENTY OF FLUID to replace perspiration at least once/hour.
ALTERNATE
REST
WALKING stimulates labor by helping open cervix.
PREPARE PSYCHE
• SINCERE CHEERFULNESS.
• CONFIDENT ENCOURAGEMENT.
• FEAR, HATE, UNHAPPINESS WILL TIGHTEN MUSCLES, PROLONG AND PAIN LABOUR.
• CALMNESS, TOGETHERNESS
• EXPLAIN ALL THAT CAN BE DETERMINED.
CERVIX DILATION: Contraction of uterine muscles gradually open cervix. (Determine by RECTAL EXAM.)
(cervix is a general term referring to the constricted part of an organ; in this instance, the uterus.
BEGINNING LABOR
uterus
Uterus Mouth
Vagina
50% STRETCHED
Mouth not dilated
100% STRETCHED
Mouth not dilated
STAGE TWO BEGINS
MOUTH 100% Dilated

10. Thus architecture verifies the necessary complimentarity of being and nonbeing.

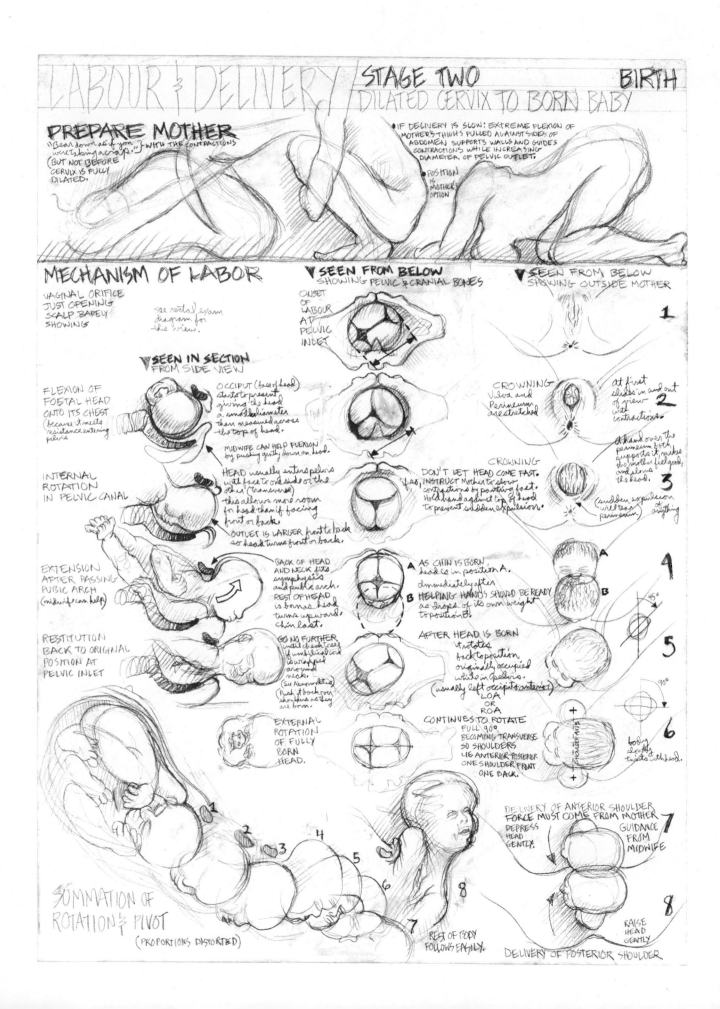

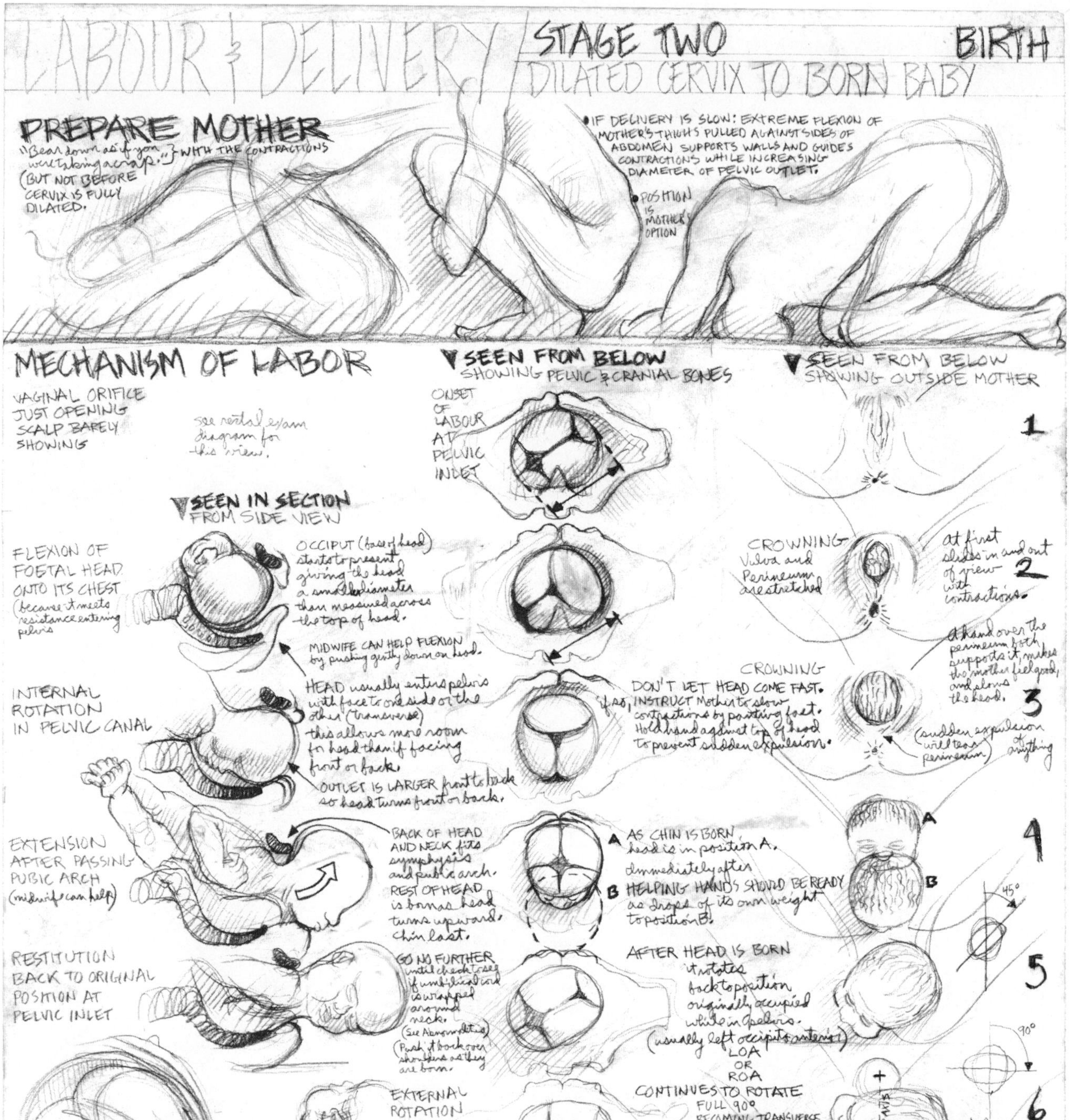

LABOUR & DELIVERY
STAGE TWO
BIRTH
DILATED CERVIX TO BORN BABY
PREPARE MOTHER
"Bear down as if you were taking a crap." WITH THE CONTRACTIONS
(BUT NOT BEFORE CERVIX IS FULLY DILATED.
IF DELIVERY IS SLOW: EXTREME FLEXION OF MOTHER'S THIGHS PULLED AGAINST SIDES OF ABDOMEN SUPPORTS WALLS AND GUIDES CONTRACTIONS WHILE INCREASING DIAMETER OF PELVIC OUTLET.
POSITION IS MOTHER'S OPTION
MECHANISM OF LABOR
SEEN FROM BELOW
SHOWING PELVIC & CRANIAL BONES
SEEN FROM BELOW
SHOWING OUTSIDE MOTHER
VAGINAL ORIFICE JUST OPENING SCALP BARELY SHOWING
see rectal exam diagram for this view.
ONSET OF LABOUR AT PELVIC INLET
1
SEEN IN SECTION FROM SIDE VIEW
FLEXION OF FOETAL HEAD ONTO ITS CHEST (because it meets resistance entering pelvis
OCCIPUT (base of head) starts to present giving the head a smaller diameter than measured across the top of head.
MIDWIFE CAN HELP FLEXION by pushing gently down on head.
CROWNING Vulva and Perineum are stretched
at first slides in and out of view with contractions.
2
A hand over the perineum both supports it, makes the mother feel good, and slows the head.
INTERNAL ROTATION IN PELVIC CANAL
HEAD usually enters pelvis with face to one side or the other (transverse) this allows more room for head than if facing front or back.
OUTLET IS LARGER front to back so head turns front or back.
CROWNING
DON'T LET HEAD COME FAST. if so, INSTRUCT Mother to slow contractions by panting fast. Hold hand against top of head to prevent sudden expulsion.
(sudden expulsion of anything will tear perineum)
3
EXTENSION AFTER PASSING PUBIC ARCH (midwife can help)
BACK OF HEAD AND NECK fits symphysis and pubic arch. REST OF HEAD is born as head turns upward. Chin last.
A AS CHIN IS BORN, head is in position A.
Immediately after
B HELPING HANDS SHOULD BE READY as drops of its own weight to position B.
A
B
4
45°
RESTITUTION BACK TO ORIGINAL POSITION AT PELVIC INLET
GO NO FURTHER until check to see if umbilical cord is wrapped around neck.
AFTER HEAD IS BORN it rotates back to position originally occupied while in pelvis. (usually left occiput anterior) LOA OR ROA
5
90°
EXTERNAL ROTATION OF FULLY BORN HEAD.
CONTINUES TO ROTATE FULL 90° BECOMING TRANSVERSE SO SHOULDERS LIE ANTERIOR POSTERIOR ONE SHOULDER FRONT ONE BACK.
SHOULDER AXIS
body slowly twists with head.
6
DELIVERY OF ANTERIOR SHOULDER
FORCE MUST COME FROM MOTHER
DEPRESS HEAD GENTLY.
GUIDANCE FROM MIDWIFE
7
1 2 3 4 5 6 7 8
SUMMATION OF ROTATION & PIVOT
(PROPORTIONS DISTORTED)
REST OF BODY FOLLOWS EASILY.
RAISE HEAD GENTLY
8
DELIVERY OF POSTERIOR SHOULDER

11. The wonder of the birth process, as in design, is that something larger than the aperture comes through to see the light. The sun is the lantern of architecture.

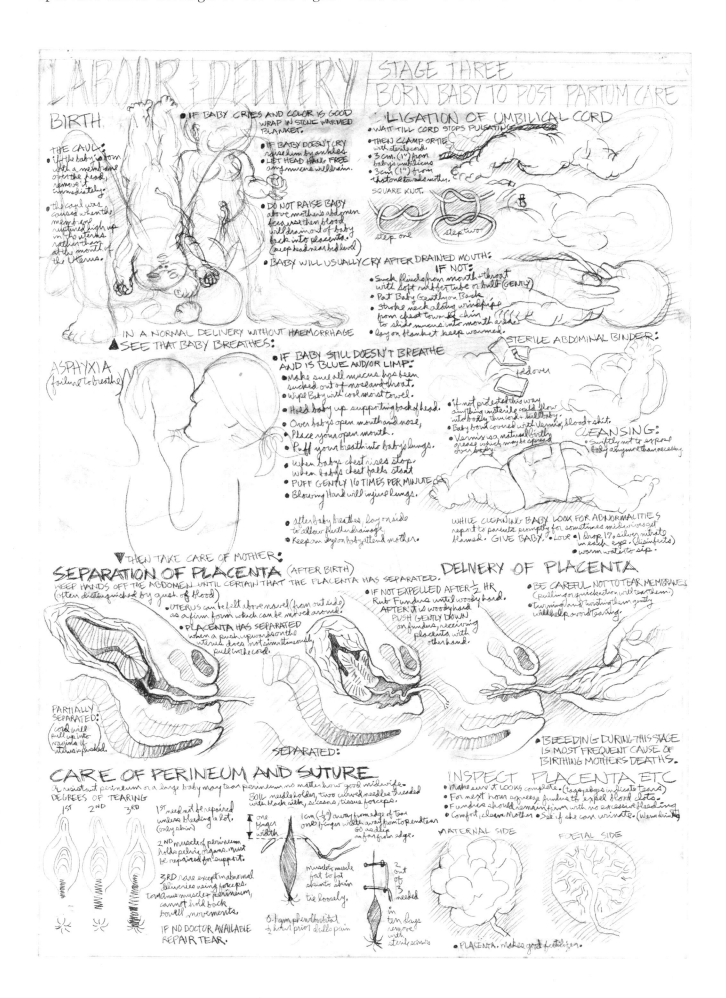

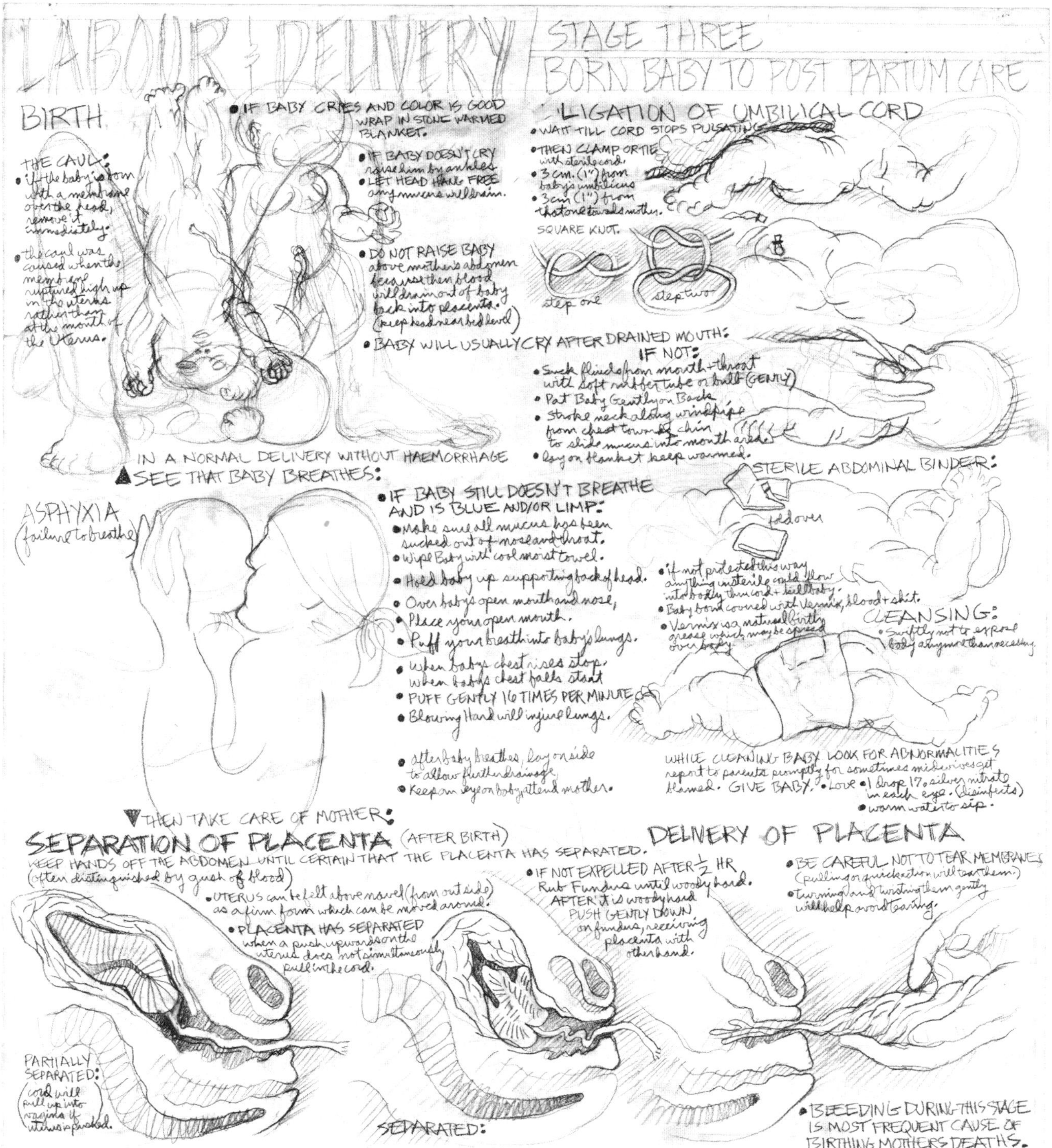

LABOUR & DELIVERY
STAGE THREE
BORN BABY TO POST PARTUM CARE
BIRTH
THE CAUL:
• If the baby is born with a membrane over the head, remove it immediately.
• the caul was caused when the membrane ruptured high up in the uterus rather than at the mouth of the Uterus.
• IF BABY CRIES AND COLOR IS GOOD WRAP IN STONE WARMED BLANKET.
• IF BABY DOESN'T CRY raise him by ankles
• LET HEAD HANG FREE any mucus will drain.
• DO NOT RAISE BABY above mother's abdomen because then blood will drain out of baby back into placenta. (keep head near bed level)
• BABY WILL USUALLY CRY AFTER DRAINED MOUTH:
LIGATION OF UMBILICAL CORD
• WAIT TILL CORD STOPS PULSATING
• THEN CLAMP OR TIE with sterile cord:
• 3 cm. (1") from baby's umbilicus
• 3 cm (1") from that one towards mother.
SQUARE KNOT.
step one
step two
IF NOT:
• Suck fluids from mouth + throat with soft rubber tube or bulb (GENTLY)
• Pat Baby Gently on Back
• Stroke neck along windpipe from chest towards chin to slide mucus into mouth area.
• lay on blanket keep warmed.
IN A NORMAL DELIVERY WITHOUT HAEMORRHAGE
SEE THAT BABY BREATHES:
ASPHYXIA (failure to breathe)
• IF BABY STILL DOESN'T BREATHE AND IS BLUE AND/OR LIMP:
• Make sure all mucus has been sucked out of nose and throat.
• Wipe Baby with cool moist towel.
• Hold baby up supporting back of head.
• Over baby's open mouth and nose,
• Place your open mouth.
• Puff your breath into baby's lungs.
• When baby's chest rises stop, when baby's chest falls start
• PUFF GENTLY 16 TIMES PER MINUTE
• Blowing Hard will injure lungs.
• after baby breathes, lay on side to allow further drainage.
• Keep an eye on baby, attend mother.
STERILE ABDOMINAL BINDER:
fold over
CLEANSING:
• Swiftly not to expose baby any more than necessary.
WHILE CLEANING BABY LOOK FOR ABNORMALITIES & report to parents promptly
• 1 drop 1% silver nitrate in each eye. (disinfects)
• warm water to sip.
THEN TAKE CARE OF MOTHER:
SEPARATION OF PLACENTA (AFTER BIRTH)
KEEP HANDS OFF THE ABDOMEN UNTIL CERTAIN THAT THE PLACENTA HAS SEPARATED.
(often distinguished by gush of blood)
• UTERUS can be felt above navel (from outside) as a firm form which can be moved around.
• PLACENTA HAS SEPARATED when a push upwards on the uterus does not simultaneously pull in the cord.
PARTIALLY SEPARATED:
SEPARATED:
DELIVERY OF PLACENTA
• IF NOT EXPELLED AFTER 1/2 HR Rub Fundus until woody hard.
AFTER it is woody hard PUSH GENTLY DOWN on fundus, receiving placenta with other hand.
• BE CAREFUL NOT TO TEAR MEMBRANES
• BLEEDING DURING THIS STAGE IS MOST FREQUENT CAUSE OF BIRTHING MOTHERS DEATHS.
CARE OF PERINEUM AND SUTURE
DEGREES OF TEARING
1ST
2ND
3RD
1ST need not be repaired unless bleeding a lot. (only skin)
2ND muscle of perineum holds pelvic organs. must be repaired for support.
3RD rare except in abnormal deliveries using forceps.
IF NO DOCTOR AVAILABLE REPAIR TEAR.
one finger width
tie loosely.
INSPECT PLACENTA ETC
• Make sure it LOOKS complete.
• For next hour squeeze fundus to expel blood clots.
• Fundus should remain firm with no excessive bleeding
MATERNAL SIDE
FOETAL SIDE
• PLACENTA. Makes good fertilizer.

12. The earth goes around the sun.

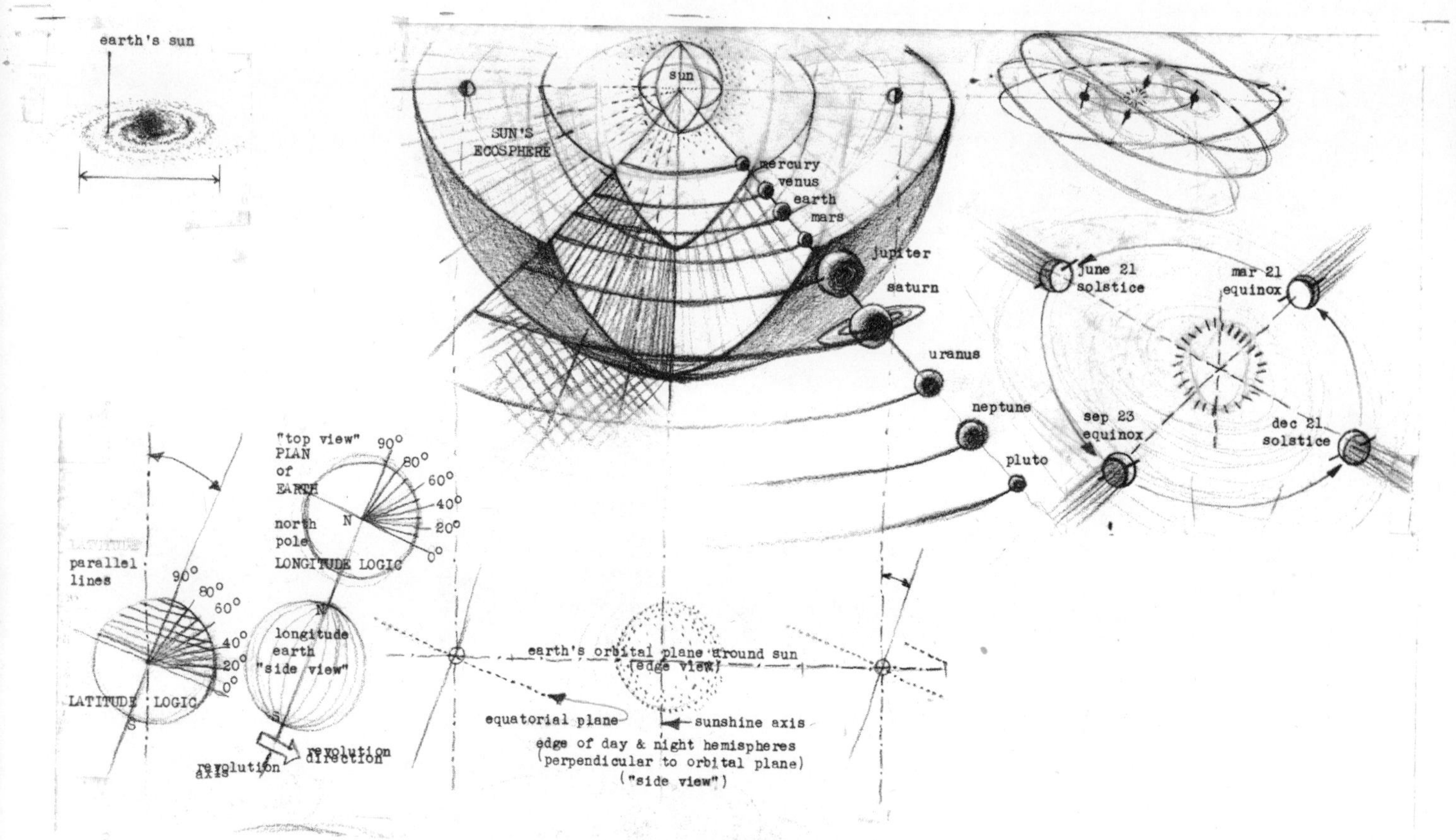

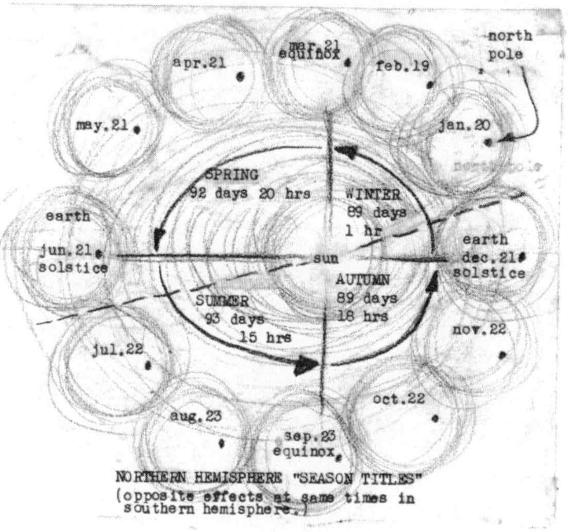

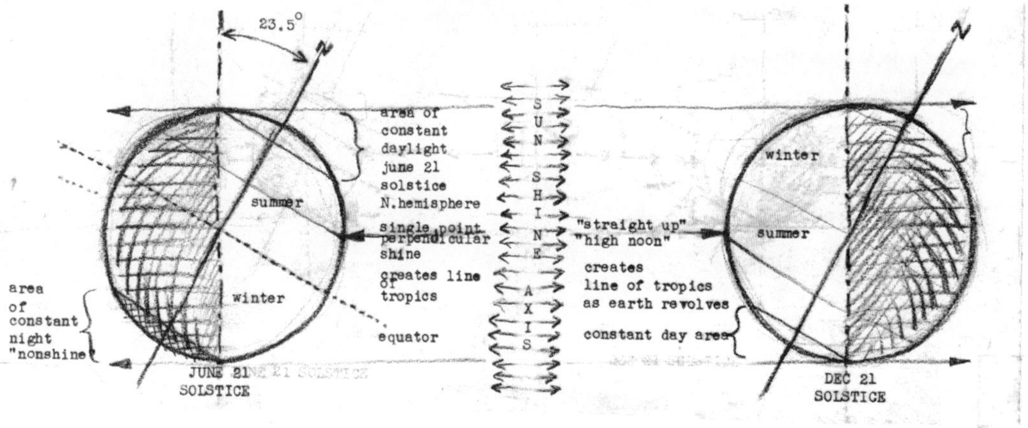

13. The sun goes around the earth.

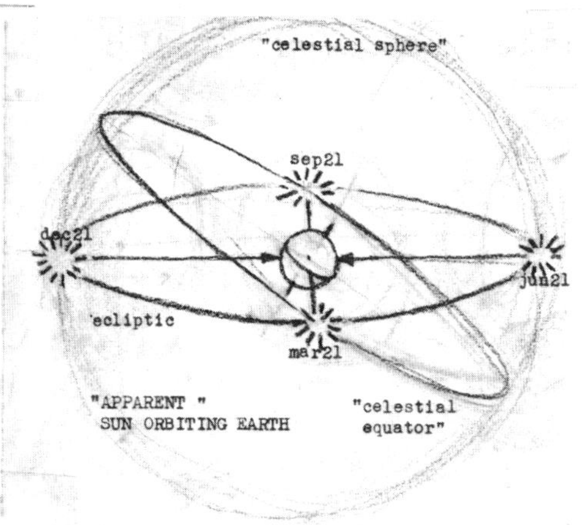

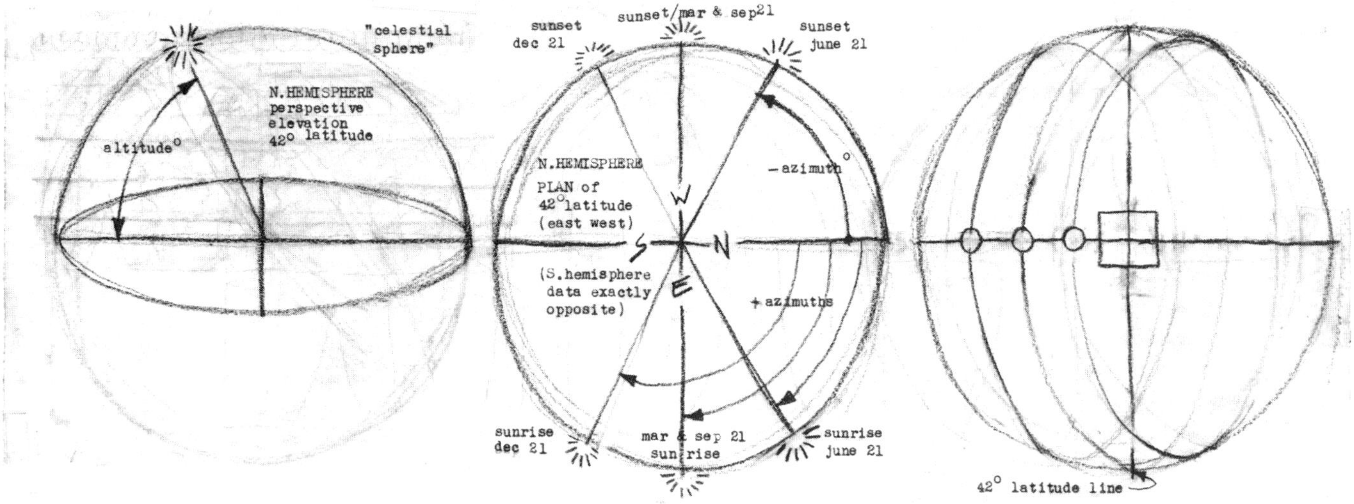

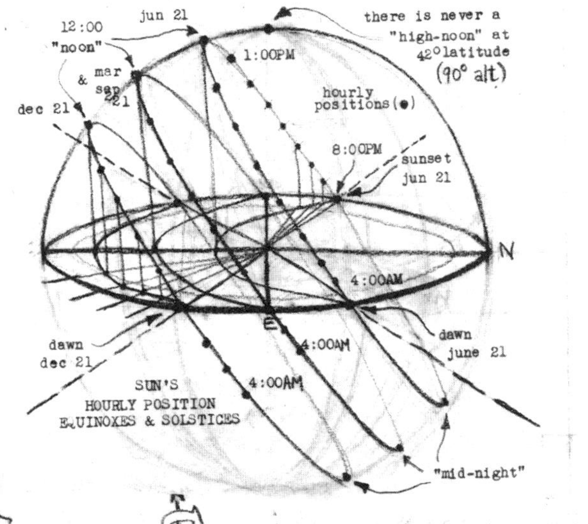

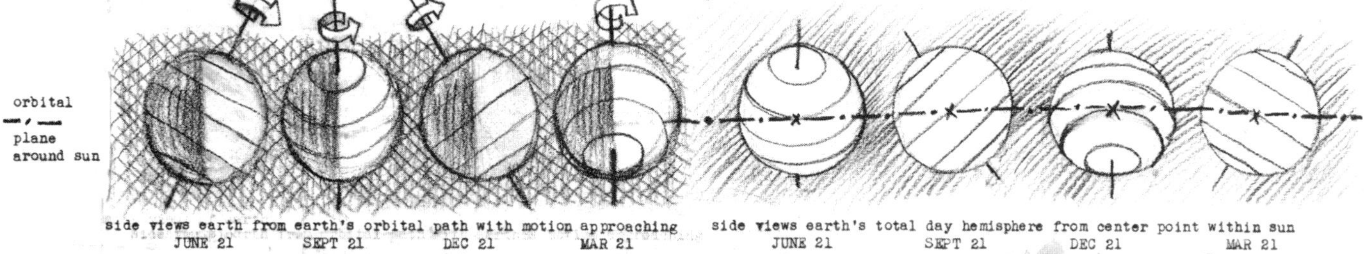

14. A shadow is cast by the sun onto the earth whenever any visible thing is between them.

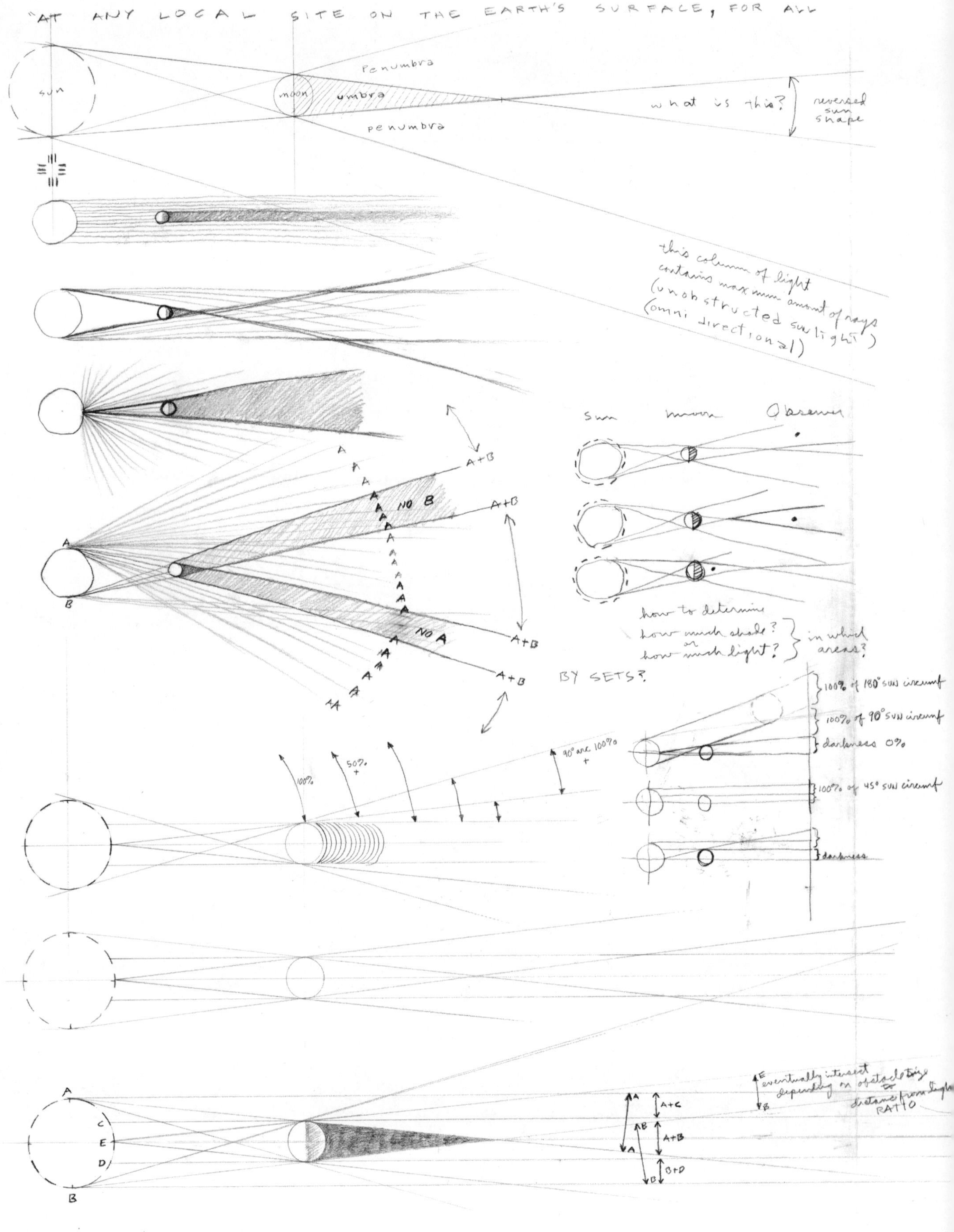

15. For all practical purposes, the sun's rays are considered parallel when they strike the earth.

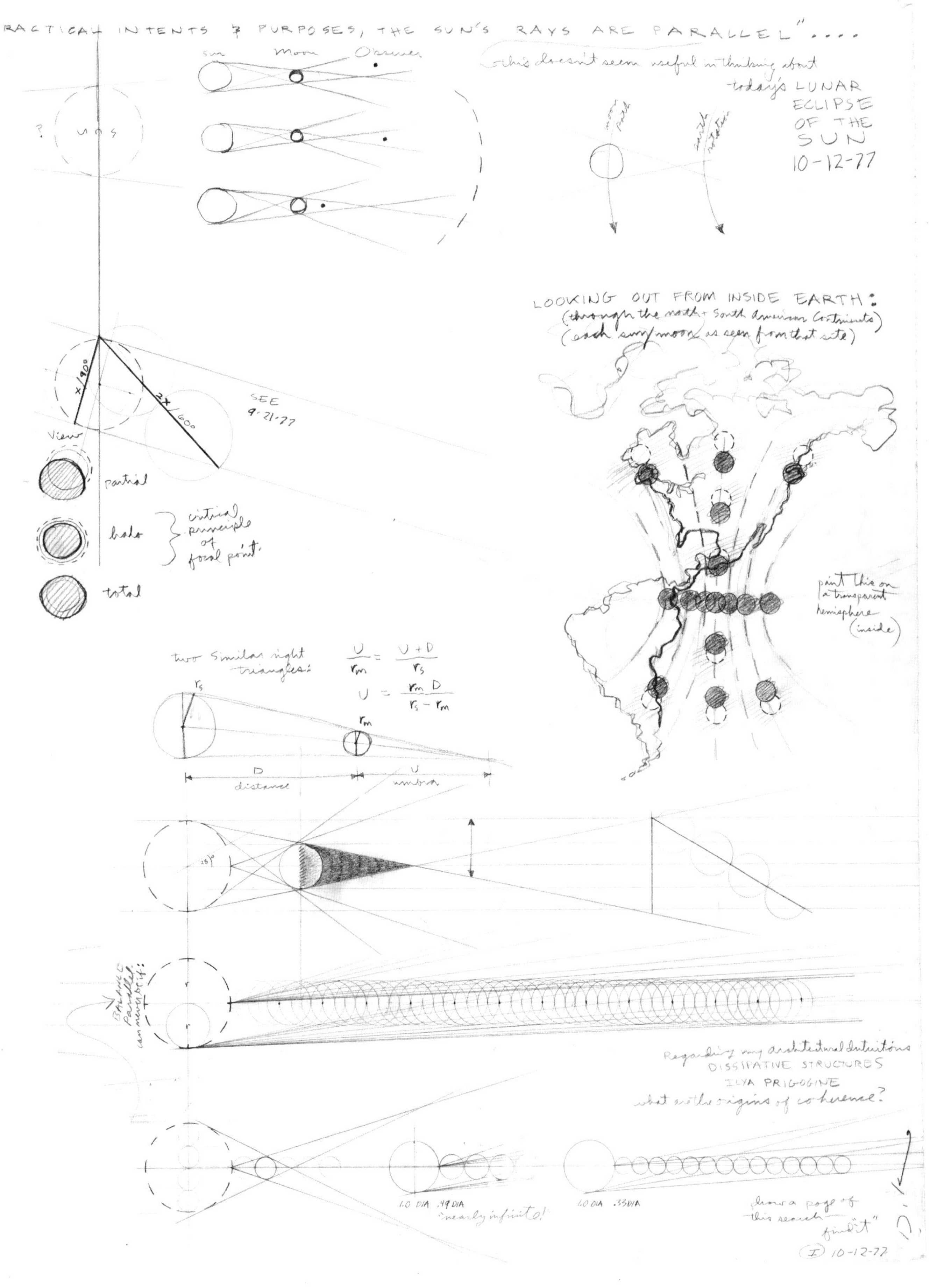

"RACTICAL INTENTS & PURPOSES, THE SUN'S RAYS ARE PARALLEL"....
Sun Moon Observer
this doesn't seem useful in thinking about today's LUNAR ECLIPSE OF THE SUN 10-12-77
SEE 9-21-77
View
partial
halo
total
critical principle of focal point.
LOOKING OUT FROM INSIDE EARTH:
(through the North + South American Continents)
(each sun/moon as seen from that site)
paint this on a transparent hemisphere (inside)
two similar right triangles:
$\frac{U}{r_m} = \frac{U+D}{r_s}$
$U = \frac{r_m D}{r_s - r_m}$
r_s
r_m
D distance
U umbra
Regarding my architectural intuitions
DISSIPATIVE STRUCTURES
ILYA PRIGOGINE
what are the origins of coherence?
1.0 DIA .49 DIA
"nearly infinite!
1.0 DIA .33 DIA
draw a page of this search — "find it"
(I) 10-12-77

16. Shadows can deter people like walls. Diagramming the possibilities of shadows is an infinitely tedious process. The journal of our yearly cycles can be drawn on our surroundings by tracing shadow edges at commemorable times.

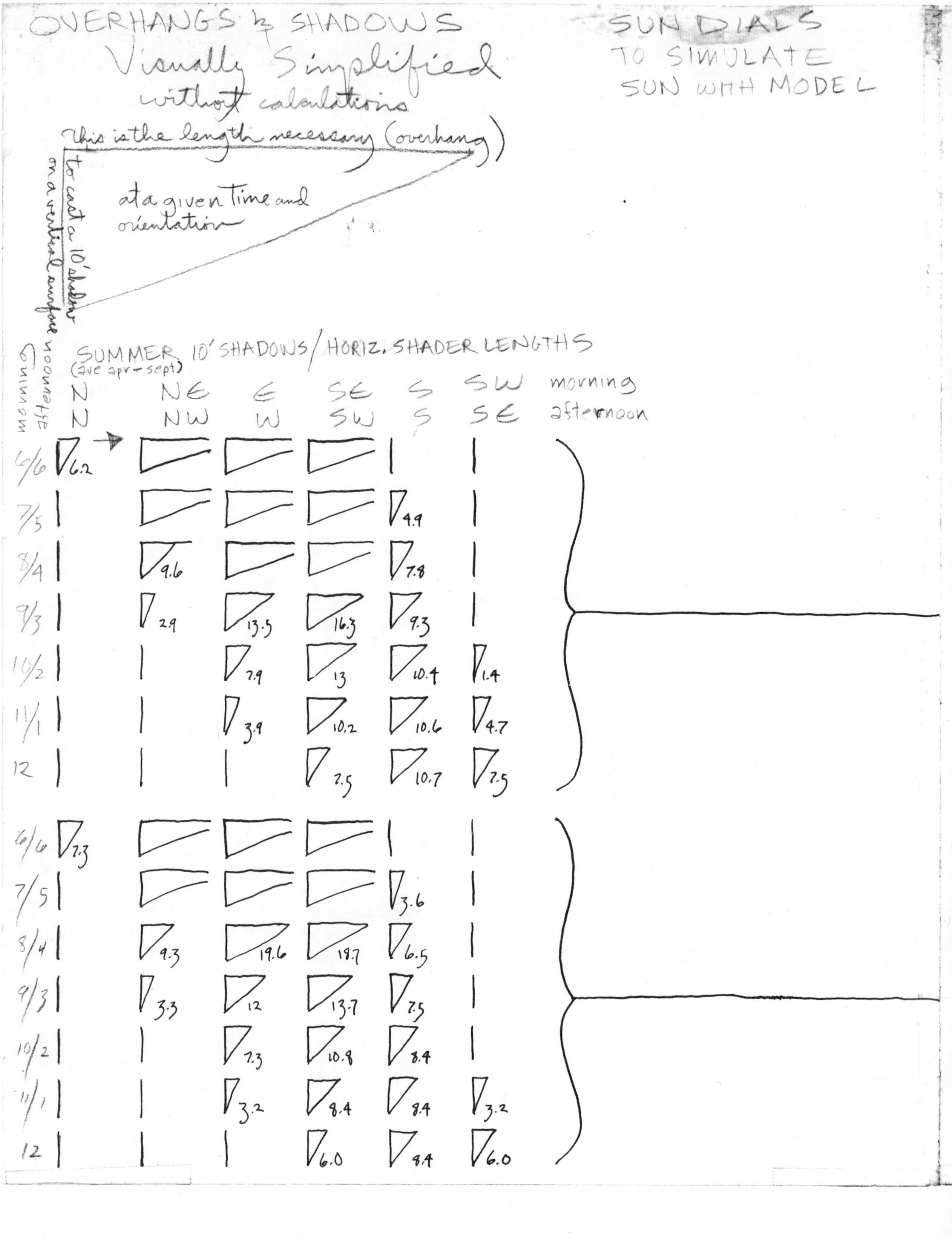

17. Shadow path diagrams for the base of a homemade solar-sextant (shadowlabe) are given on the following pages for six latitudes, 30°, 32°, 40°, 48°, 50°, 56°. Any moment in the year can then be modeled or predicted.

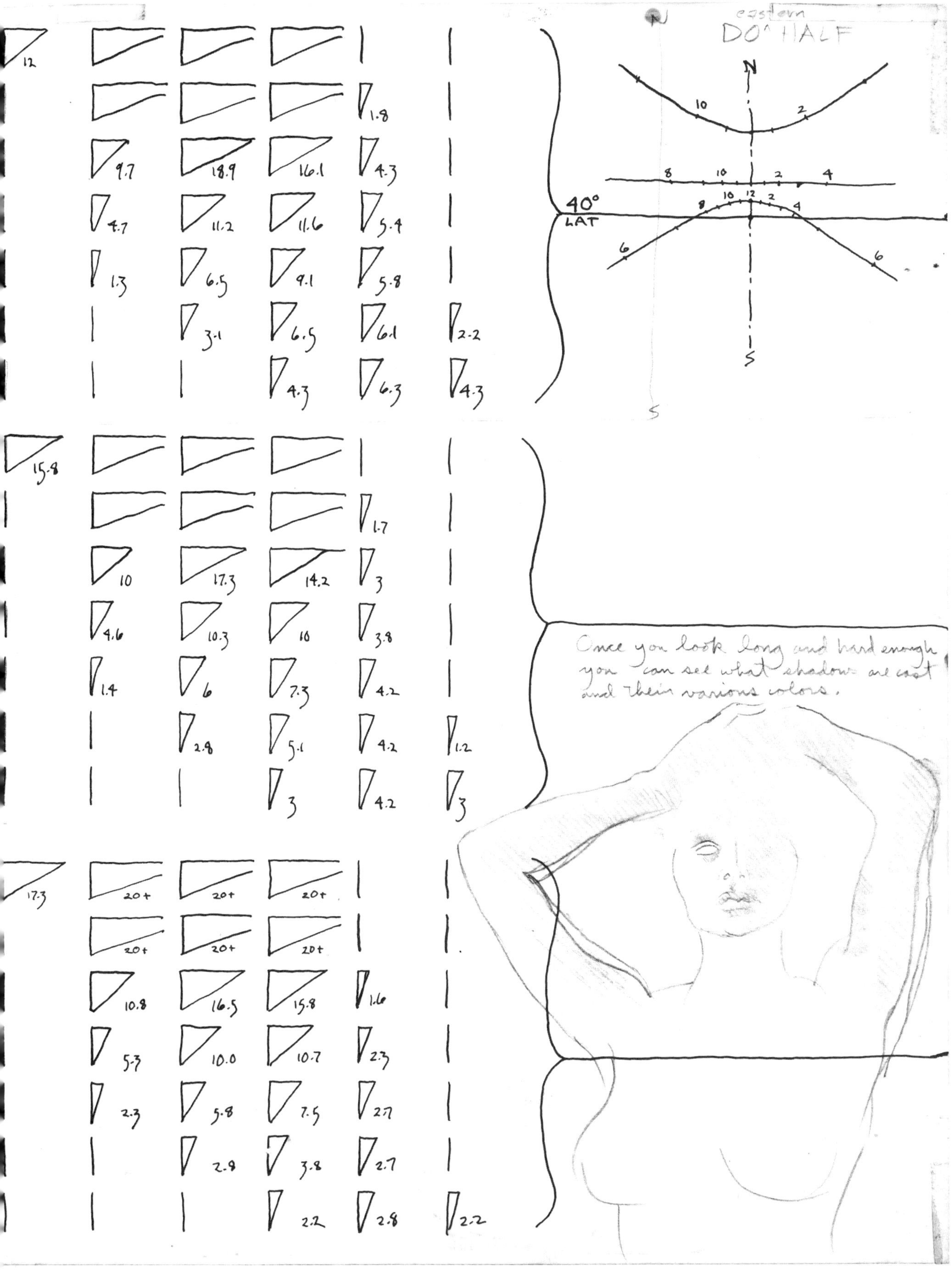
eastern
N
40°
LAT
S
Once you look long and hard enough you can see what shadows are cast and their various colors.

18. Climate is the ratio of sun to slope. Both the earth's atmosphere and the surface of the earth cause the parallel light rays to be changed.

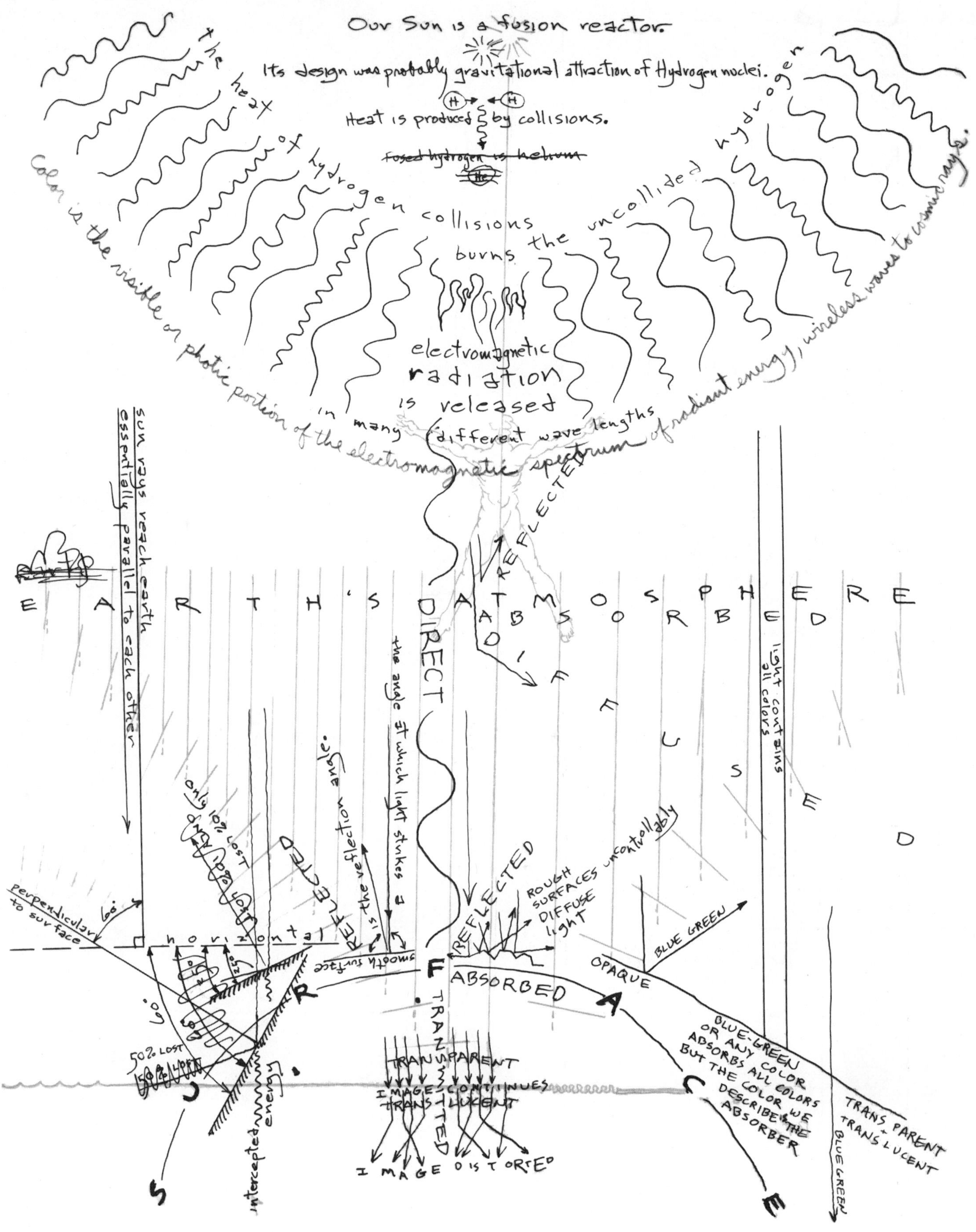

Our Sun is a fusion reactor.
Its design was probably gravitational attraction of Hydrogen nuclei.
Heat is produced by collisions.
the heat of hydrogen collisions burns the uncollided hydrogen
electromagnetic radiation is released in many different wave lengths
Color is the visible or photic portion of the electromagnetic spectrum of radiant energy, wireless waves to cosmic rays.
sun rays reach earth essentially parallel to each other
EARTH'S ATMOSPHERE
REFLECTED
ABSORBED
DIFFUSED
DIRECT
light contains all colors
the angle at which light strikes is the reflection angle
only 10% lost
REFLECTED
perpendicular to surface
60°
horizontal
smooth surface
REFLECTED
ROUGH SURFACES DIFFUSE LIGHT uncontrollably
BLUE GREEN
OPAQUE
ABSORBED
TRANSMITTED
TRANSPARENT
IMAGE CONTINUES
TRANSLUCENT
IMAGE DISTORTED
50% LOST
intercepted energy
BLUE-GREEN OR ANY COLOR ABSORBS ALL COLORS BUT THE COLOR WE DESCRIBE THE ABSORBER
TRANSPARENT + TRANSLUCENT
BLUE GREEN

19. Architectural climates are created by various conversions of light, but first the designer must know where the sun is.

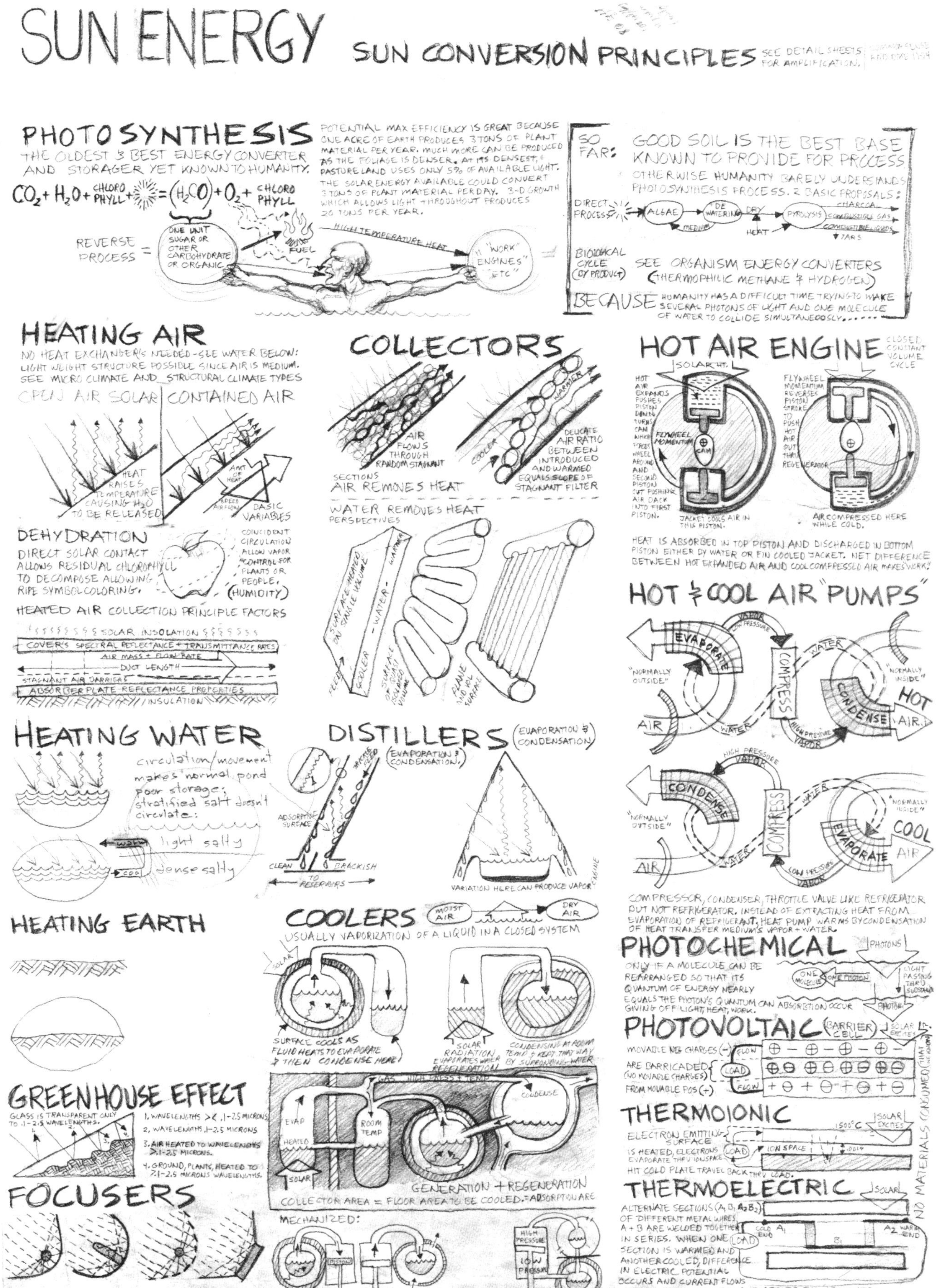
SUN ENERGY
SUN CONVERSION PRINCIPLES
SEE DETAIL SHEETS FOR AMPLIFICATION.
PHOTOSYNTHESIS
THE OLDEST & BEST ENERGY CONVERTER AND STORAGER YET KNOWN TO HUMANITY.
POTENTIAL MAX EFFICIENCY IS GREAT BECAUSE ONE ACRE OF EARTH PRODUCES 3 TONS OF PLANT MATERIAL PER YEAR. MUCH MORE CAN BE PRODUCED AS THE FOLIAGE IS DENSER. AT ITS DENSEST, PASTURE LAND USES ONLY 5% OF AVAILABLE LIGHT. THE SOLAR ENERGY AVAILABLE COULD CONVERT 3 TONS OF PLANT MATERIAL PER DAY. 3-D GROWTH WHICH ALLOWS LIGHT THROUGHOUT PRODUCES 20 TONS PER YEAR.
CO2 + H2O + CHLORO PHYLL + = (H2CO) + O2 + CHLORO PHYLL
REVERSE PROCESS =
ONE UNIT SUGAR OR OTHER CARBOHYDRATE OR ORGANIC
FUEL
HIGH TEMPERATURE HEAT
"WORK" ENGINES "ETC"
SO FAR:
GOOD SOIL IS THE BEST BASE KNOWN TO PROVIDE FOR PROCESS
OTHERWISE HUMANITY BARELY UNDERSTANDS PHOTOSYNTHESIS PROCESS. 2 BASIC PROPOSALS:
DIRECT PROCESS
ALGAE
DE WATERING
DRY
PYROLYSIS
CHARCOAL
COMBUSTIBLE GAS
COMBUSTIBLE LIQUIDS
TARS
MEDIUM
HEAT
BIOLOGICAL CYCLE (BY PRODUCT)
SEE ORGANISM ENERGY CONVERTERS (THERMOPHILIC METHANE & HYDROGEN)
BECAUSE HUMANITY HAS A DIFFICULT TIME TRYING TO MAKE SEVERAL PHOTONS OF LIGHT AND ONE MOLECULE OF WATER TO COLLIDE SIMULTANEOUSLY......
HEATING AIR
NO HEAT EXCHANGERS NEEDED - SEE WATER BELOW: LIGHT WEIGHT STRUCTURE POSSIBLE SINCE AIR IS MEDIUM. SEE MICRO CLIMATE AND STRUCTURAL CLIMATE TYPES
OPEN AIR SOLAR
CONTAINED AIR
HEAT RAISES TEMPERATURE CAUSING H2O TO BE RELEASED
BASIC VARIABLES
DEHYDRATION
DIRECT SOLAR CONTACT ALLOWS RESIDUAL CHLOROPHYLL TO DECOMPOSE ALLOWING RIPE SYMBOL COLORING.
COINCIDENT CIRCULATION ALLOW VAPOR CONTROL FOR PLANTS OR PEOPLE. (HUMIDITY)
HEATED AIR COLLECTION PRINCIPLE FACTORS
SOLAR INSOLATION
COVER'S SPECTRAL REFLECTANCE + TRANSMITTANCE RATES
AIR MASS + FLOW RATE
DUCT LENGTH
STAGNANT AIR BARRIERS
ABSORBER PLATE REFLECTANCE PROPERTIES
INSULATION
COLLECTORS
AIR FLOWS THROUGH RANDOM STAGNANT
WARMER
COOLER
DELICATE AIR RATIO BETWEEN INTRODUCED AND WARMED EQUALS SLOPE OF STAGNANT FILTER
SECTIONS
AIR REMOVES HEAT
WATER REMOVES HEAT
PERSPECTIVES
SURFACE HEATED ON SINGLE VOLUME
WATER
WARMER
COOLER
FEED
PLANE NO VOL SURFACE
HOT AIR ENGINE
CLOSED CONSTANT VOLUME CYCLE
SOLAR HT.
HOT AIR EXPANDS PUSHES PISTON DOWN TURNS CAM WHICH FORCES WHEEL AROUND AND SECOND PISTON CUT PUSHING AIR BACK INTO FIRST PISTON.
FLYWHEEL MOMENTUM
CAM
JACKET COOLS AIR IN THIS PISTON.
FLYWHEEL MOMENTUM REVERSES PISTON STROKE TO PUSH HOT AIR OUT THRU REGENERATOR
AIR COMPRESSED HERE WHILE COLD.
HEAT IS ABSORBED IN TOP PISTON AND DISCHARGED IN BOTTOM PISTON EITHER BY WATER OR FIN COOLED JACKET. NET DIFFERENCE BETWEEN HOT EXPANDED AIR AND COOL COMPRESSED AIR MAKES WORK!
HOT & COOL AIR "PUMPS"
EVAPORATE
LOW PRESSURE
VAPOR
COMPRESS
WATER
"NORMALLY OUTSIDE"
"NORMALLY INSIDE"
CONDENSE
HOT AIR
AIR
HIGH PRESSURE
CONDENSE
COMPRESS
COOL AIR
EVAPORATE
COMPRESSOR, CONDENSER, THROTTLE VALVE LIKE REFRIGERATOR BUT NOT REFRIGERATOR. INSTEAD OF EXTRACTING HEAT FROM EVAPORATION OF REFRIGERANT, HEAT PUMP WARMS BY CONDENSATION OF HEAT TRANSFER MEDIUM'S VAPOR + WATER.
HEATING WATER
circulation/movement makes normal pond poor storage; stratified salt doesn't circulate:
light salty
dense salty
DISTILLERS (EVAPORATION & CONDENSATION)
ABSORPTIVE SURFACE
CLEAN
BRACKISH
TO RESERVOIRS
VARIATION HERE CAN PRODUCE VAPOR
HEATING EARTH
COOLERS
MOIST AIR
DRY AIR
USUALLY VAPORIZATION OF A LIQUID IN A CLOSED SYSTEM
SOLAR
SURFACE COOLS AS FLUID HEATS TO EVAPORATE & THEN CONDENSE HERE
SOLAR RADIATION EVAPORATES WATER REGENERATION
CONDENSING AT ROOM TEMP & KEPT THAT WAY BY SURROUNDING WATER
CONDENSE
EVAP
HEATED
ROOM TEMP
GENERATION + REGENERATION
COLLECTOR AREA = FLOOR AREA TO BE COOLED - ABSORPTION ARE
MECHANIZED:
HIGH PRESSURE
LOW PRESSURE
COOLING
REGENERATION
GREENHOUSE EFFECT
GLASS IS TRANSPARENT ONLY TO .1-2.5 WAVELENGTHS.
1. WAVELENGTHS > .1-2.5 MICRONS
2. WAVELENGTHS .1-2.5 MICRONS
3. AIR HEATED TO WAVELENGTHS >.1-2.5 MICRONS.
4. GROUND, PLANTS, HEATED TO 7.1-2.5 MICRONS WAVELENGTHS.
FOCUSERS
PHOTOCHEMICAL
PHOTONS
ONLY IF A MOLECULE CAN BE REARRANGED SO THAT ITS QUANTUM OF ENERGY NEARLY EQUALS THE PHOTON'S QUANTUM CAN ABSORPTION OCCUR GIVING OFF LIGHT, HEAT, WORK.
ONE MOLECULE
ONE PHOTON
PHOTOVOLTAIC
BARRIER CELL
SOLAR EXCITES
MOVABLE NEG CHARGES (-)
ARE BARRICADED (NO MOVABLE CHARGES)
FROM MOVABLE POS (+)
FLOW
LOAD
THERMOIONIC
SOLAR EXCITES
ELECTRON EMITTING SURFACE IS HEATED, ELECTRONS EVAPORATE THRU IONSPACE HIT COLD PLATE TRAVEL BACK THRU LOAD.
ION SPACE
THERMOELECTRIC
SOLAR
ALTERNATE SECTIONS (A1 B1 A2 B2) OF DIFFERENT METAL WIRES A + B ARE WELDED TOGETHER IN SERIES. WHEN ONE SECTION IS WARMED AND ANOTHER COOLED, DIFFERENCE IN ELECTRIC POTENTIAL OCCURS AND CURRENT FLOWS IF CIRCUITED FROM ONE END OF SERIES TO THE OTHER.
COLD END
WARM END
NO MATERIALS CONSUMED

20. Any point in the celestial sphere (the sky) can be located by two numbers, the altitude and the azimuth.

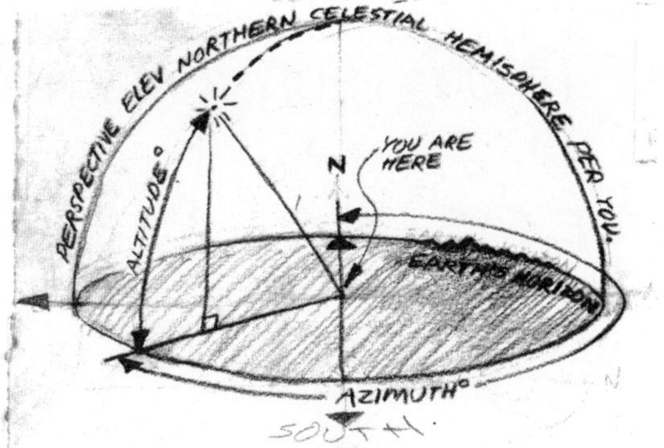

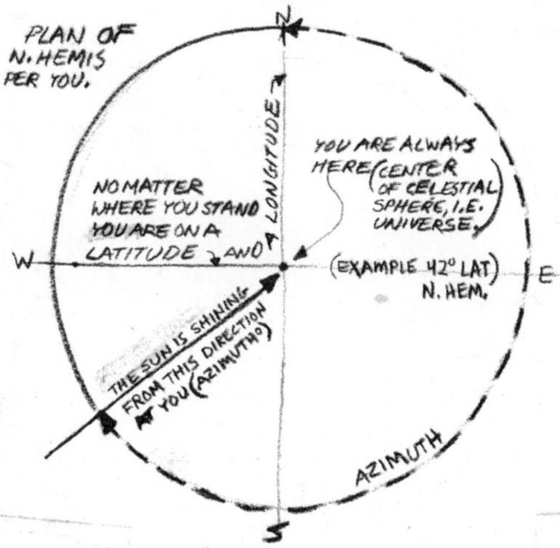

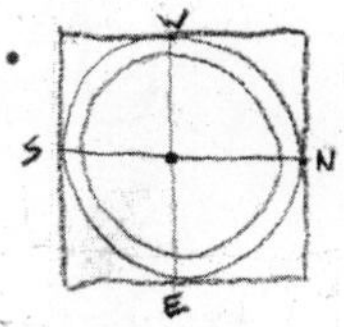
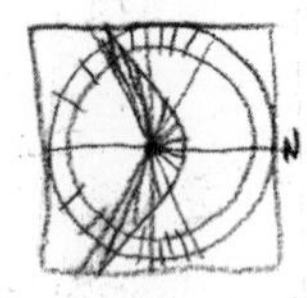
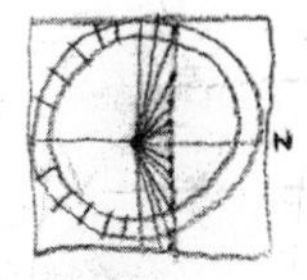
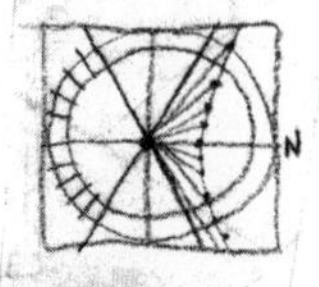

The apparent path of the sun (from any viewpoint on earth) is that it "rises" in the east and "sets" in the west.

These sun paths together with the stars & planets, create earth's Celestial Sphere.

Any point can be described (located) on that sphere by two factors,

ALTITUDE°

which is how high the sun is perpendicular to the horizon.

AZIMUTH°

which is how far the sun is clockwise from north on horizon. (from south in southern hemisphere)

The sun's path is different at every

LATITUDE°

At any given latitude the sun's path is the same at every

LONGITUDE°

These known facts of sun paths can be modelled with a light source and a shadow caster.

The opposite page shows how to diagram a plan of shadow lengths for a 10 UNIT vertical shadowcaster.

Altitude in degrees / tangent

1	.017	588
2	.035	286
3	.052	192
4	.070	143
5	.087	115
6	.105	952
7	.123	81.3
8	.141	70.9
9	.158	63.3
10	.176	56.8
11	.194	51.5
12	.213	47
13	.231	43.3
14	.249	40.2
15	.268	37.3
16	.287	35
17	.306	32.7
18	.325	30.8
19	.344	29
20	.364	27.5
21	.384	26
22	.404	24.7
23	.424	23.6
24	.445	22.5
25	.466	21.5
26	.488	20.5
27	.510	19.6
28	.538	18.7
29	.554	18
30	.577	17.3
31	.601	16.6
32	.625	16.0
33	.649	15.4
34	.675	14.8
35	.700	14.3
36	.727	13.8
37	.754	13.2
38	.781	12.8
39	.810	12.3
40	.839	11.9
41	.869	11.5
42	.900	11.1
43	.933	10.7
44	.966	10.3
45	1,000	10

IN the north hemisphere,

LONGEST DAY, SHORTEST NIGHT
JUNE 21 SOLSTICE

EQUAL DAY AND NIGHT
MAR & SEP 21 EQUINOXES

SHORTEST DAY, LONGEST NIGHT
DEC 21 SOLSTICE

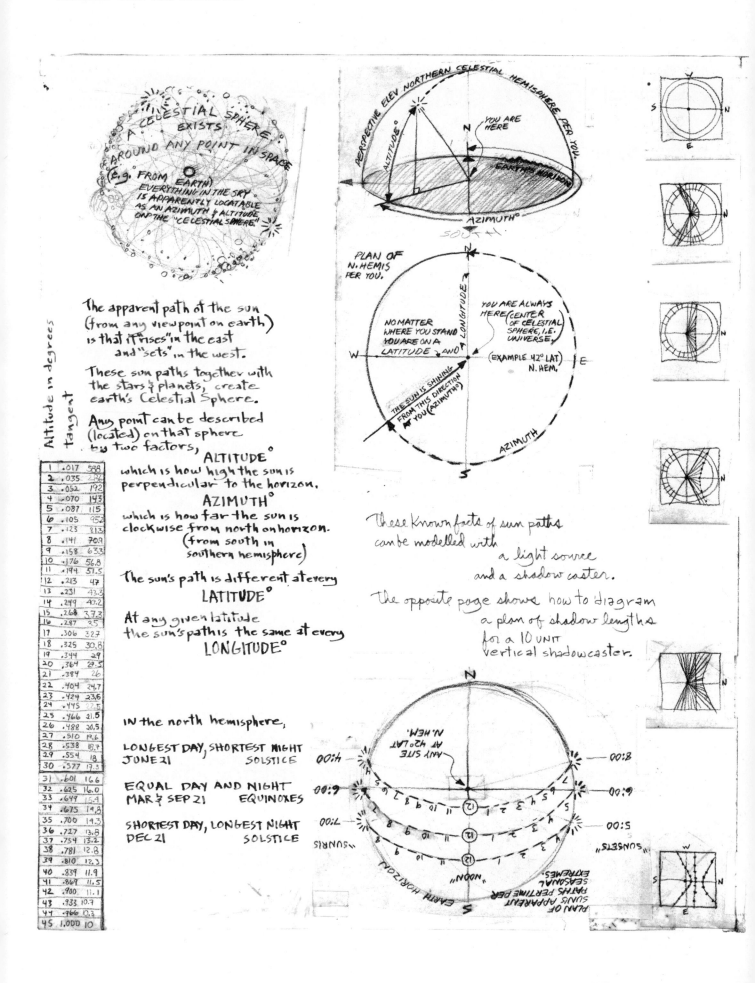

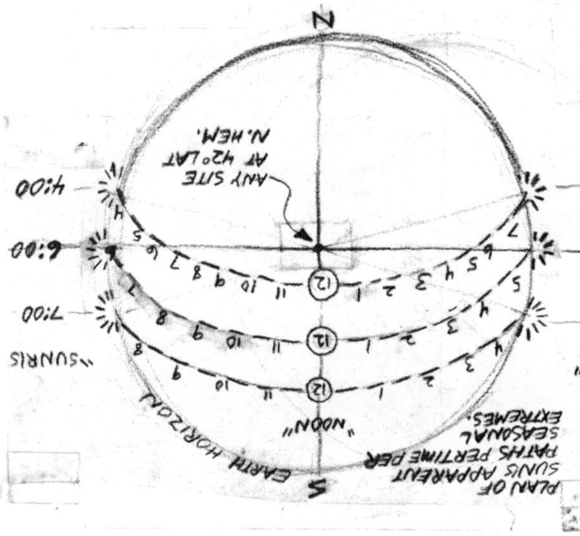

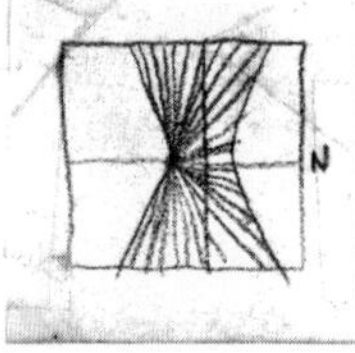
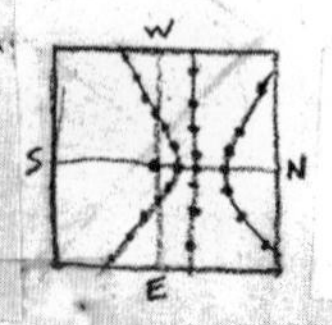

21. These two numbers can be used to make a shadowlabe to predict shadows at any time and angle.

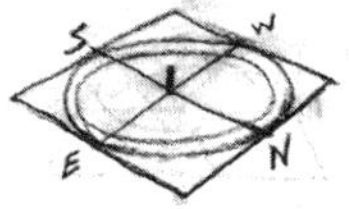

sheet of cardboard with pin, geometrically prepared to locate shadow lengths of pin.

If shadow lengths are known for certain times for a certain shadow caster, then for those times... sunshine can be simulated.

TO CALCULATE SHADOW LENGTH

DIVIDE: 10 UNIT HT. SHADOW CASTER

BY: TAN. SUN'S ALTITUDE/TIME

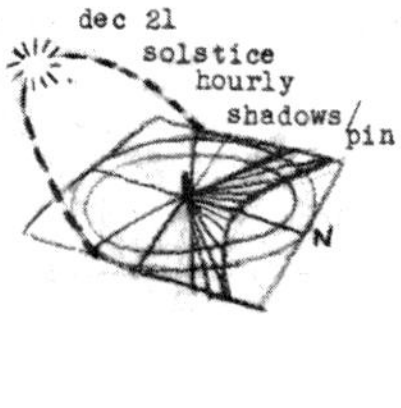

dec 21 solstice hourly shadows/pin

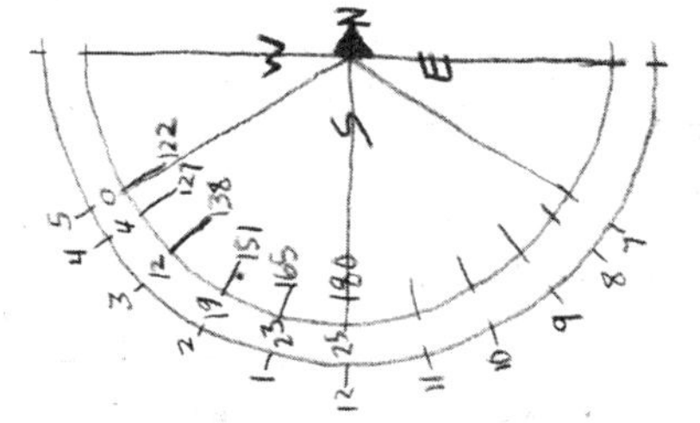

TO CONSTRUCT SHADOW CASTER

Find a solar ephemeris.
For the particular site latitude, locate, on circumferences, azimuths per hour per at least solstices & equinoxes.

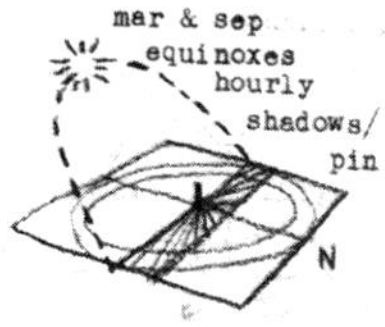

mar & sep equinoxes hourly shadows/pin

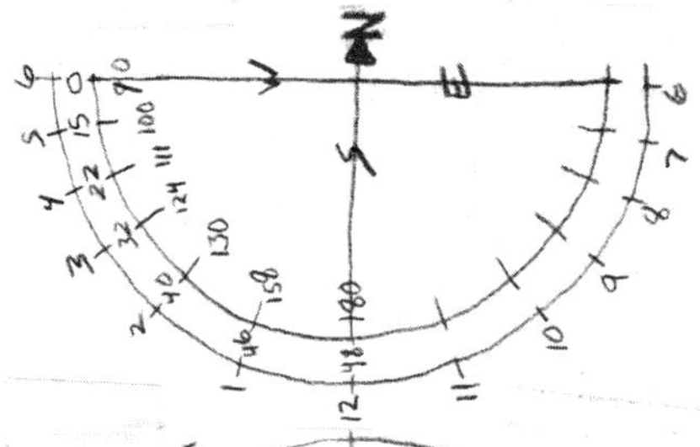

Morning & afternoon data will mirror each other.
Use the altitude data to calculate shadow lengths.
Overlay the plotted azimuths and from the center of the circle away from the direction to the sun, plot the shadow-length end-points.
I recommend drawing only the "ends" of the shadow lengths, then connecting those of a given day by a continous line. Rough visual interpolations may be made between these accurate sun paths. The more accuracy required of the interpolations, the more frequently the end-shadow paths should be plotted (at least monthly instead of quarterly)

june 21 solstice hourly shadows/pin

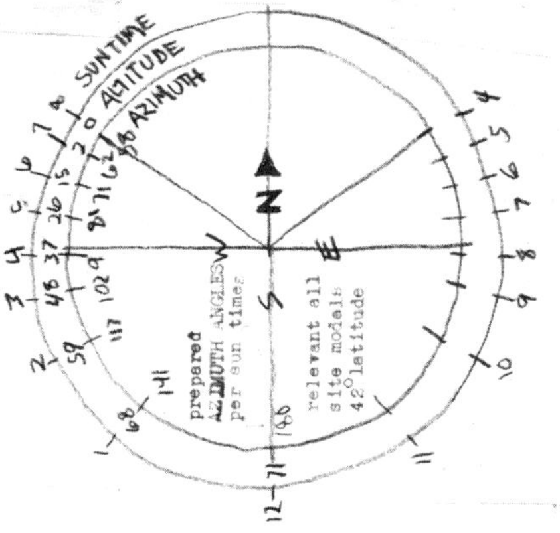

46	1.036	9.6
47	1.072	9.3
48	1.111	9.0
49	1.150	8.7
50	1.192	8.4
51	1.235	8.1
52	1.280	7.8
53	1.327	7.5
54	1.376	7.3
55	1.428	7
56	1.483	6.7
57	1.540	6.5
58	1.600	6.3
59	1.664	6.0
60	1.732	5.8
61	1.804	5.5
62	1.881	5.3
63	1.963	5.1
64	2.050	4.9
65	2.145	4.7
66	2.246	4.5
67	2.356	4.2
68	2.475	4.0
69	2.605	3.8
70	2.748	3.6
71	2.904	3.4
72	3.078	3.25
73	3.271	3
74	3.487	2.9
75	3.732	2.7
76	4.011	2.5
77	4.332	2.3
78	4.705	2.1
79	5.145	1.9
80	5.671	1.8
81	6.314	1.6
82	7.115	1.4
83	8.144	1.2
84	9.514	1.1
85	11.43	.9
86	14.30	.7
87	19.08	.5
88	28.64	.35
89	57.29	.17
90		

Shadow Lengths.

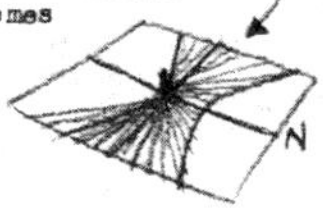

all possible shadows between solstice extremes

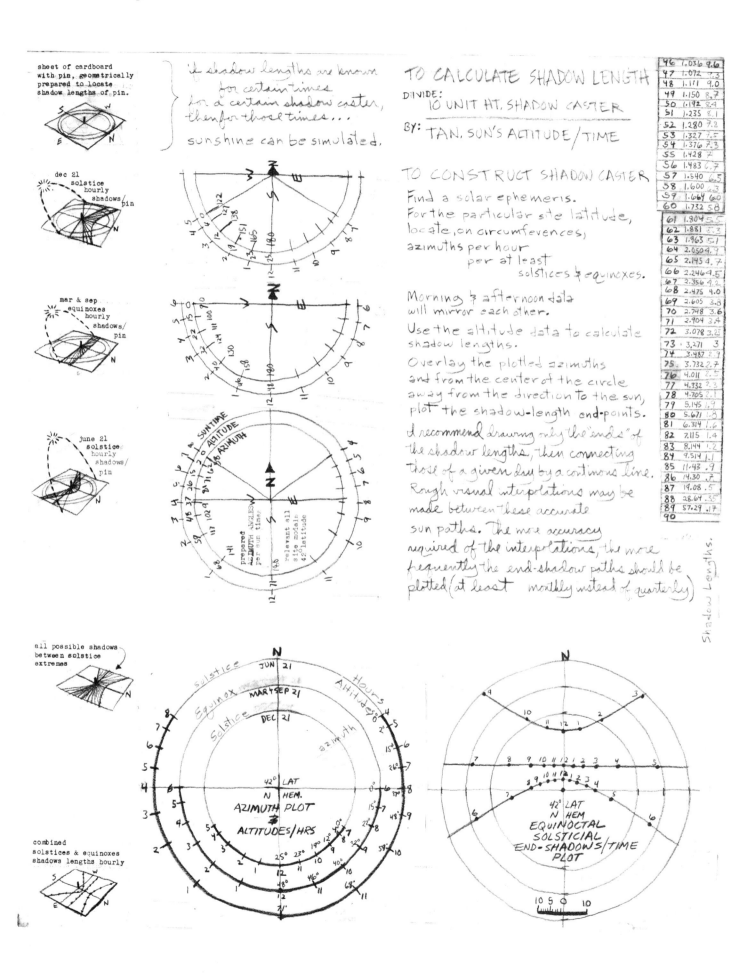

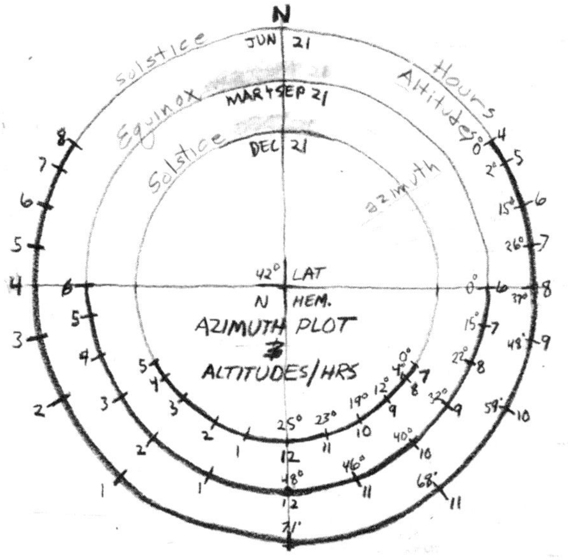

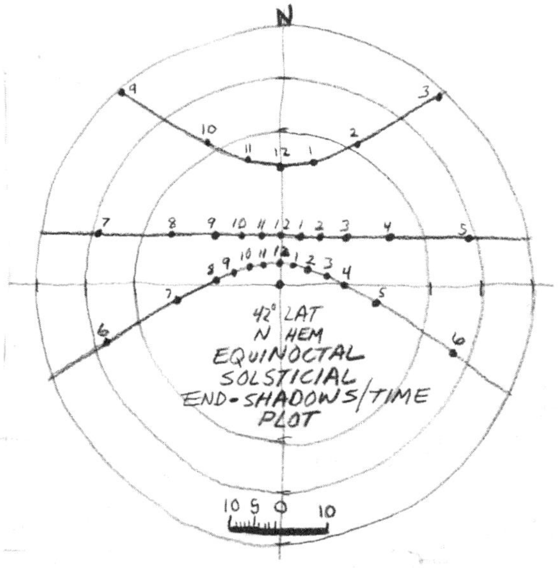

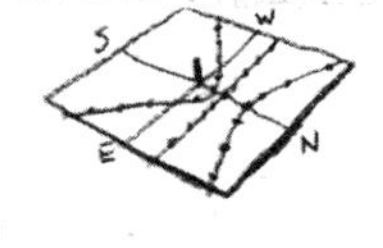

combined solstices & equinoxes shadows lengths hourly

22. Shadow and sun area predictions can be made with the shadowlabe on the site or on a model of the site.

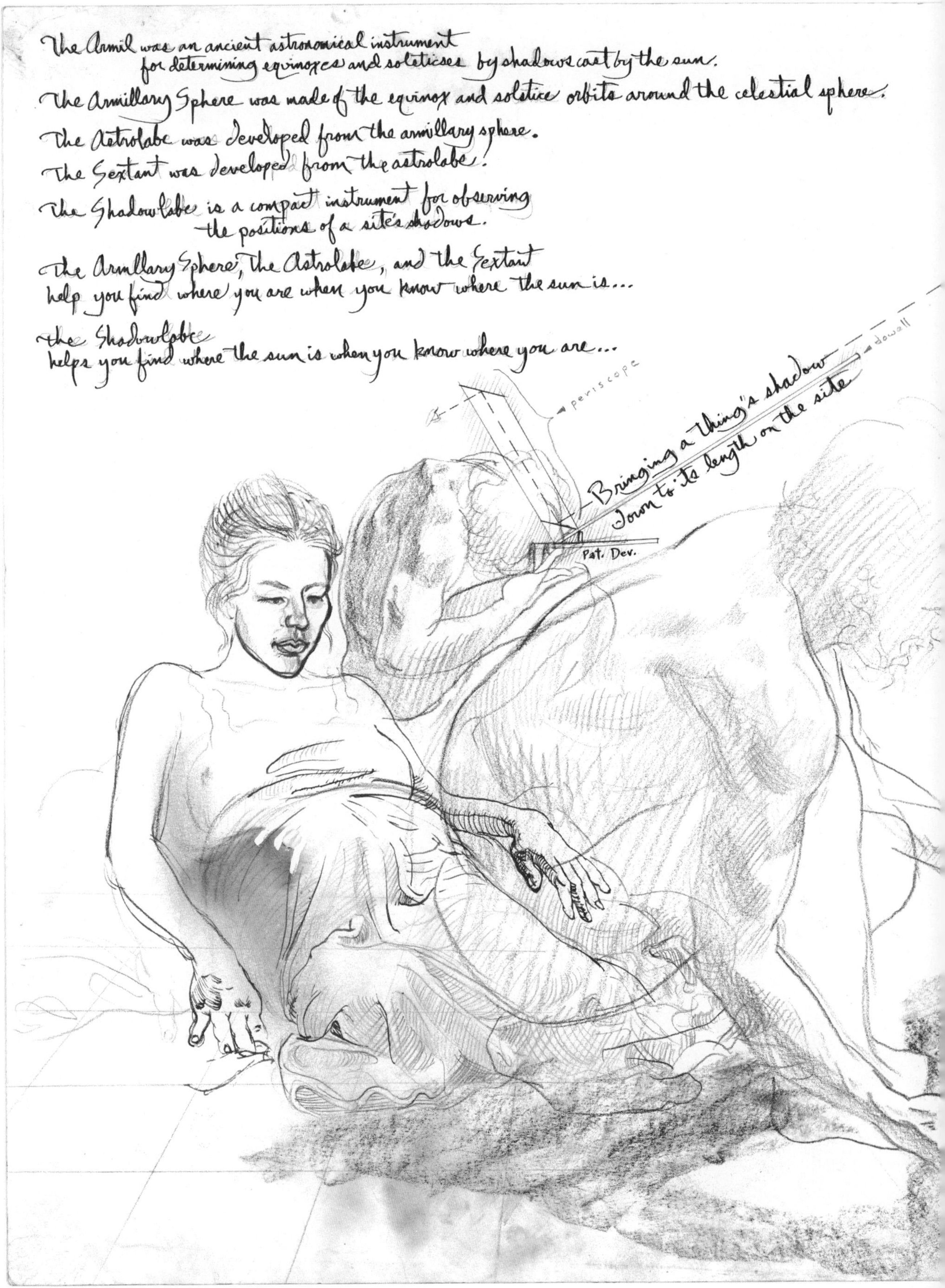

The Armil was an ancient astronomical instrument
for determining equinoxes and solsticises by shadows cast by the sun.
The Armillary Sphere was made of the equinox and solstice orbits around the celestial sphere.
The Astrolabe was developed from the armillary sphere.
The Sextant was developed from the astrolabe.
The Shadowlabe is a compact instrument for observing
the positions of a site's shadows.
The Armillary Sphere, the Astrolabe, and the Sextant
help you find where you are when you know where the sun is...
the Shadowlabe
helps you find where the sun is when you know where you are...
dowall
periscope
Bringing a thing's shadow
down to its length on the site
Pat. Dev.

23. These sun and shadow patterns offer form possibilities of interesting and useful configurations, especially in a complex landscape such as the city.

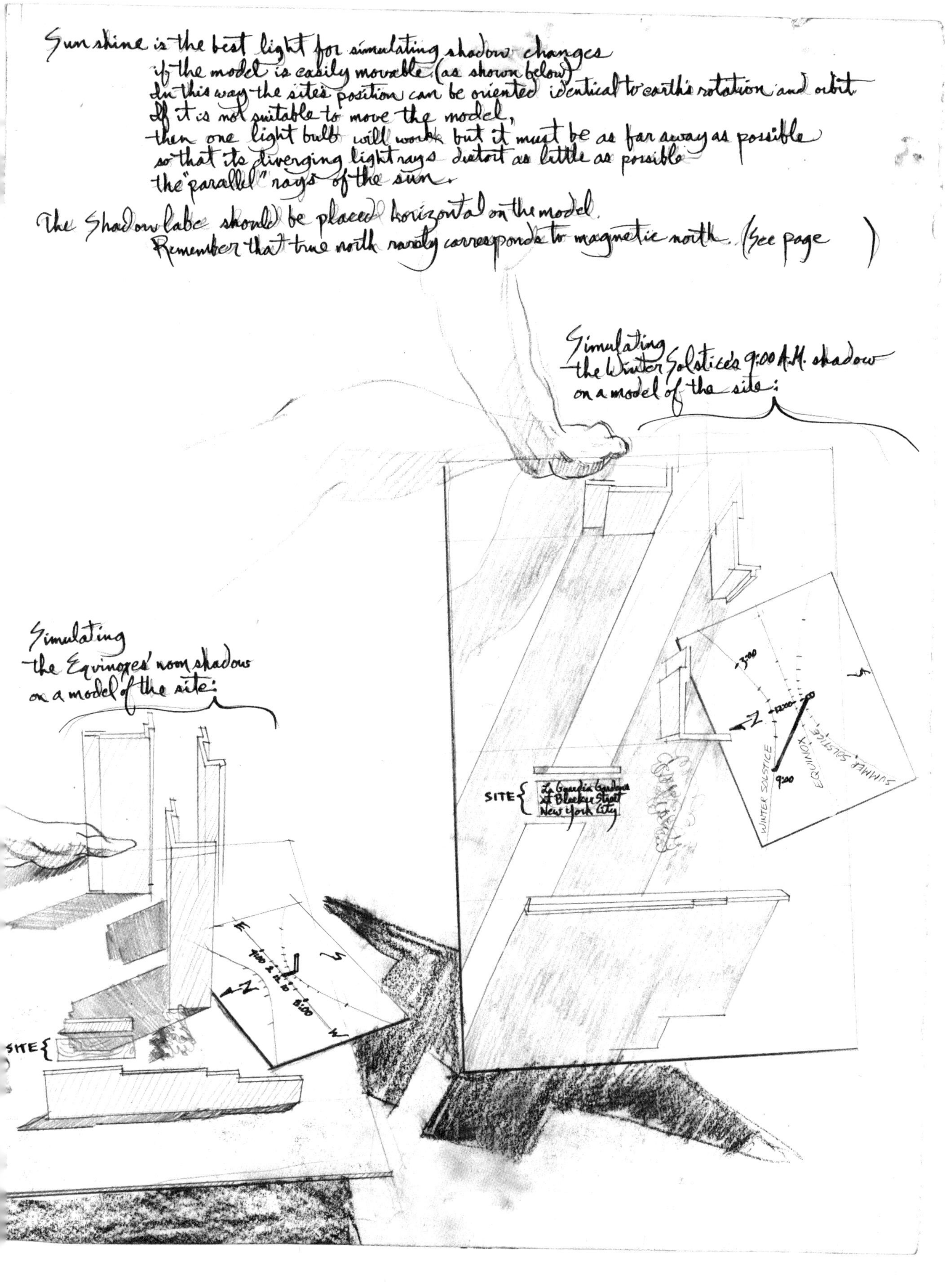

Sunshine is the best light for simulating shadow changes
if the model is easily movable (as shown below)
In this way the site's position can be oriented identical to earth's rotation and orbit
If it is not suitable to move the model,
then one light bulb will work but it must be as far away as possible
so that its diverging light rays distort as little as possible
the "parallel" rays of the sun.
The Shadow labe should be placed horizontal on the model.
Remember that true north rarely corresponds to magnetic north. (See page)
Simulating the Winter Solstice's 9:00 A.M. shadow on a model of the site:
SITE
La Guardia Gardens
at Bleeker Street
New York City
3:00
12:00
9:00
N
S
WINTER SOLSTICE
EQUINOX
SUMMER SOLSTICE
Simulating the Equinoxes' noon shadow on a model of the site:
SITE
E
W
S
N

24. The greater the latitude, the longer the shadowcaster's shadows.

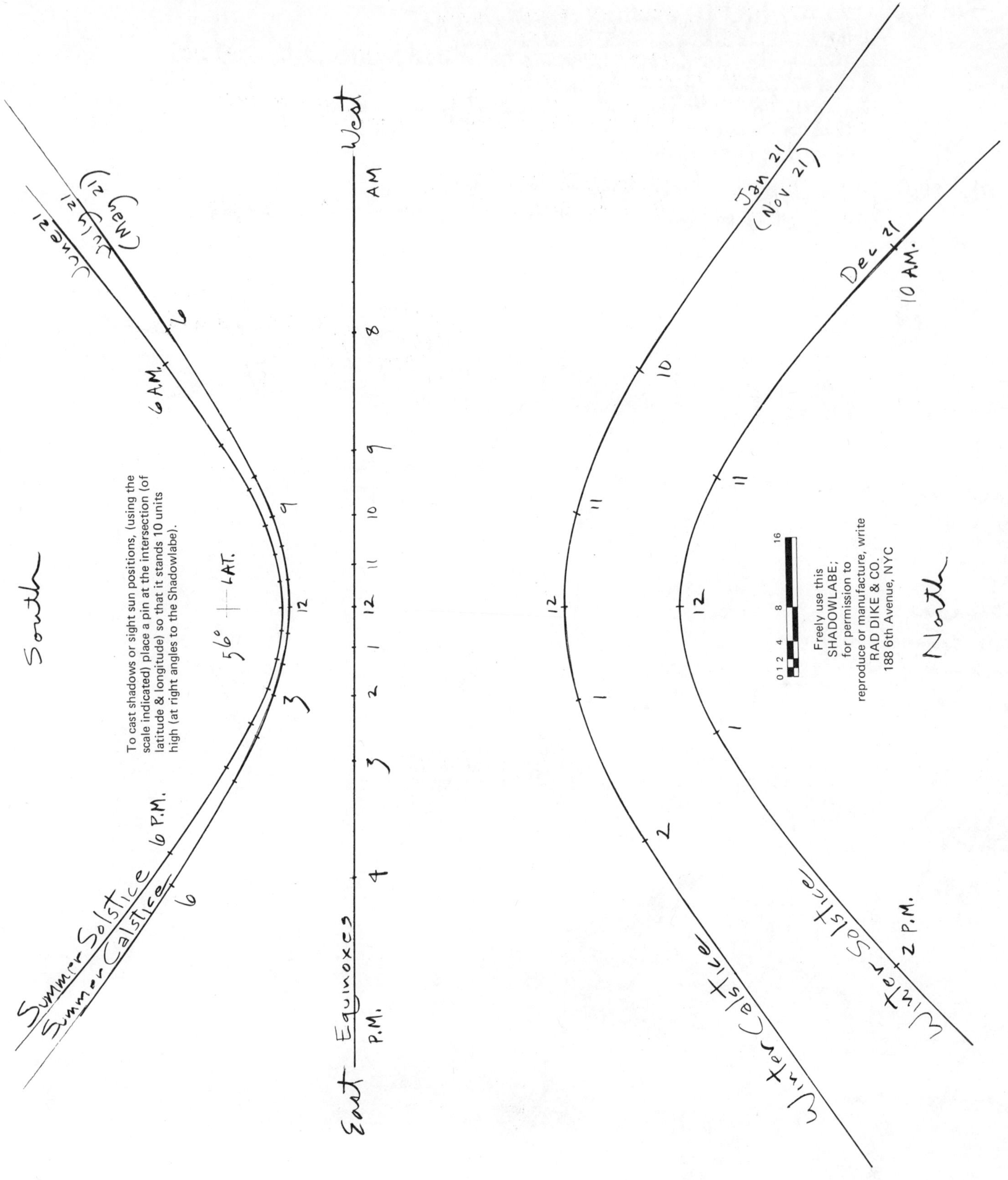

25. The lesser the latitude, the shorter the shadowcaster's shadows.

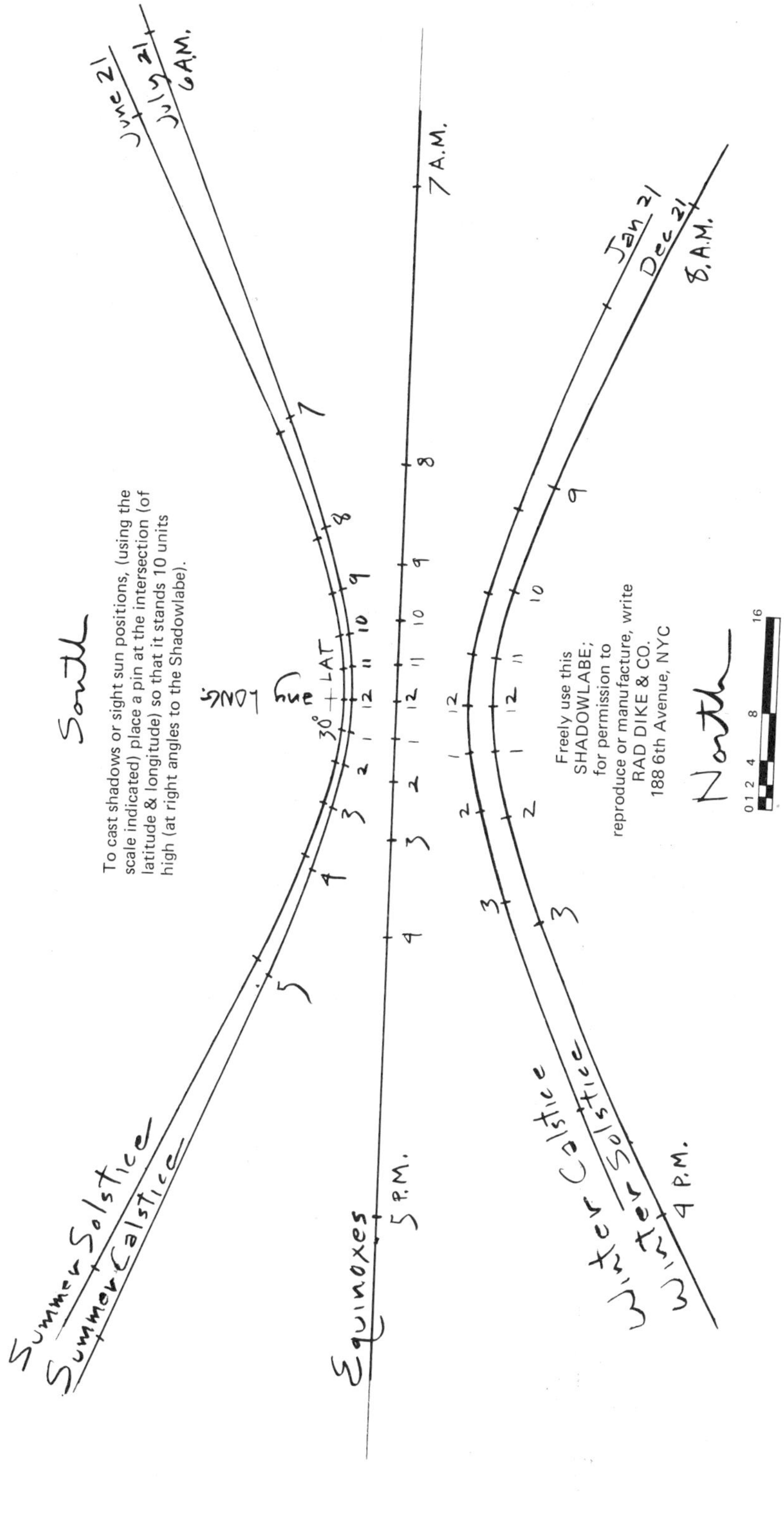

26. The calstices are the extreme heat points of the year, just as the solstices are the extreme angle points of the year. The equicals are the midpoints of the heat extremes, just as the equinoxes are the midpoints of the sun angle extremes.

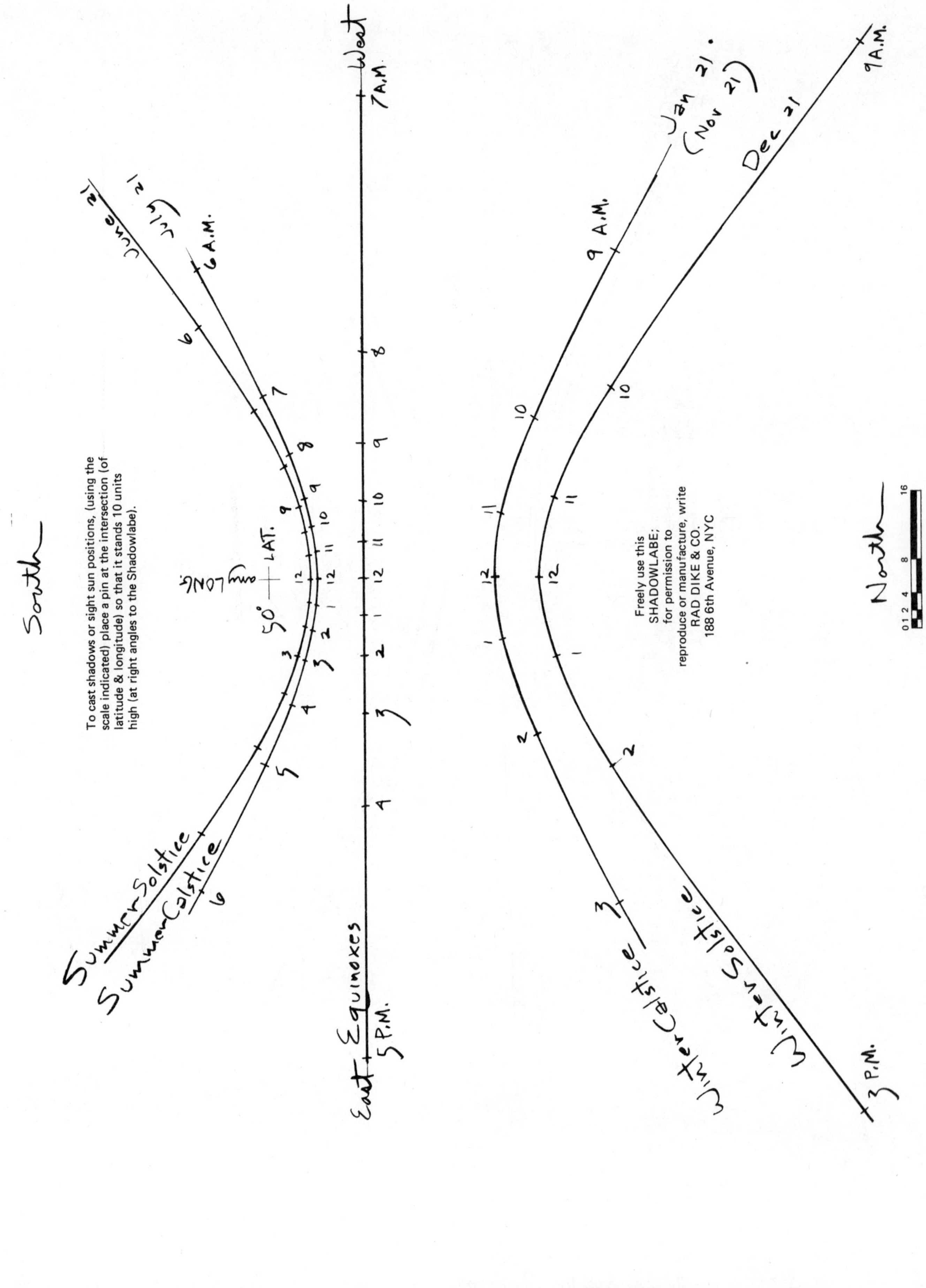

27. Five lines on the shadowlabe show the shadowcaster's lengths through six of the most architecturally significant days of the year. (2 equinoxes = one line, 2 solstices, and 2 calstices)

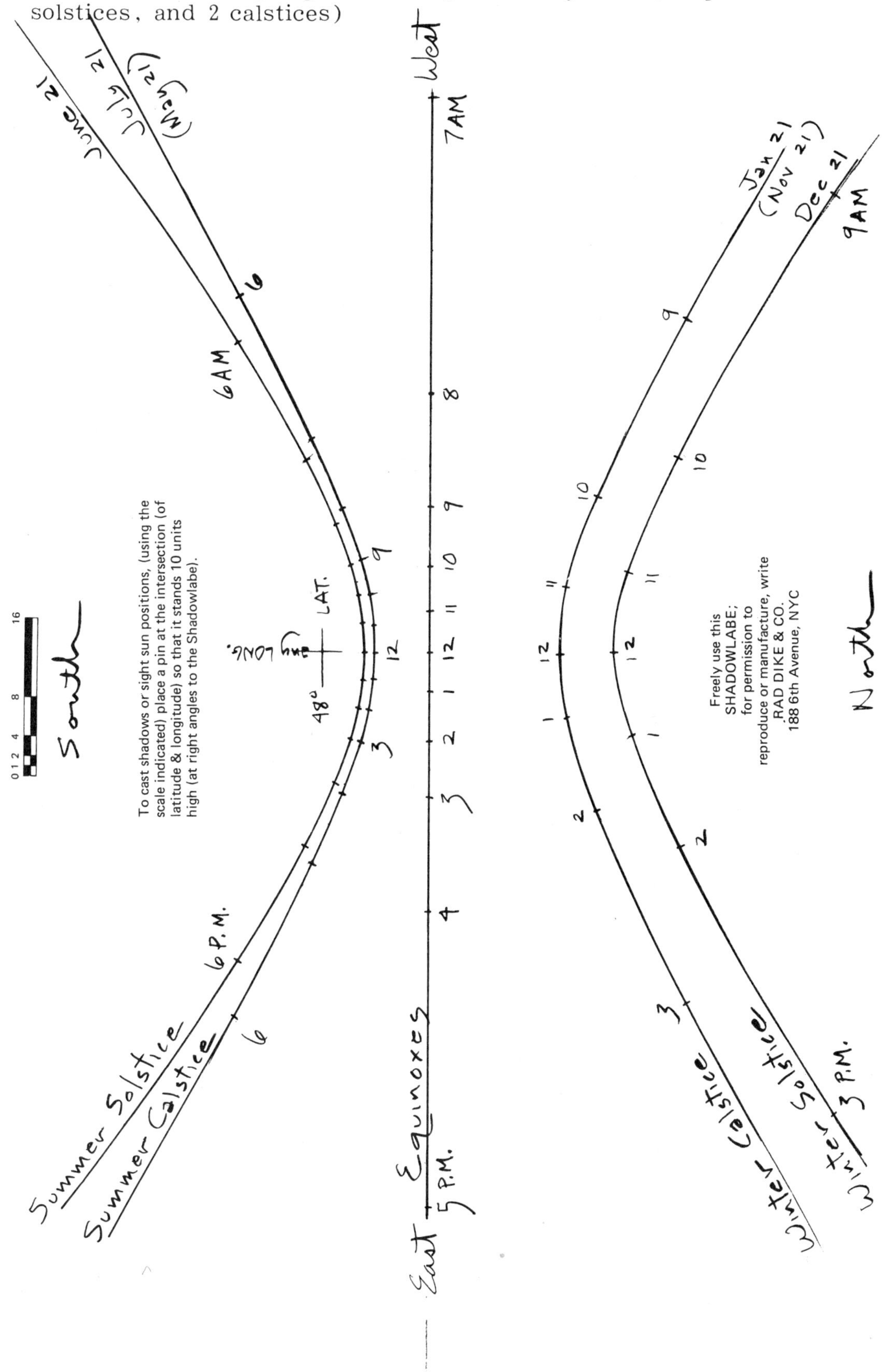

28. The summer sun moves around the shadowcaster in an arc from the northeast to the northwest.

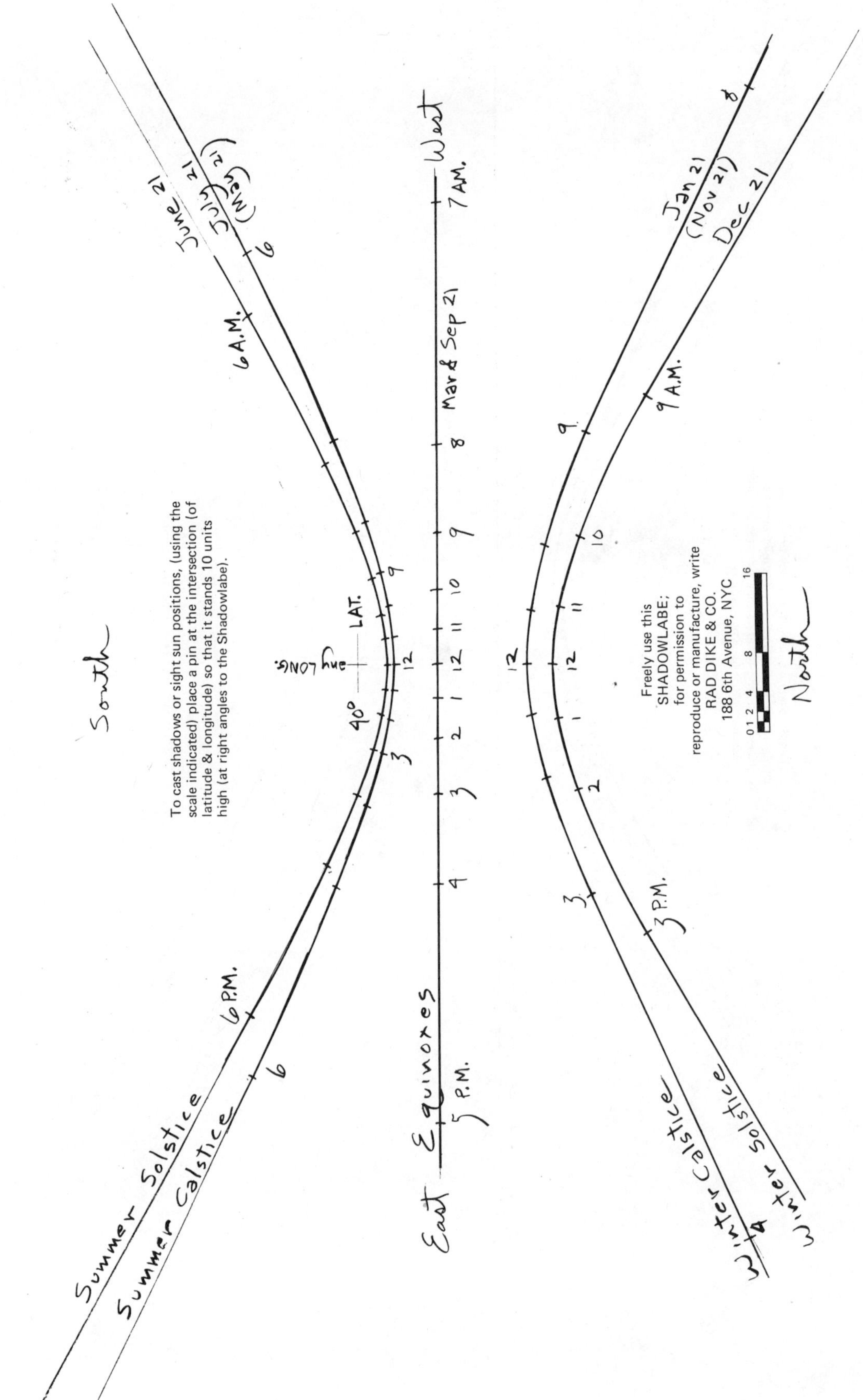

29. The winter sun's arc moves towards and away from the shadowcaster from the southeast to the southwest.

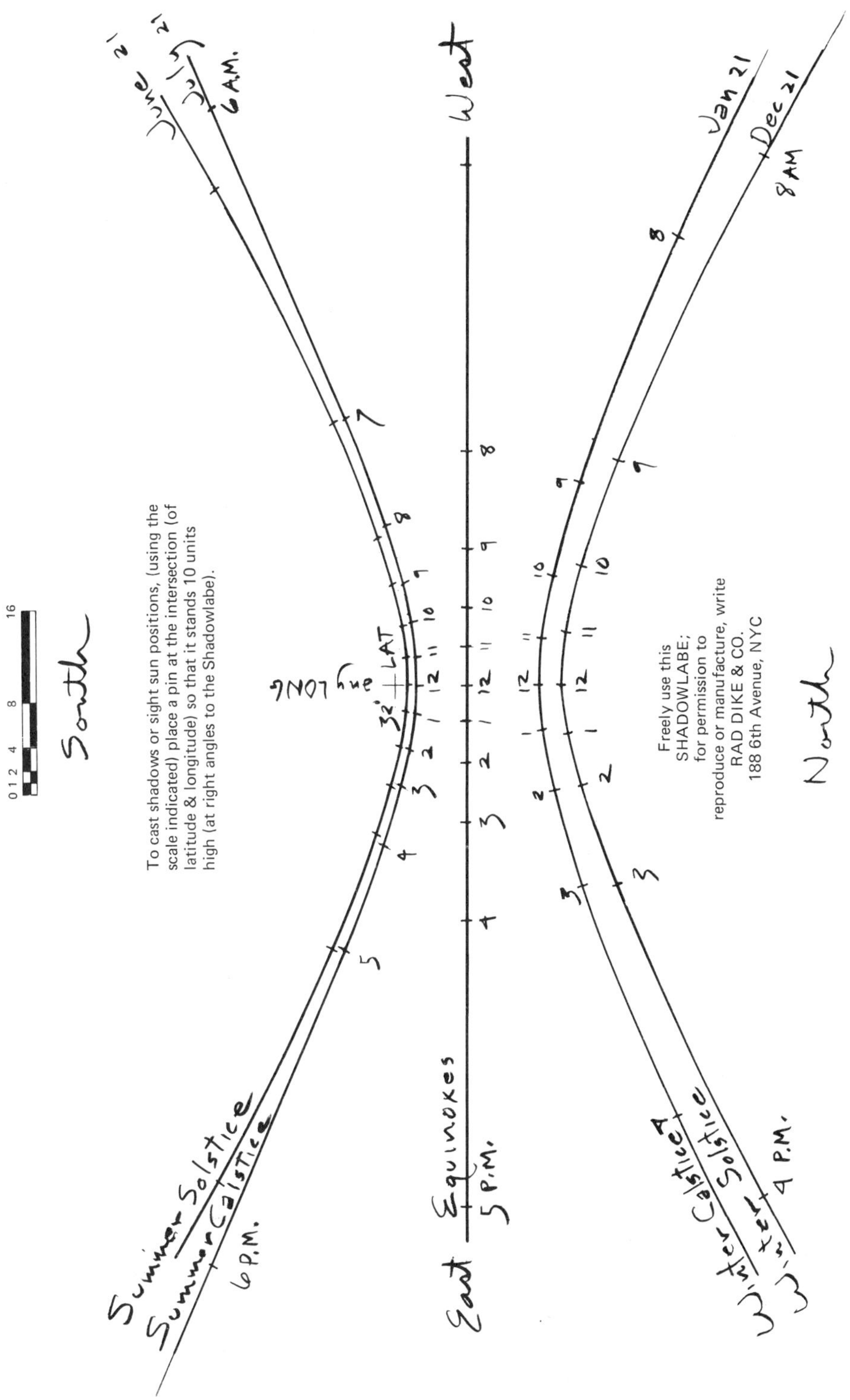

30. Sunlight and the angle at which it strikes the earth are measurable as heat potential to change temperature.

SOLAR RADIATION

is primarily measured in LANGELEYS/MINUTE

ROUGH DISTRIBUTION OF SOLAR ENERGY IN 100'S OF HOURS/YEAR

I love these kinds of forms of invisible information. Every local site has them and a responsive design will be shaped with them.

ONE LANGELY = ONE CALORIE OF RADIATION ENERGY PER ONE SQUARE CENTIMETER

MEASUREMENT BASICS

ONE CALORIE IS THE AMOUNT OF HEAT WHICH RAISES THE TEMPERATURE OF ONE GRAM OF WATER ONE CENTIGRADE DEGREE.

(need a map of BTU's given off by body heat not just pop. densities.)

aerial thermal photos of N.Y.C. exist?

ONE GRAM OF WATER IS ABOUT THIS MUCH OF A SPHERE:

ONE SQUARE CENTIMETER IS THIS MUCH:
(30 cm = 1 FT 900 sq cm = 1 sq ft.)

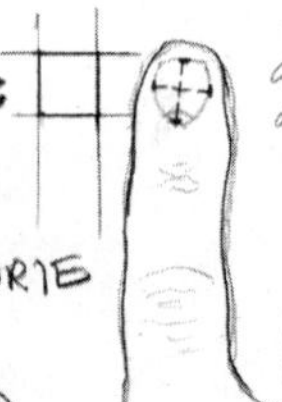

call Koefoed/Swithness re-research assistance without contract...? with contract.

ONE KILOCALORIE OR ONE KILOGRAM CALORIE IS 1000 CALORIES.

ANNUAL # Sunshine Hours

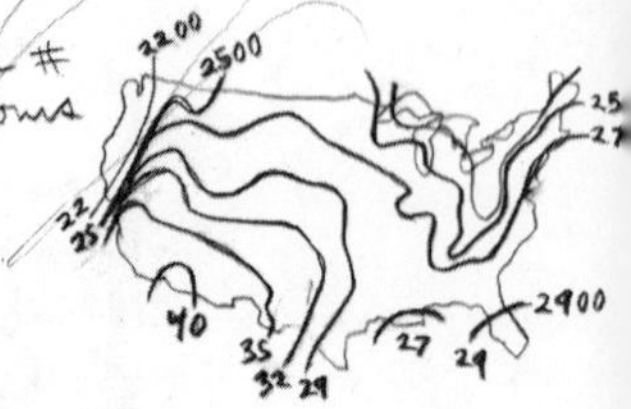

ONE BTU (BRITISH THERMAL UNIT) IS THE AMOUNT OF HEAT WHICH RAISES THE TEMPERATURE OF ONE POUND OF WATER ONE DEGREE FARENHEIT.

1 BTU = 453.6 × 5/9 CAL = 252 CAL
(explain)

knowing that 1 BTU of heat raises one pound of water one farenheit degree...
how can one determine how many calories are in one BTU?
knowing that 1 cal of heat raises one gram of water one centigrade degree...
there are 454 grams in one lb.
centigrade degrees are 5/9 farenheit degrees

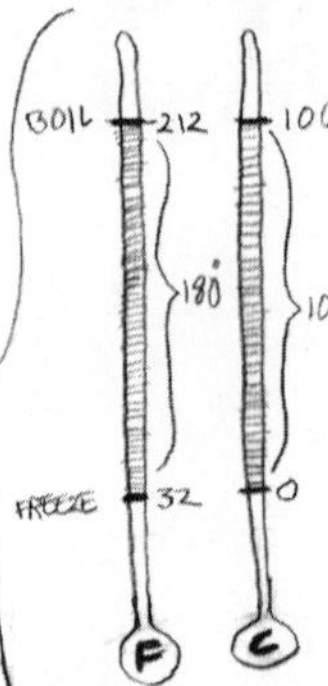

$\frac{C}{F}$ centigrade is to farenheit as 100 is to 180

$\frac{100}{180} = \frac{5}{9}$

the assumption is that both thermometers are made of the same materials used in the same universe of temperature laws.

so 1 BTU is so many calories...
BTU's are expressed in lbs per farenheit
Cals are expressed in grams per centigrade

1 BTU is so many grams per 1 pound per (1 degree F = 5/9 1 deg C)

$$1 \text{ BTU} = \frac{454 \text{ grams}}{1 \text{ pound}} \times \frac{1 \text{ pound}}{1^\circ F} \times \frac{1^\circ C}{1 \text{ gram}}$$

$$= 454 \times \frac{1^\circ C}{1^\circ F}\left(\frac{5}{9}\right)$$

$$1 \text{ BTU} = 455 \times \frac{5}{9} = 252 \text{ what?}$$

(calories, but where does label come from equation?)

drawing Alves) Do a geometric

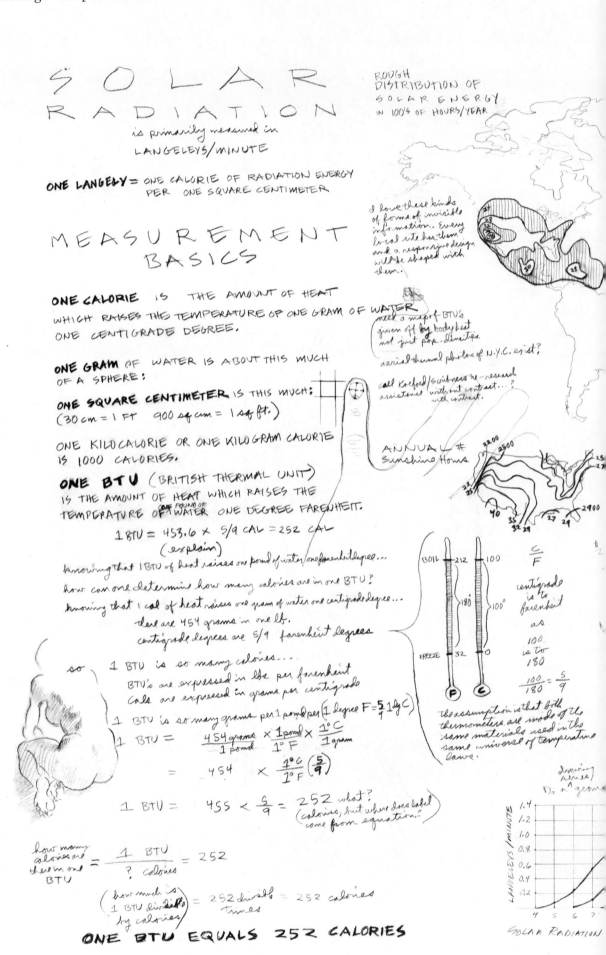

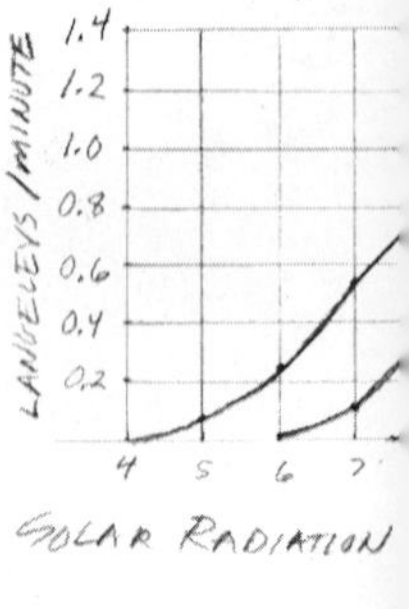

$$\text{how many calories are there in one BTU} = \frac{1 \text{ BTU}}{? \text{ calories}} = 252$$

(how much is 1 BTU divisible by calories) = 252 divisible times = 252 calories

ONE BTU EQUALS 252 CALORIES

31. The area of sunlight and its angle define the maximum energy (in this case, heat as BTUs) available on a site from the sun at a certain time.

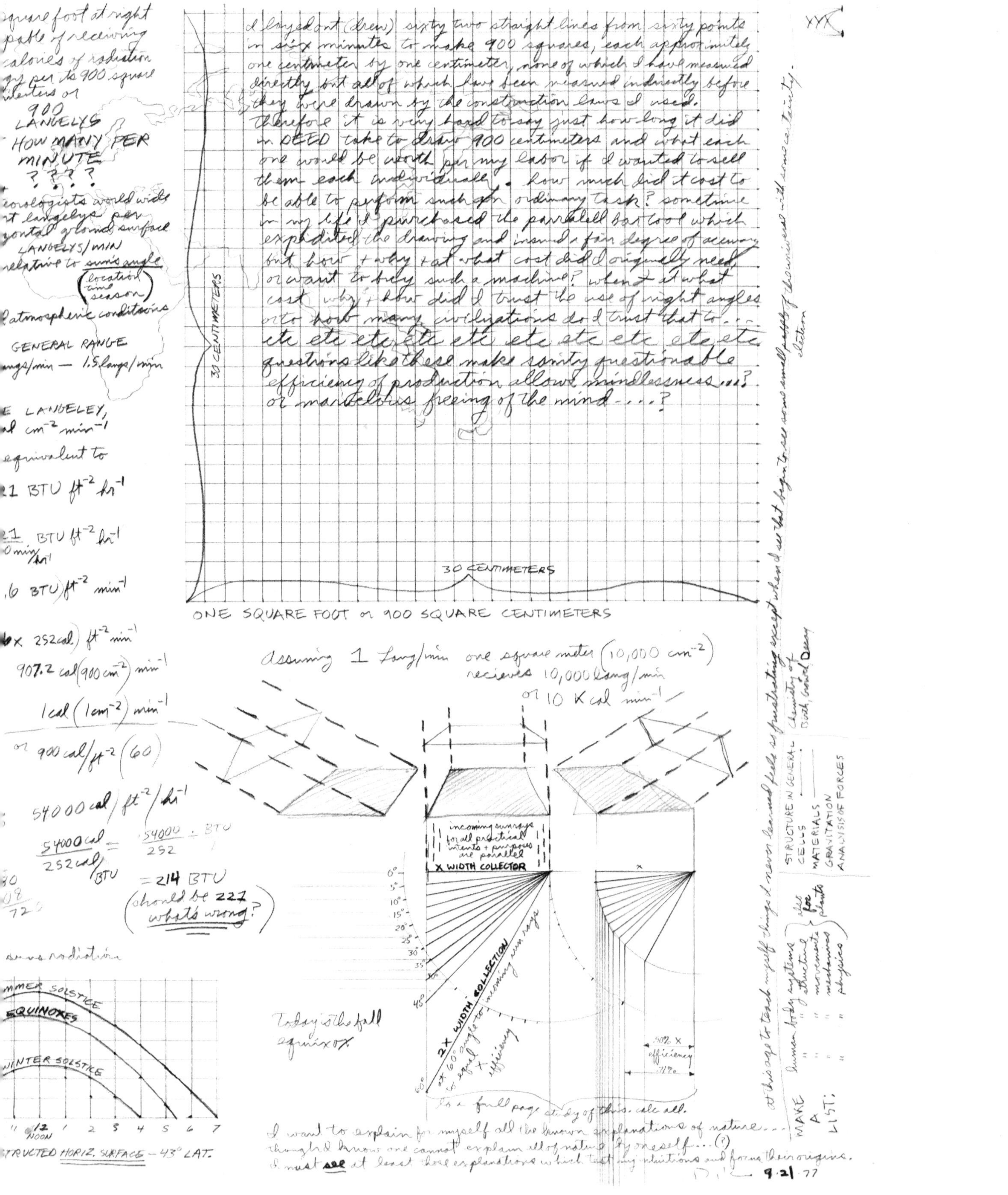
I layed out (drew) sixty two straight lines from sixty points in six minutes to make 900 squares, each approximately one centimeter by one centimeter, none of which I have measured directly but all of which have been measured indirectly before they were drawn by the construction laws I used.
Therefore it is very hard to say just how long it did in DEED take to draw 900 centimeters and what each one would be worth per my labor if I wanted to sell them each individually. how much did it cost to be able to perform such an ordinary task? sometime in my life I purchased the parallel bar tool which expedited the drawing and insured a fair degree of accuracy but how + why + at what cost did I originally need or want to buy such a machine? when + at what cost why + how did I trust the use of right angles or to how many civilizations do I trust that to...
etc etc etc etc etc etc etc etc etc etc
questions like these make sanity questionable
efficiency of production allows mindlessness...?
or marvelous freeing of the mind...?
30 CENTIMETERS
30 CENTIMETERS
ONE SQUARE FOOT or 900 SQUARE CENTIMETERS
HOW MANY PER MINUTE ????
LANGELYS/MIN
GENERAL RANGE
Assuming 1 Lang/min one square meter (10,000 cm⁻²) recieves 10,000 lang/min or 10 K cal min⁻¹
= 214 BTU (should be 227 what's wrong?)
incoming sunrays for all practical intents + purposes are parallel
X WIDTH COLLECTOR
2X WIDTH COLLECTION
Today is the fall equinox
SUMMER SOLSTICE
EQUINOXES
WINTER SOLSTICE
STRUCTURE IN GENERAL
CELLS
MATERIALS
GRAVITATION
ANALYSIS OF FORCES
MAKE A LIST:
I want to explain for myself all the known explanations of nature...
9.21.77

32. For every angle at which the sun strikes a slope, a maximum amount of BTU's has been calculated by ASHRAE, AIA, and others.

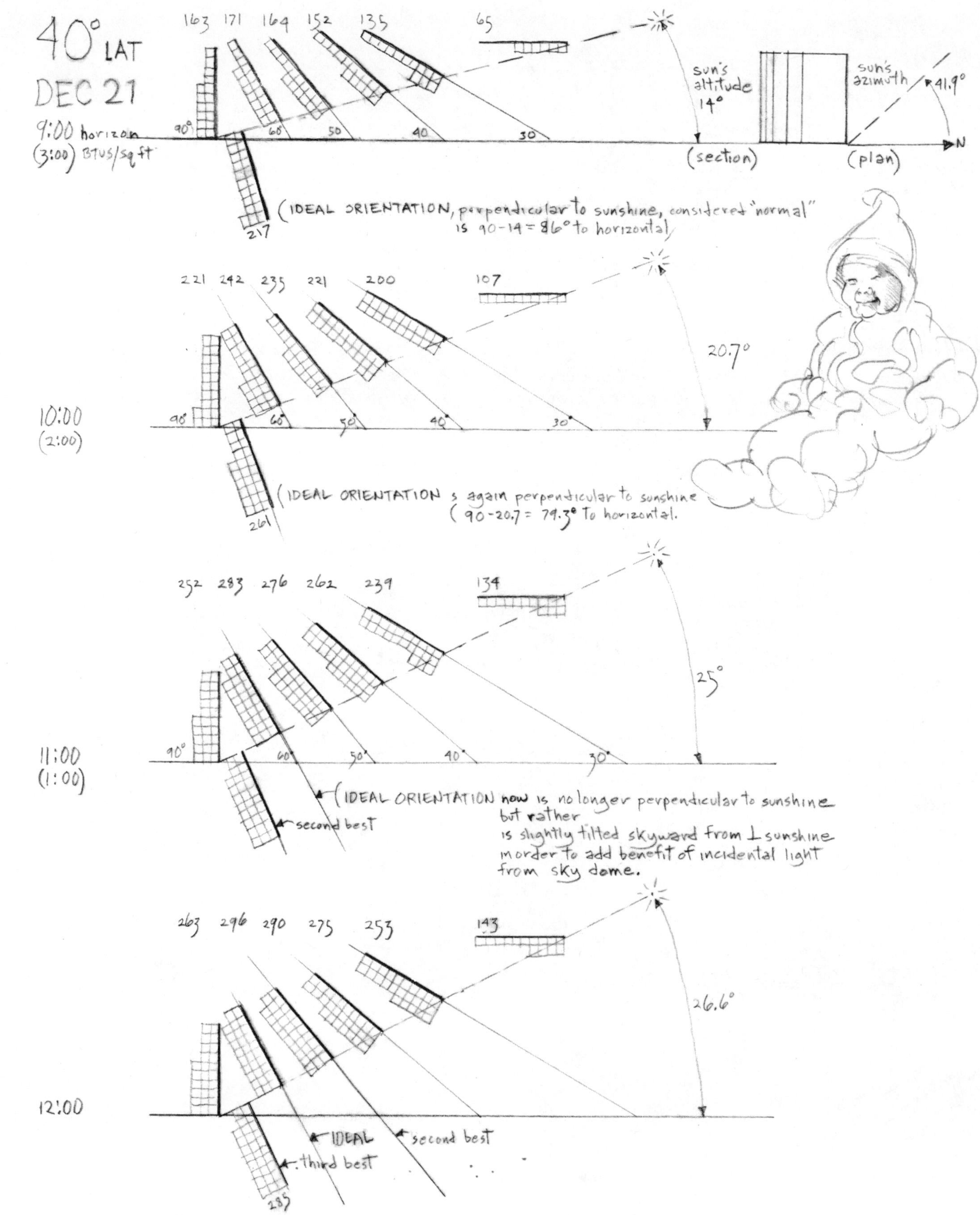

33. The diagrams at left show a variety of slopes (0°, 30°, 40°, 50°, 60°, 90°) per each hour of sunshine. Each square on the slope represents 10 BTU.

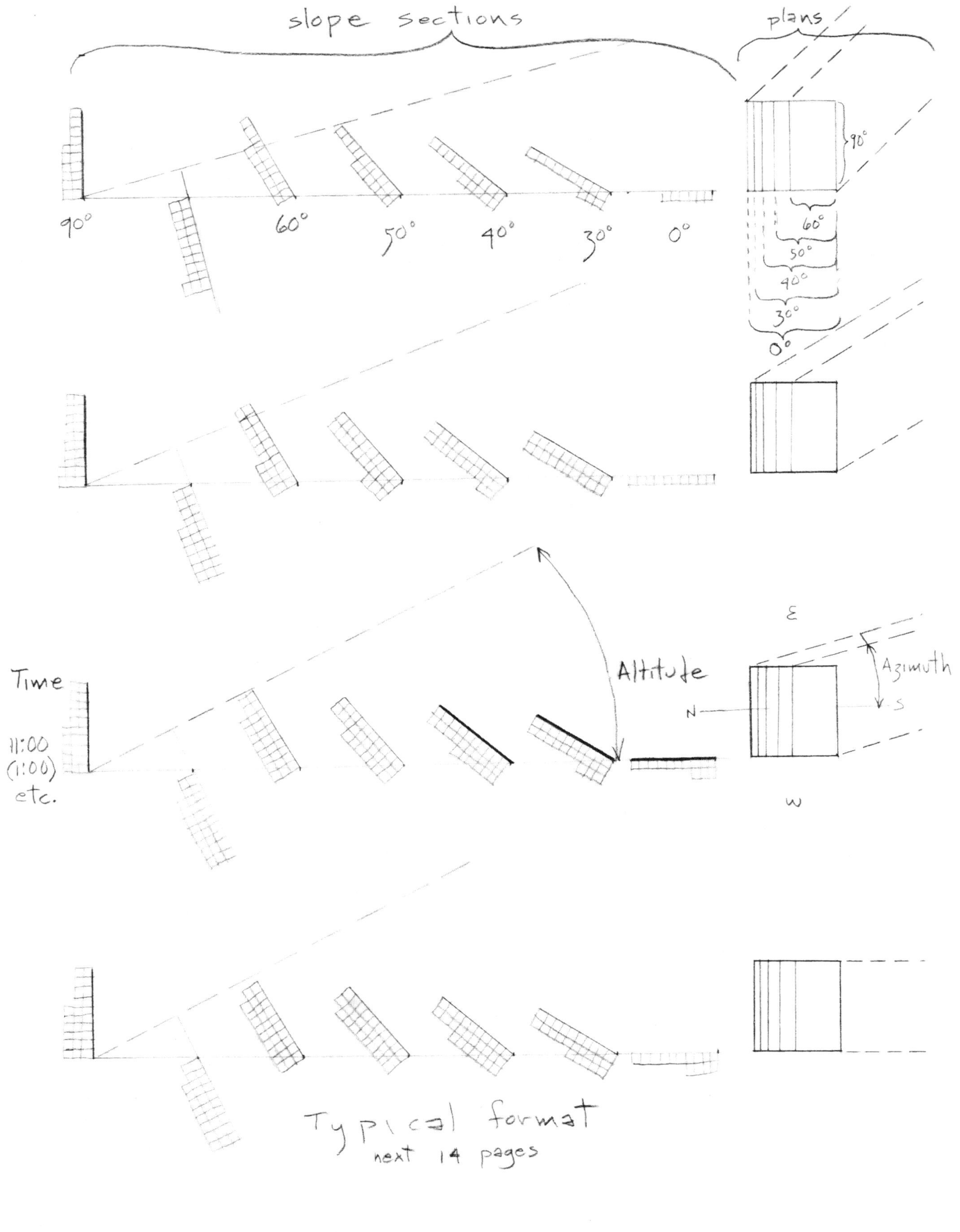

34. The plans of these slopes are shown square with true north, which is not the case, for instance, on New York's streets and avenues.

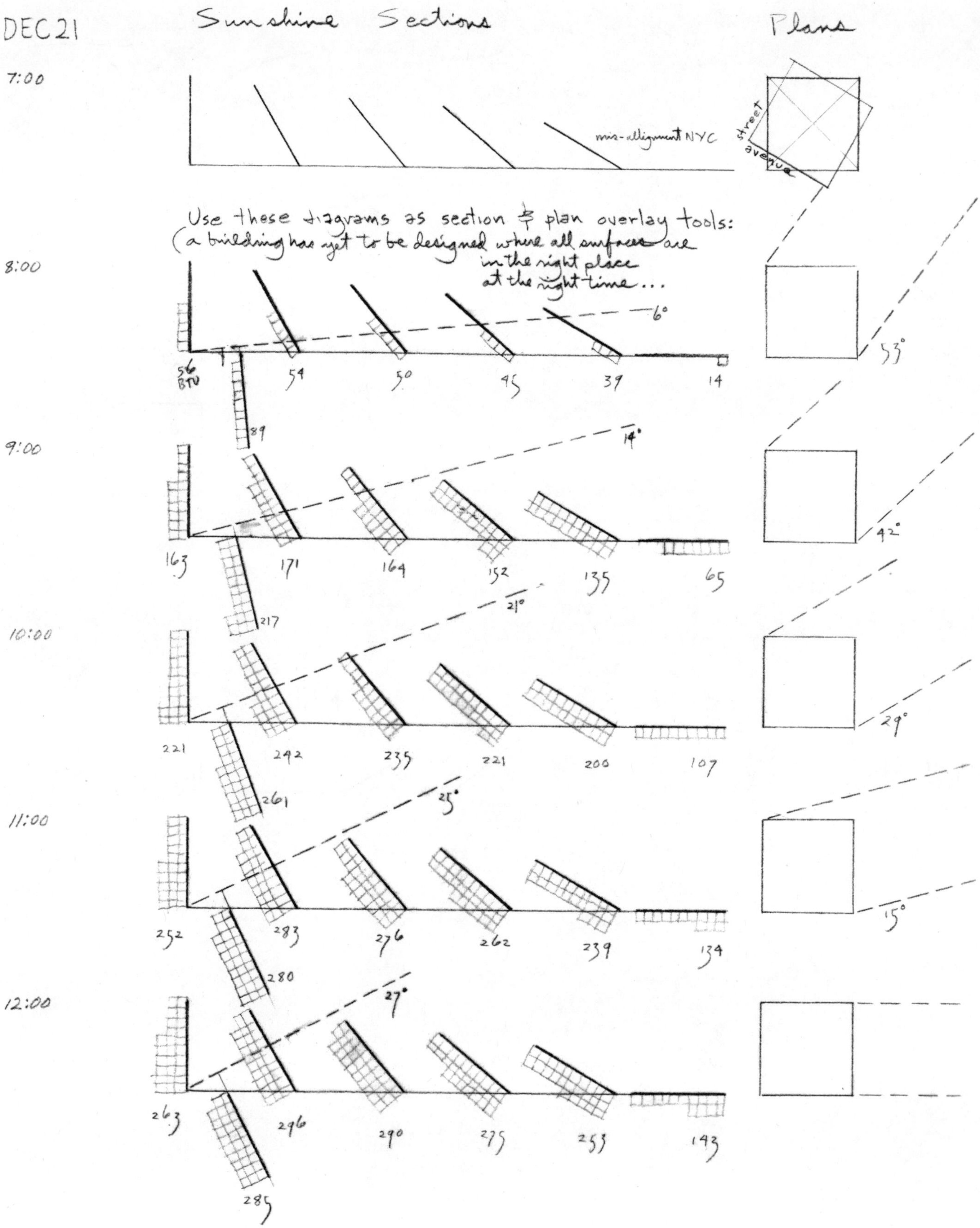

35. Imagine designs that accomodate the complexity of positions ideal for gathering sunshine.

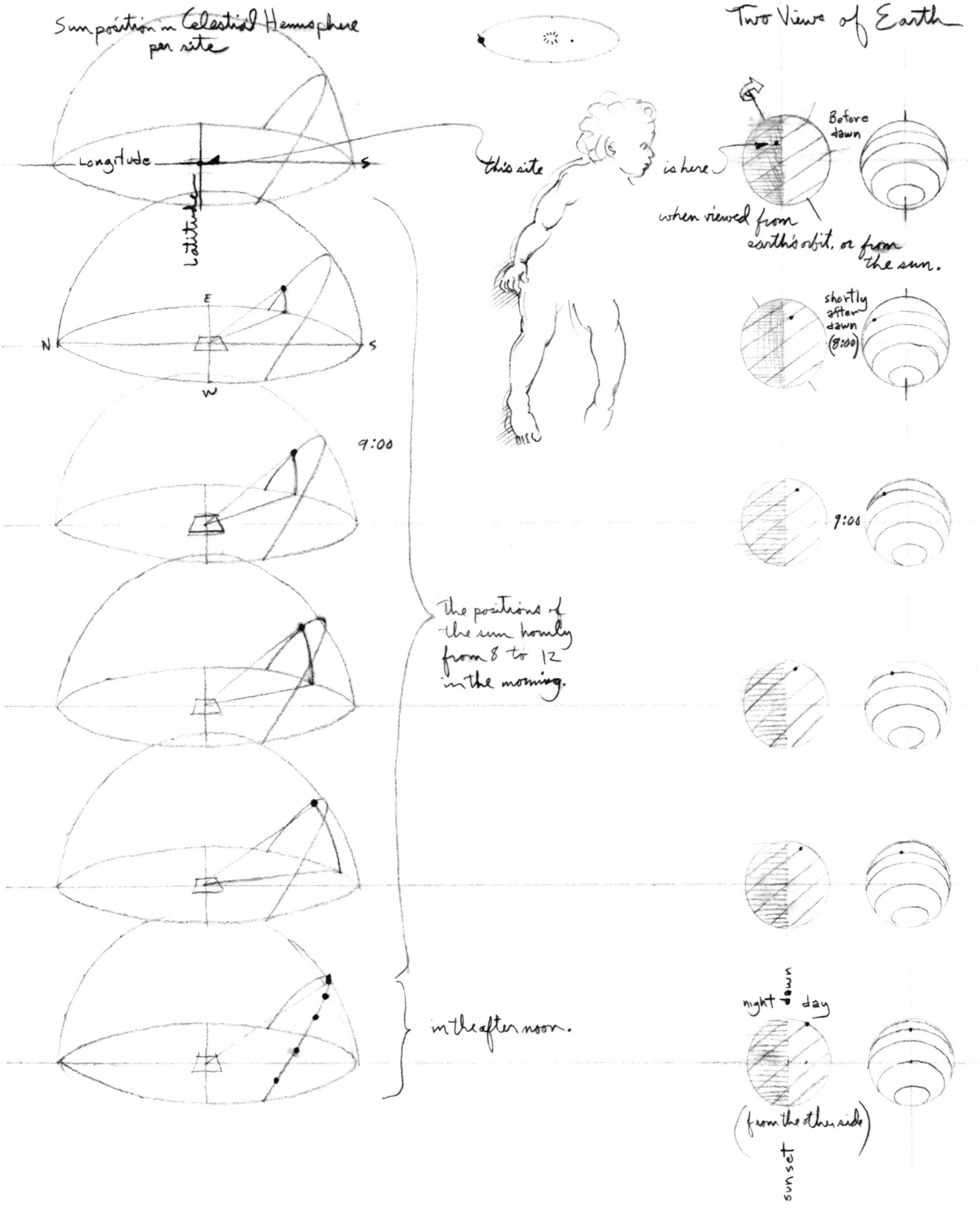

36. Though these diagrams are shown for forty degrees latitude, they can be adapted to any latitude.

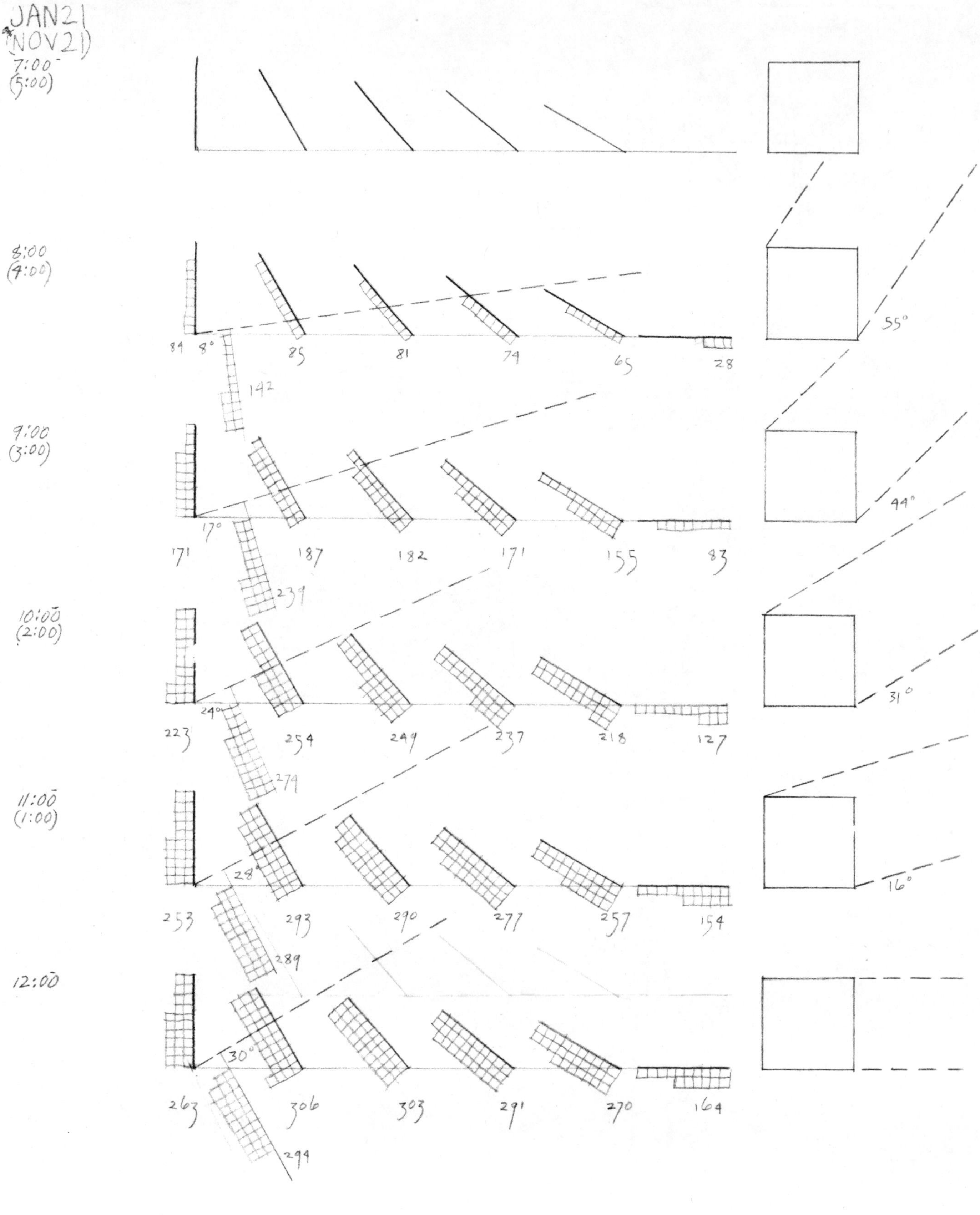

37. By filling in the uncompleted celestial hemispheres and views of the earth, the reader can familiarize himself with the sun's apparently complex relationship with earth.

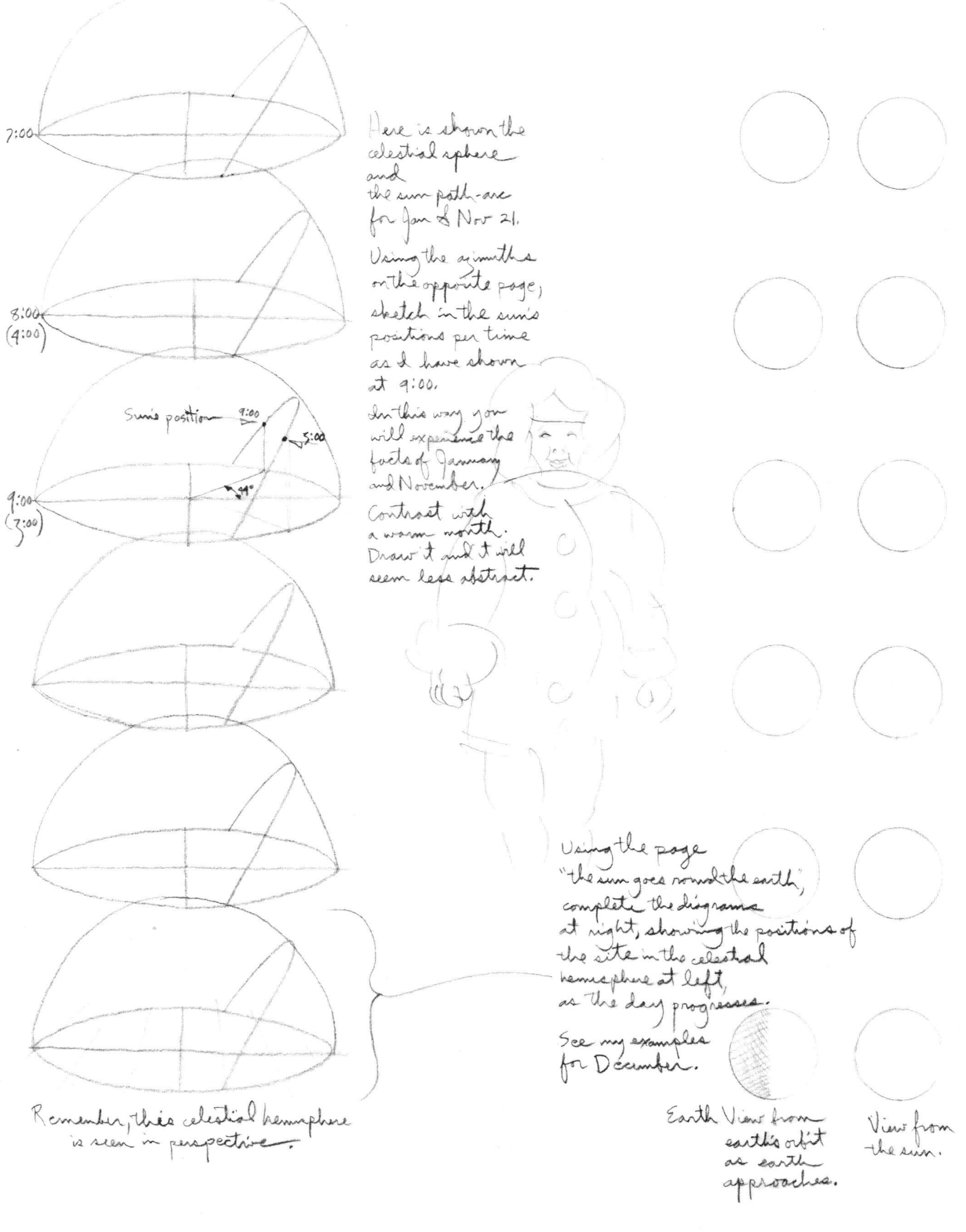
7:00
8:00
(4:00)
Sun's position 9:00
5:00
44°
9:00
(3:00)
Here is shown the celestial sphere and the sun path-arc for Jan & Nov 21.
Using the azimuths on the opposite page, sketch in the sun's positions per time as I have shown at 9:00.
In this way you will experience the facts of January and November.
Contrast with a warm month.
Draw it and it will seem less abstract.
Using the page "the sun goes round the earth", complete the diagrams at right, showing the positions of the site in the celestial hemisphere at left, as the day progresses.
See my examples for December.
Remember, this celestial hemisphere is seen in perspective.
Earth View from earth's orbit as earth approaches.
View from the sun.

38. The lowest sun angle is not exactly at the time of coldest temperatures (the winter calstice).

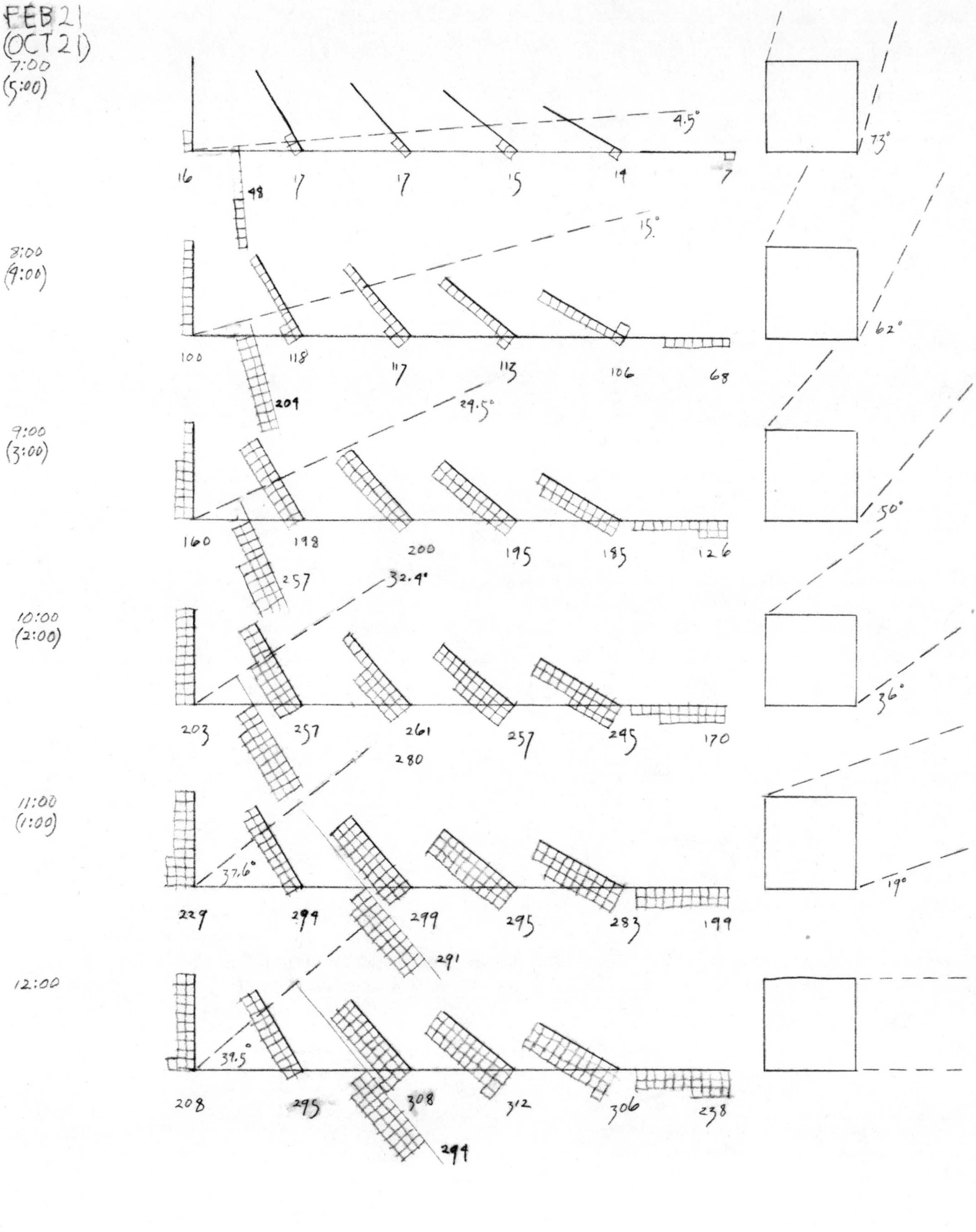

39. The calstices and equicals follow the solstices and equinoxes because of the earth's thermal lag.

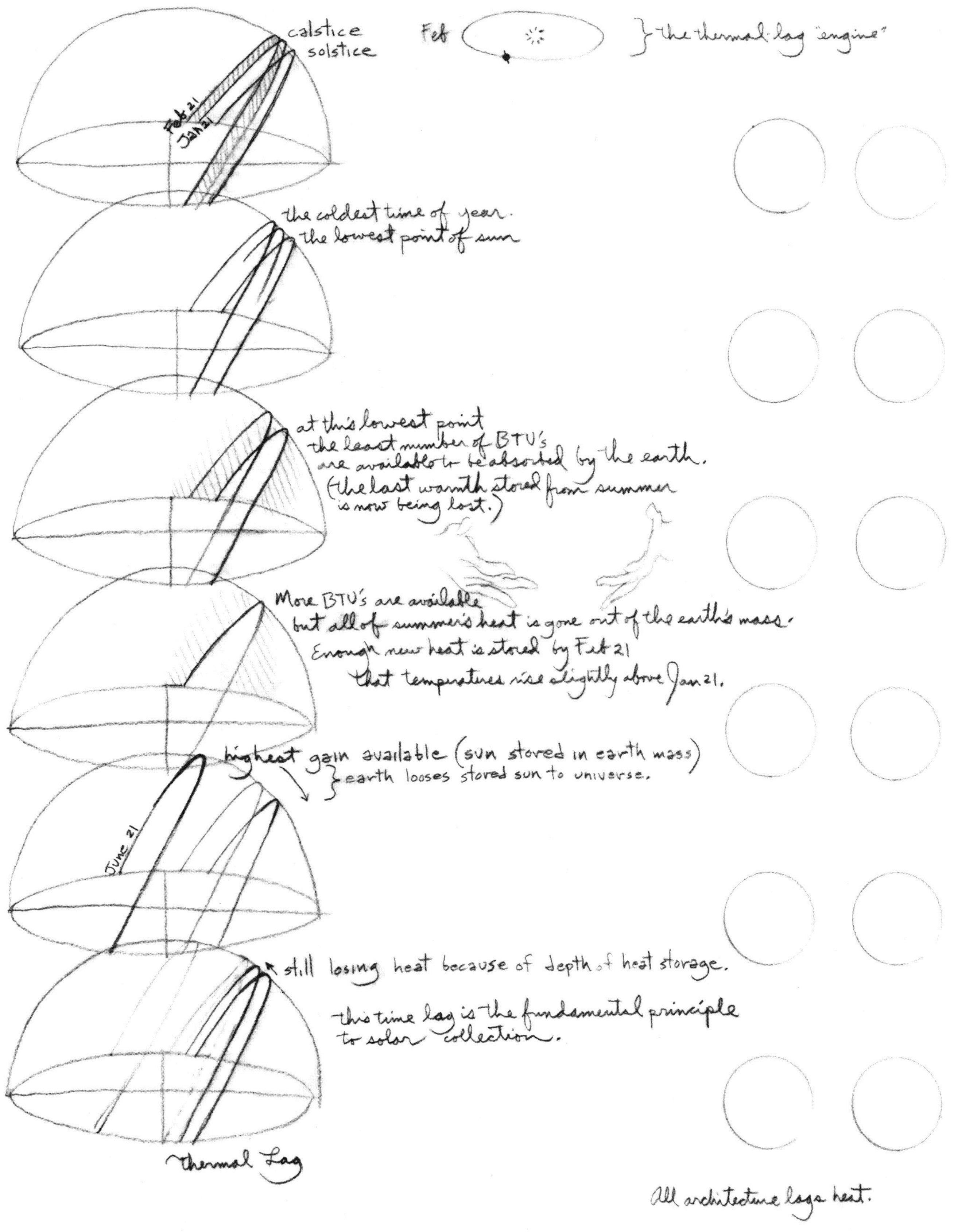

calstice
solstice
Feb
} the thermal lag "engine"
Feb 21
Jan 21
the coldest time of year.
the lowest point of sun
at this lowest point
the least number of BTU's
are available to be absorbed by the earth.
(the last warmth stored from summer
is now being lost.)
More BTU's are available
but all of summer's heat is gone out of the earth's mass.
Enough new heat is stored by Feb 21
that temperatures rise slightly above Jan 21.
highest gain available (sun stored in earth mass)
} earth looses stored sun to universe.
June 21
still losing heat because of depth of heat storage.
this time lag is the fundamental principle
to solar collection.
thermal Lag
All architecture lags heat.

40. The sun-arc angle changes quickly at the time of the equinoxes; the length of day changes noticeably from week to week.

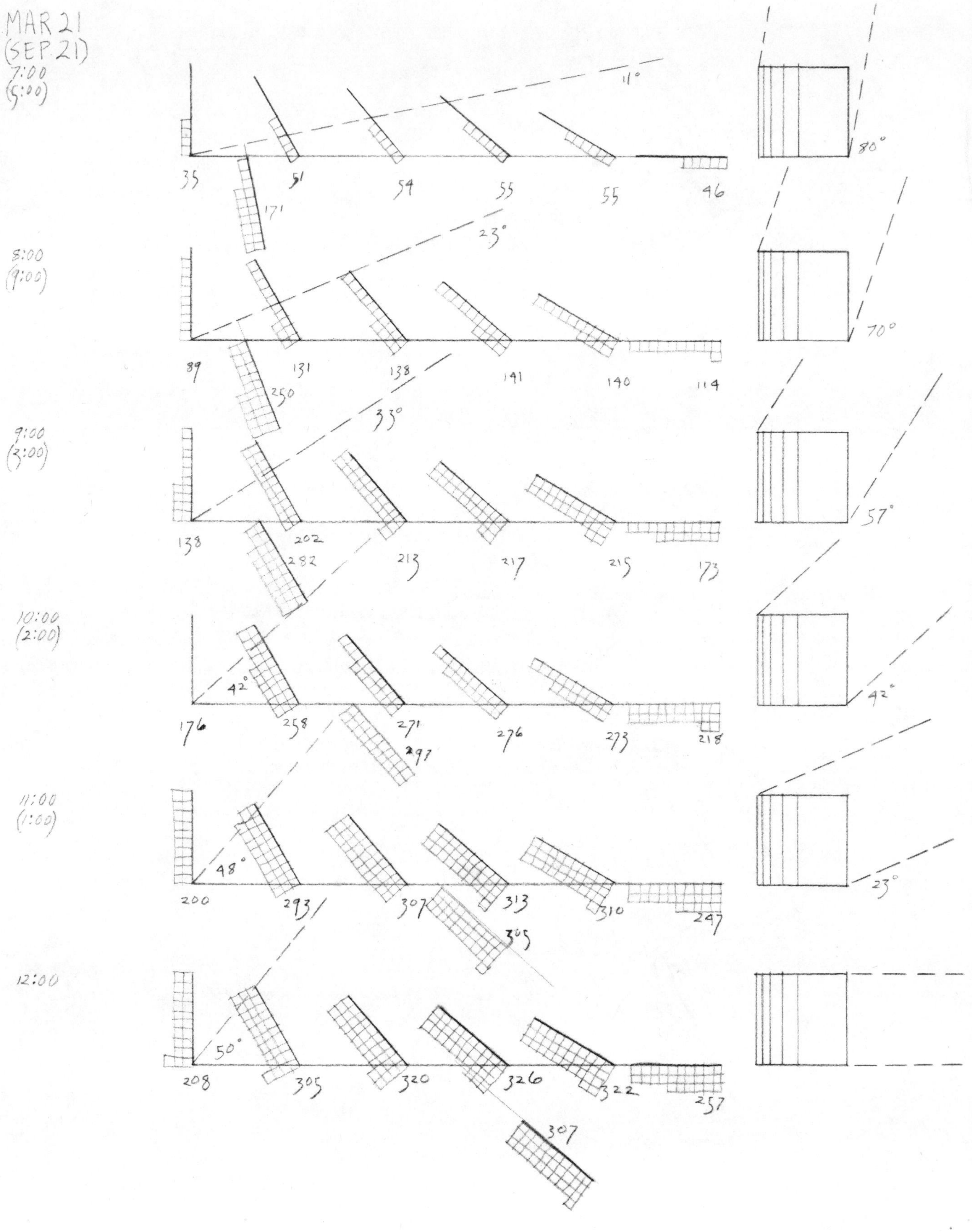

41. Any one thing casts equinox shadow lengths whose sum of endpoints through the day forms a straight line east and west.

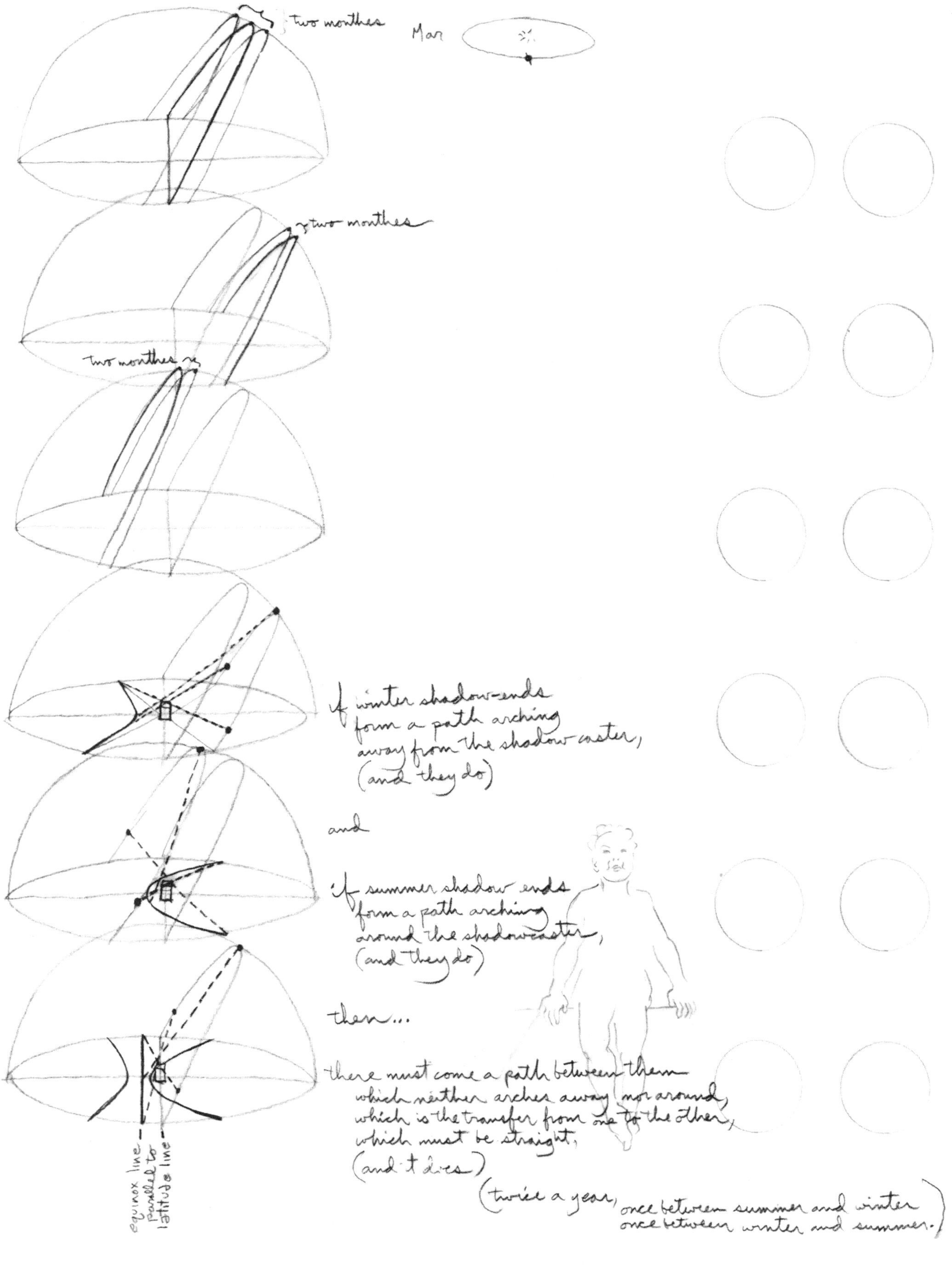
two monthes
Mar
two monthes
two monthes
if winter shadow-ends
form a path arching
away from the shadow-caster,
(and they do)
and
if summer shadow ends
form a path arching
around the shadowcaster,
(and they do)
then...
there must come a path between them
which neither arches away nor around,
which is the transfer from one to the other,
which must be straight,
(and it does)
(twice a year, once between summer and winter
once between winter and summer.)
equinox line
paralled to
latitude line

42. Sunlight is not uniform lighting.

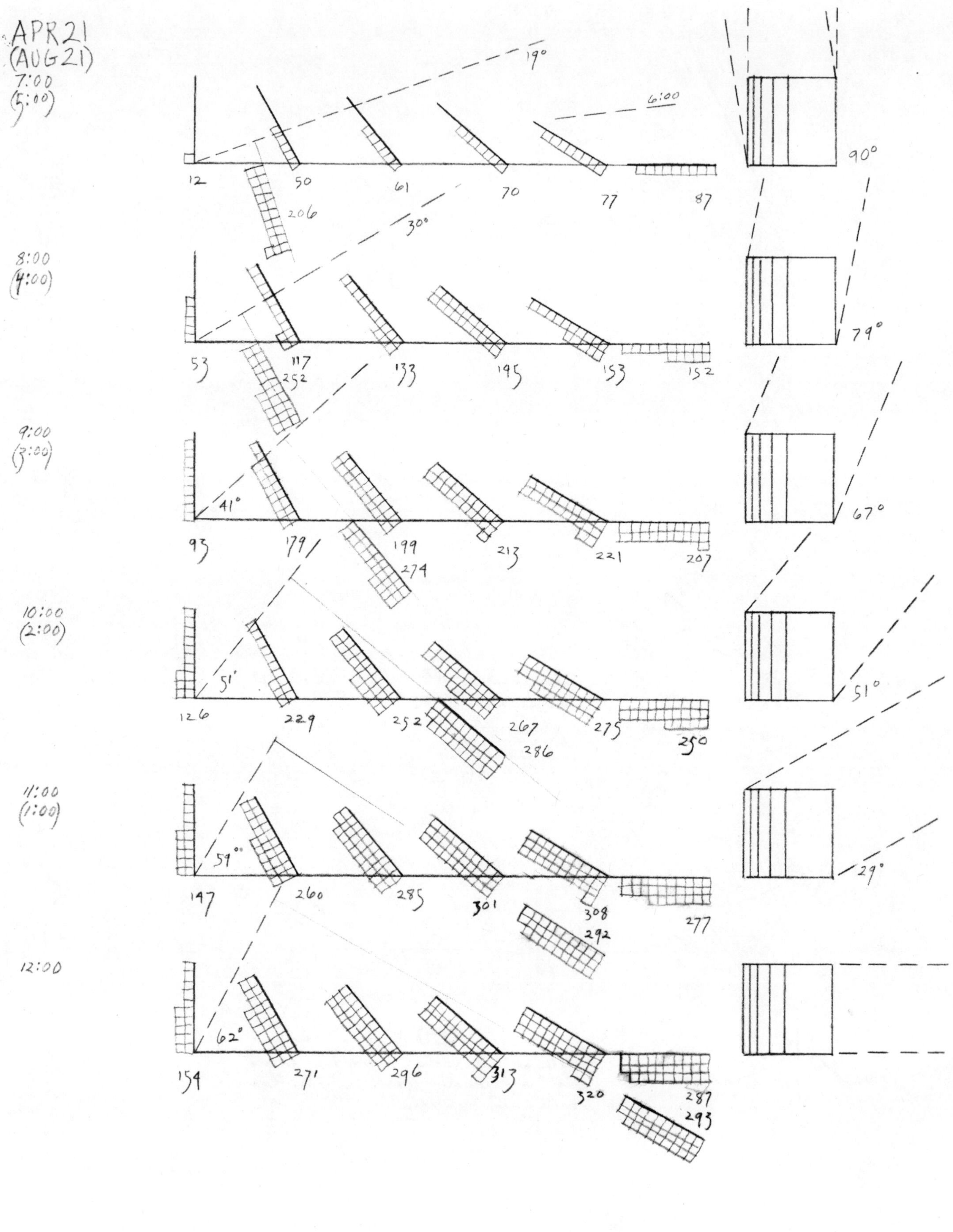

43. Orientation proportions shadows, contrasts, images, borders, diffusions, and photosynthesis.

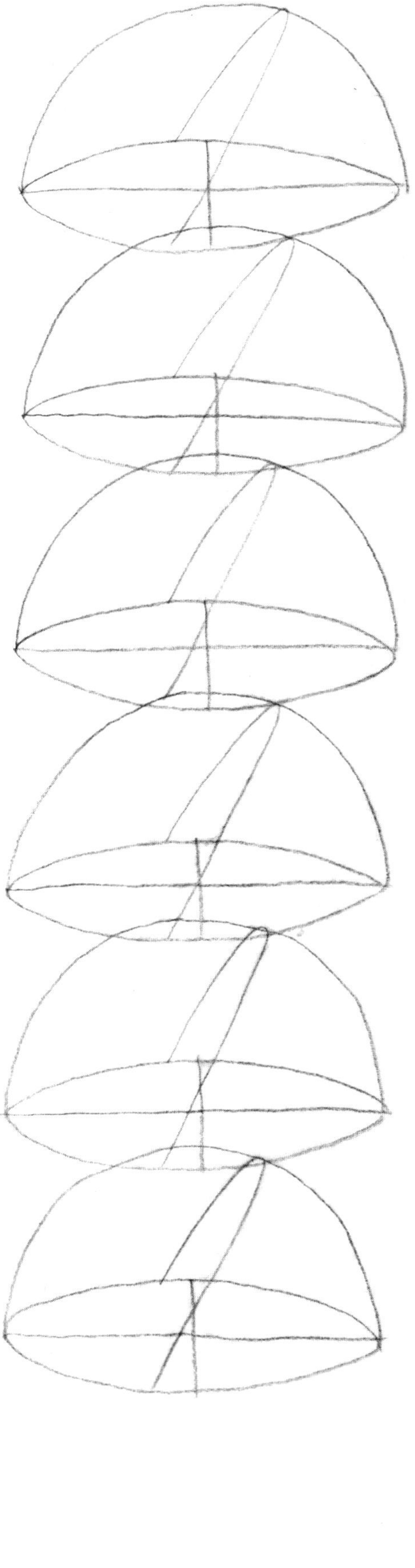

Sun light is not uniform lighting;
to this fact we may attribute
a society's non-uniformity or uniformity. (and dislike or like of evenly-bright modulor lighting)
(Even-outdoor-light is rarely seen where there is the dappling effect of leaves or ever changing clouds.

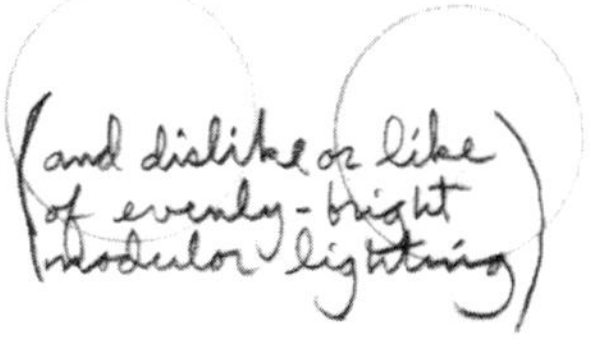

Direct light casts strong shadows,
strong contrasts and strong images.

Contrasts are reinforced by our eyes
when they fire and re-fire mechanically.
Contrasting stimuli create borders.
It is therefore our desire to visually
diffuse contrasts much as we prefer
diffused conflict.

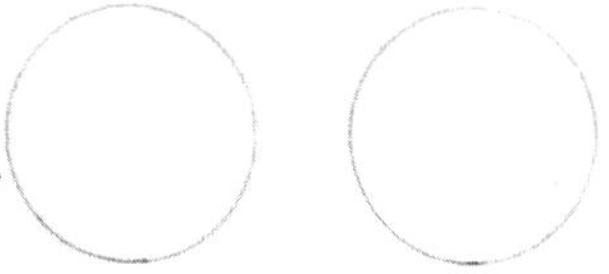

The contrast between light & darkness
results in glare at the boundary without a diffuser.

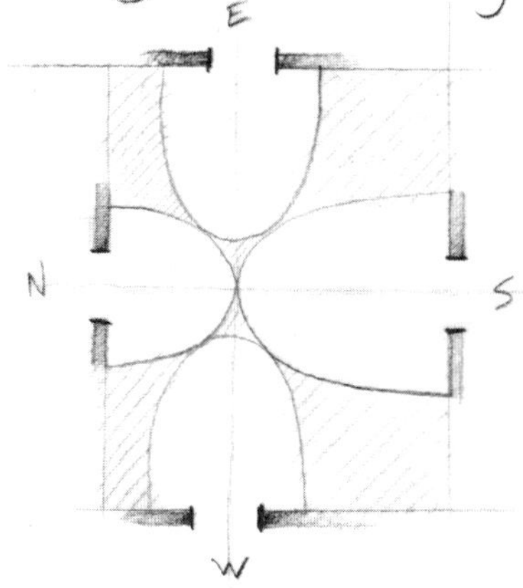

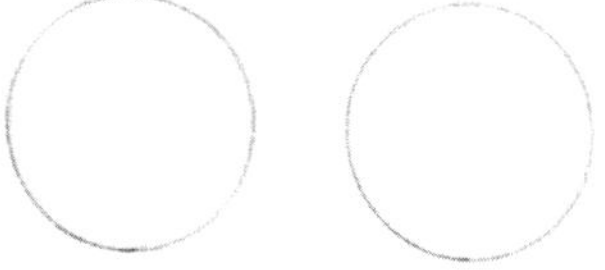

Plants will grow well in the areas
unshaded in this plan.

Leaves are natural diffusers;
Orientation naturally proportions photosynthesis.

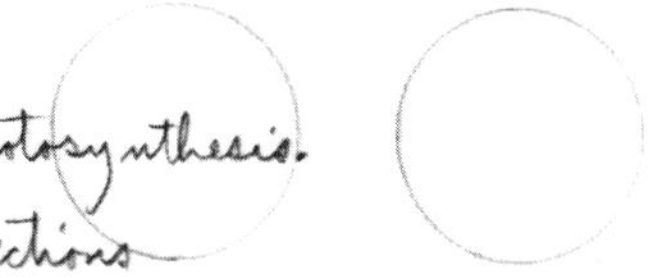

Natural light from different directions
(at least two)
automatically diffuses contrasts.

As the sun moves,
one source is always more dominant
which causes high-lights;
another source is subordinant
and gives highlights in the shade,
(the penumbra, passion in the shade)

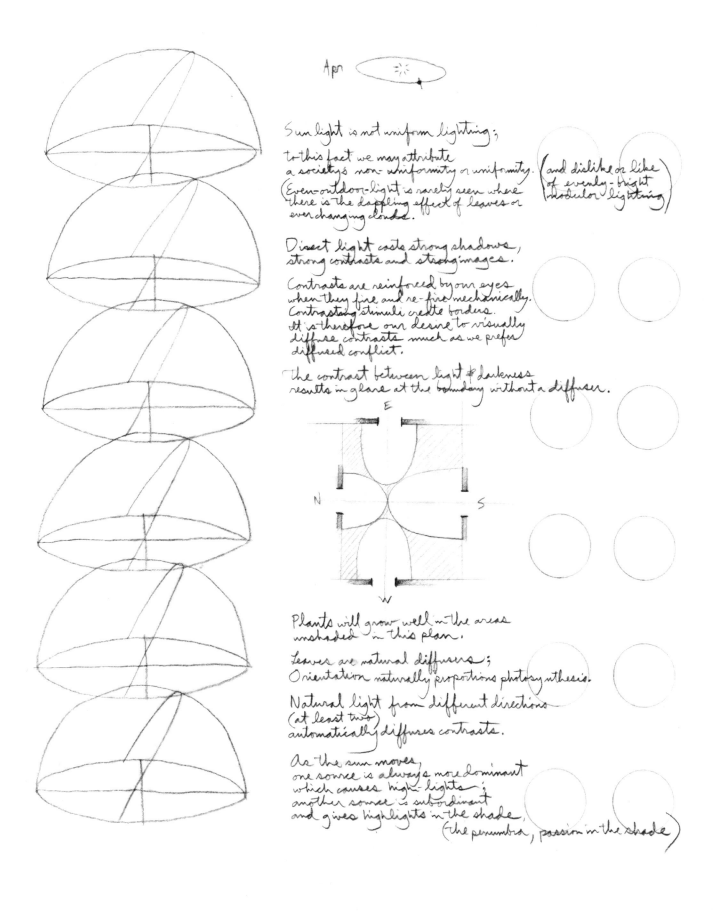

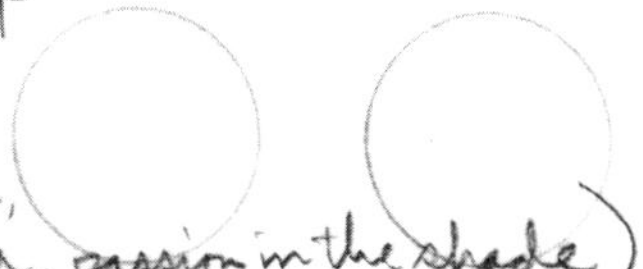

44. The highest sun angle is not exactly at the time of hottest temperatures (the summer calstice).

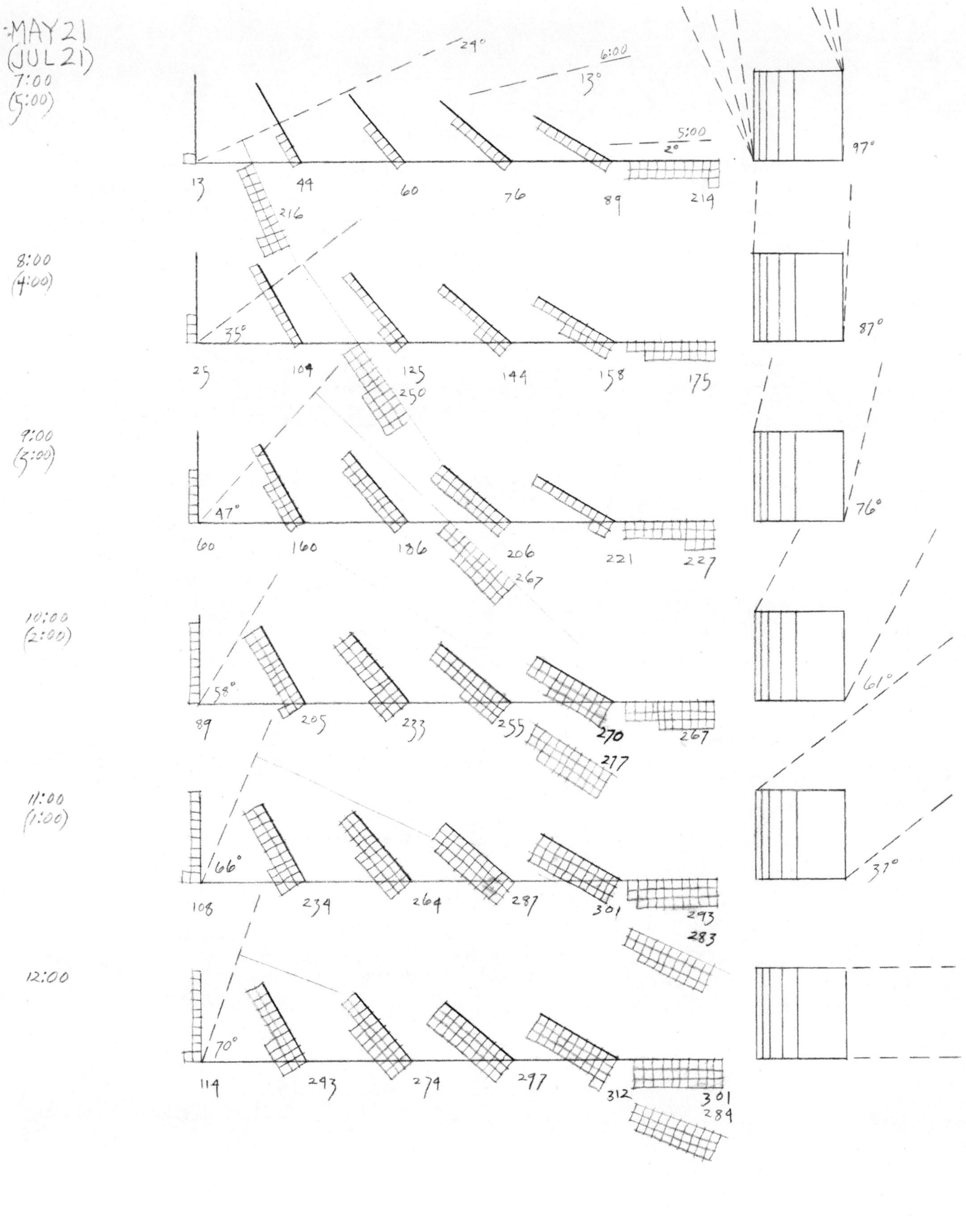

45. The calstices change their attractive polarity like magnets.

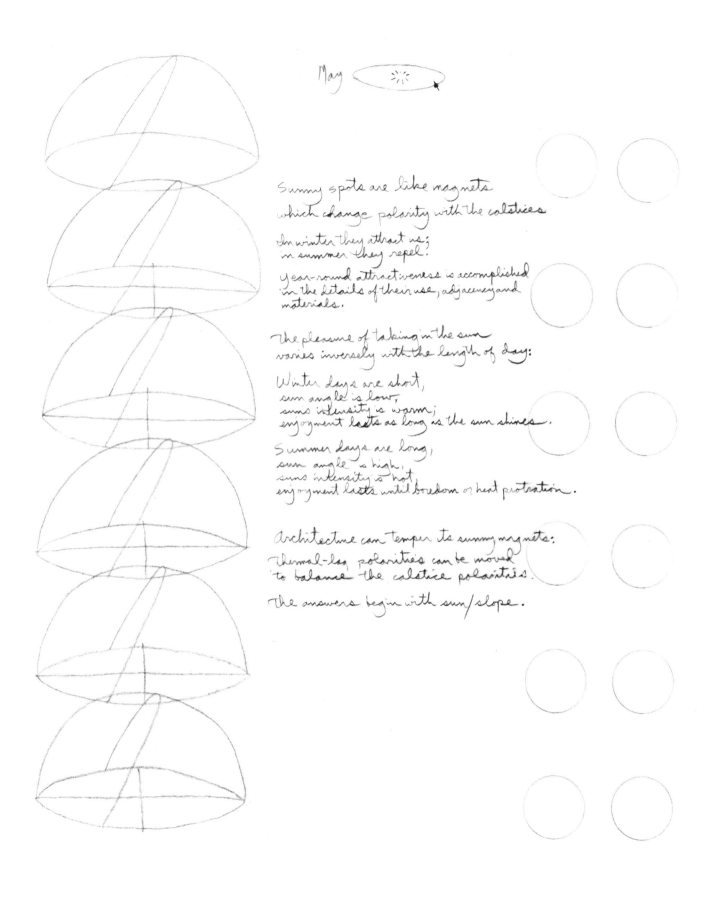

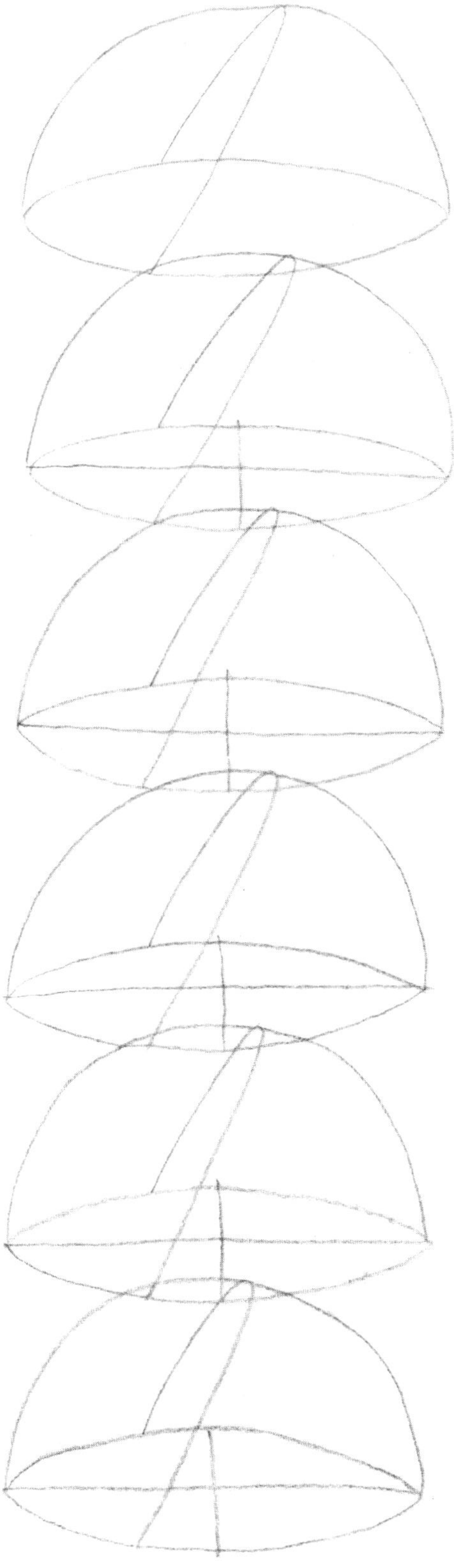

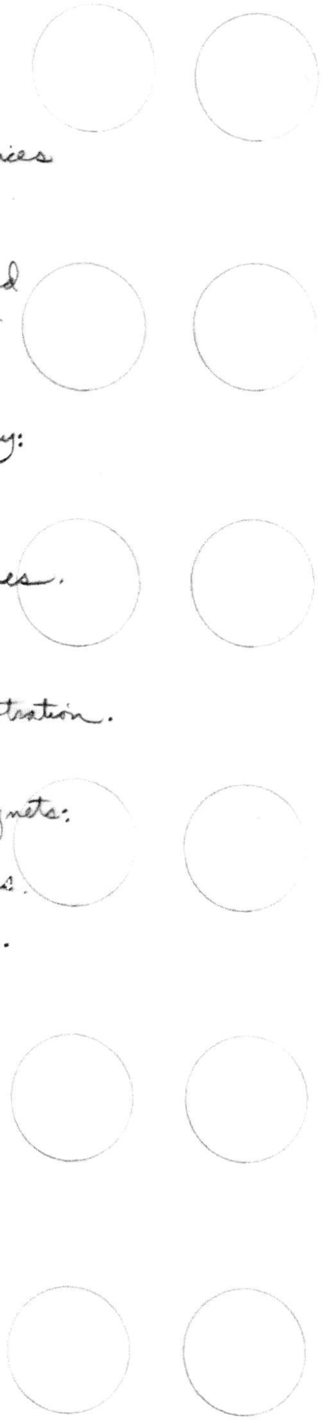

Sunny spots are like magnets
which change polarity with the calstices

In winter they attract us;
in summer they repel.

Year-round attractiveness is accomplished
in the details of their use, adjacency and
materials.

The pleasure of taking in the sun
varies inversely with the length of day:

Winter days are short,
sun angle is low,
suns intensity is warm;
enjoyment lasts as long as the sun shines.

Summer days are long,
sun angle is high,
suns intensity is hot,
enjoyment lasts until boredom or heat protration.

Architecture can temper its sunny magnets:
Thermal-lag polarities can be moved
to balance the calstice polarities.

The answers begin with sun/slope.

46. The sun-arc angle changes slowly at the time of the solstices.

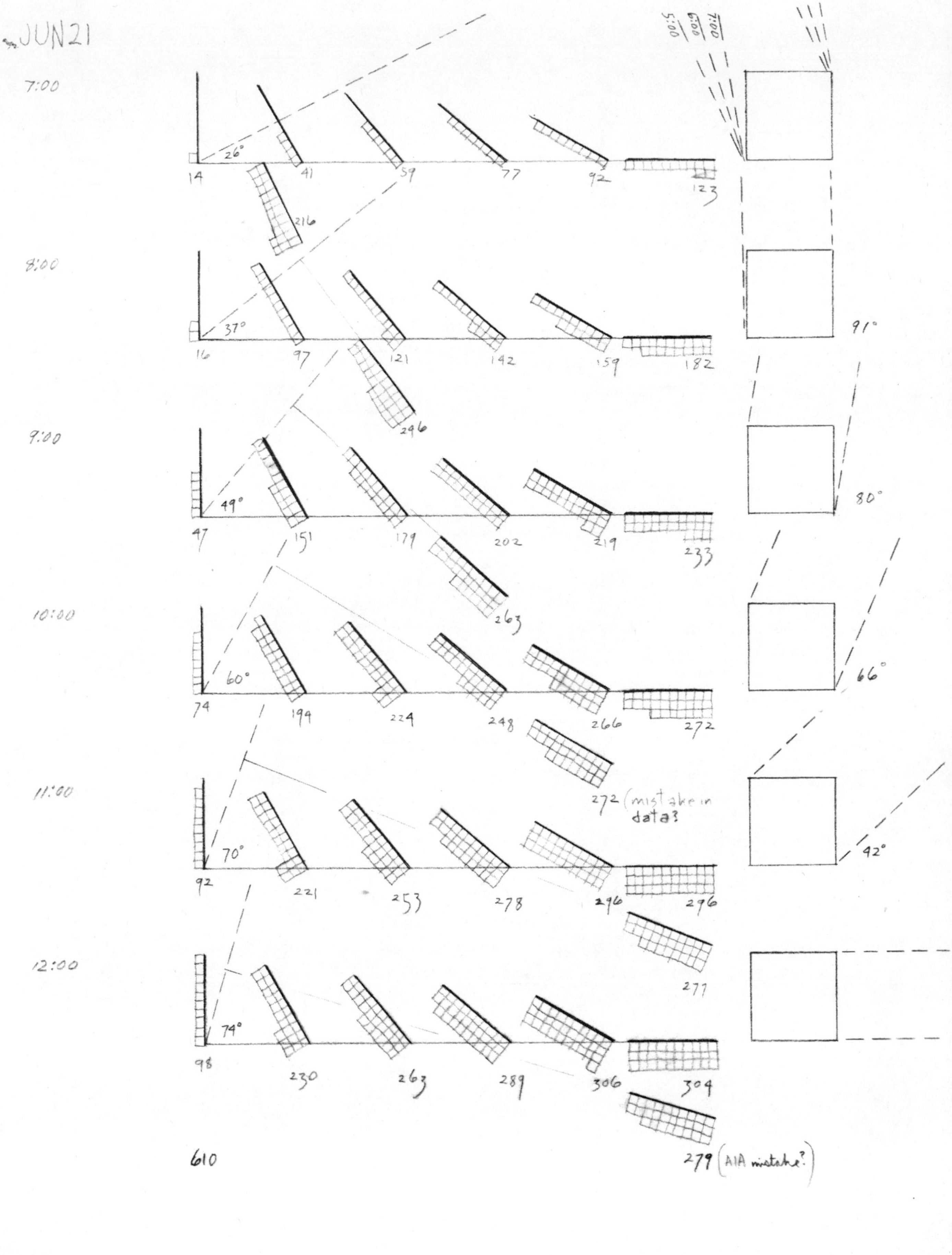

47. For a month before and after the solstice, the sun-arc does not seem to be much higher or lower; the days do not seem much longer or shorter.

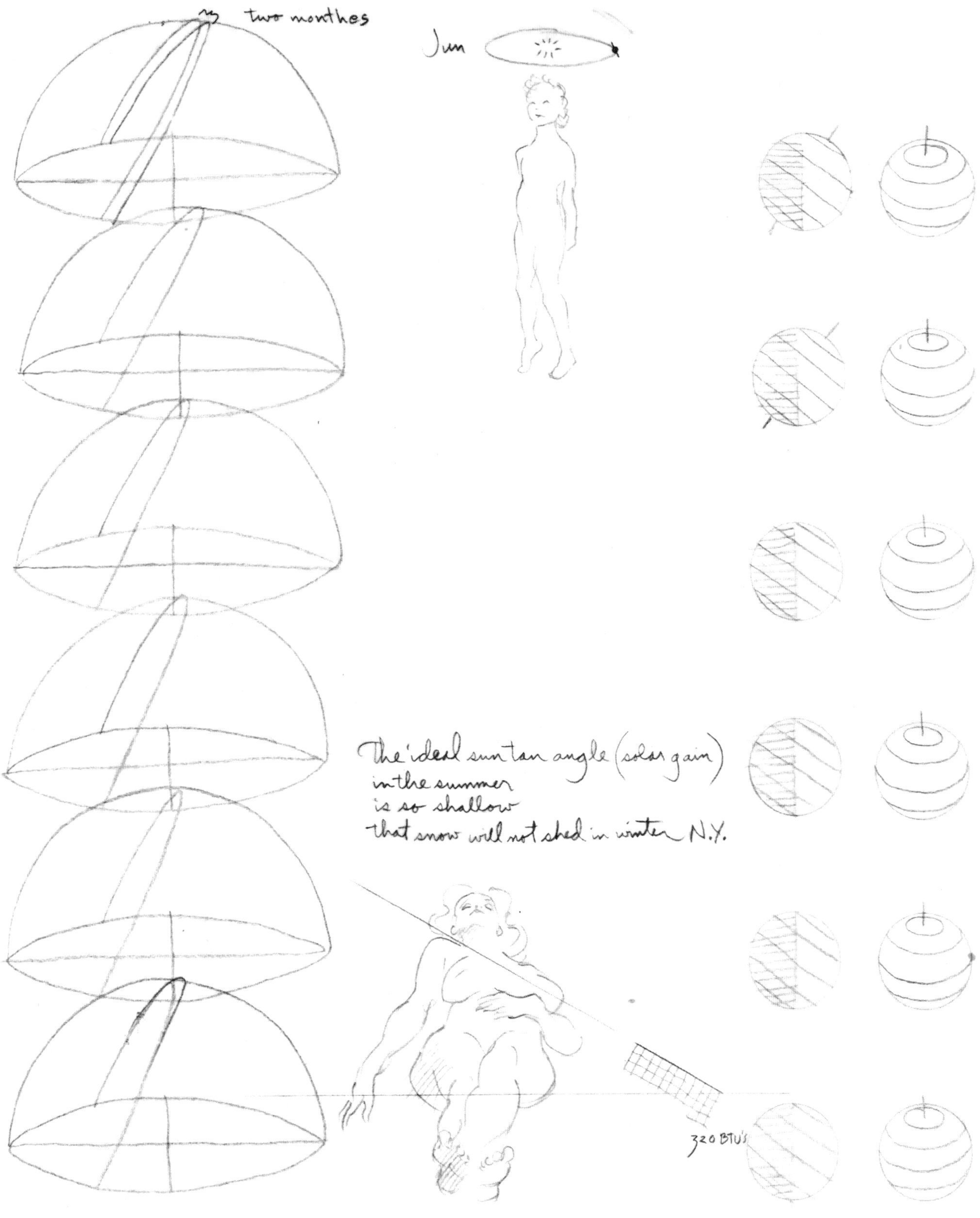
two monthes
Jun
The 'ideal sun tan angle (solar gain)
in the summer
is so shallow
that snow will not shed in winter N.Y.
320 BTU's

48. The solstices and equinoxes are the extremes of the sun angles during the year.

Winter Solstice

The tug-of-war between the warm earth and the cold universe causes frequent extreme temperature variations not midway between Autumnal Equical and Winter Solstice but simultaneously with the solstice due to earth's massive thermal lag.

Autumnal Equical

Autumnal Equinox

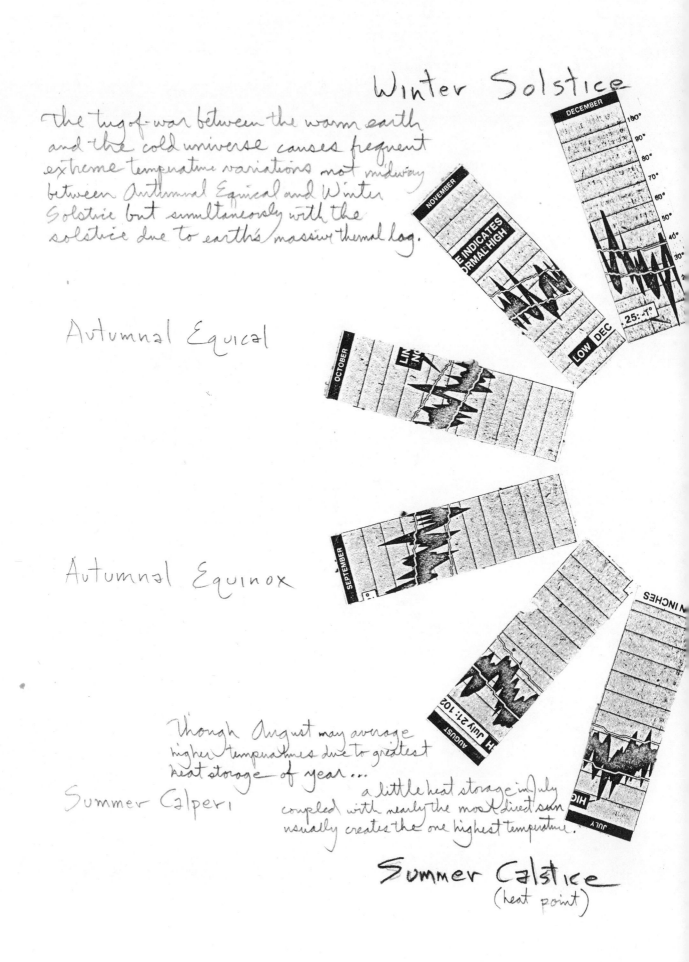

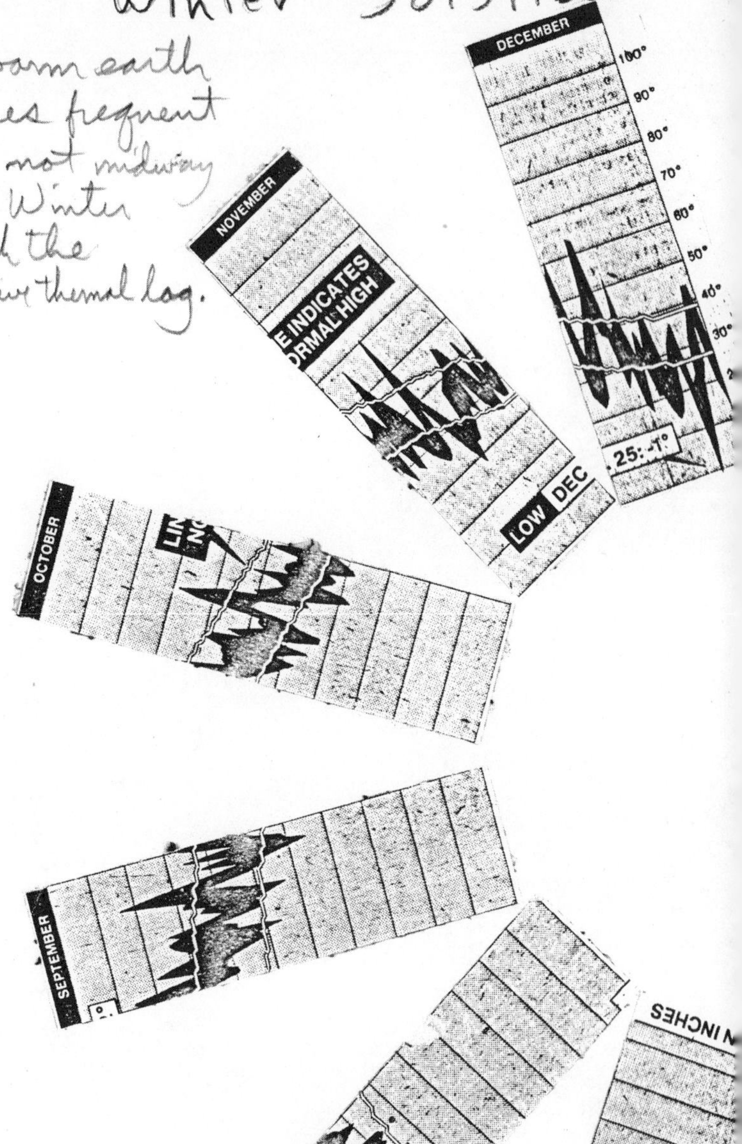

Though August may average higher temperatures due to greatest heat storage of year...

Summer Calperi

a little heat storage in July coupled with nearly the most direct sun usually creates the one highest temperature.

Summer Calstice
(heat point)

49. The calstices, equicals, and calperis are the temperature extremes of the seasons.

Winter Calstice and Calperi

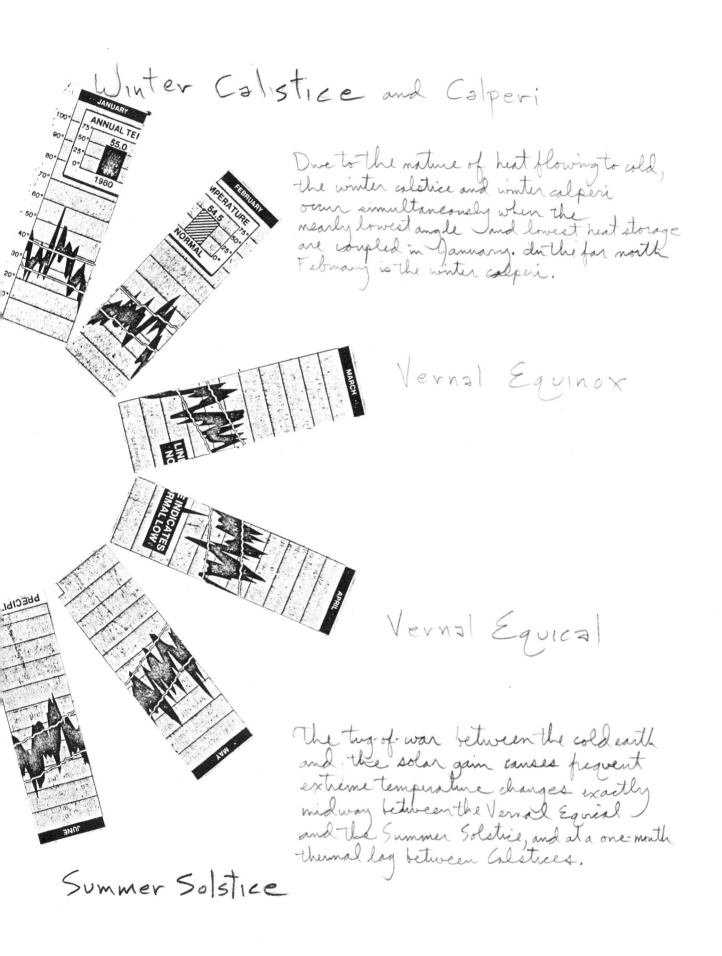

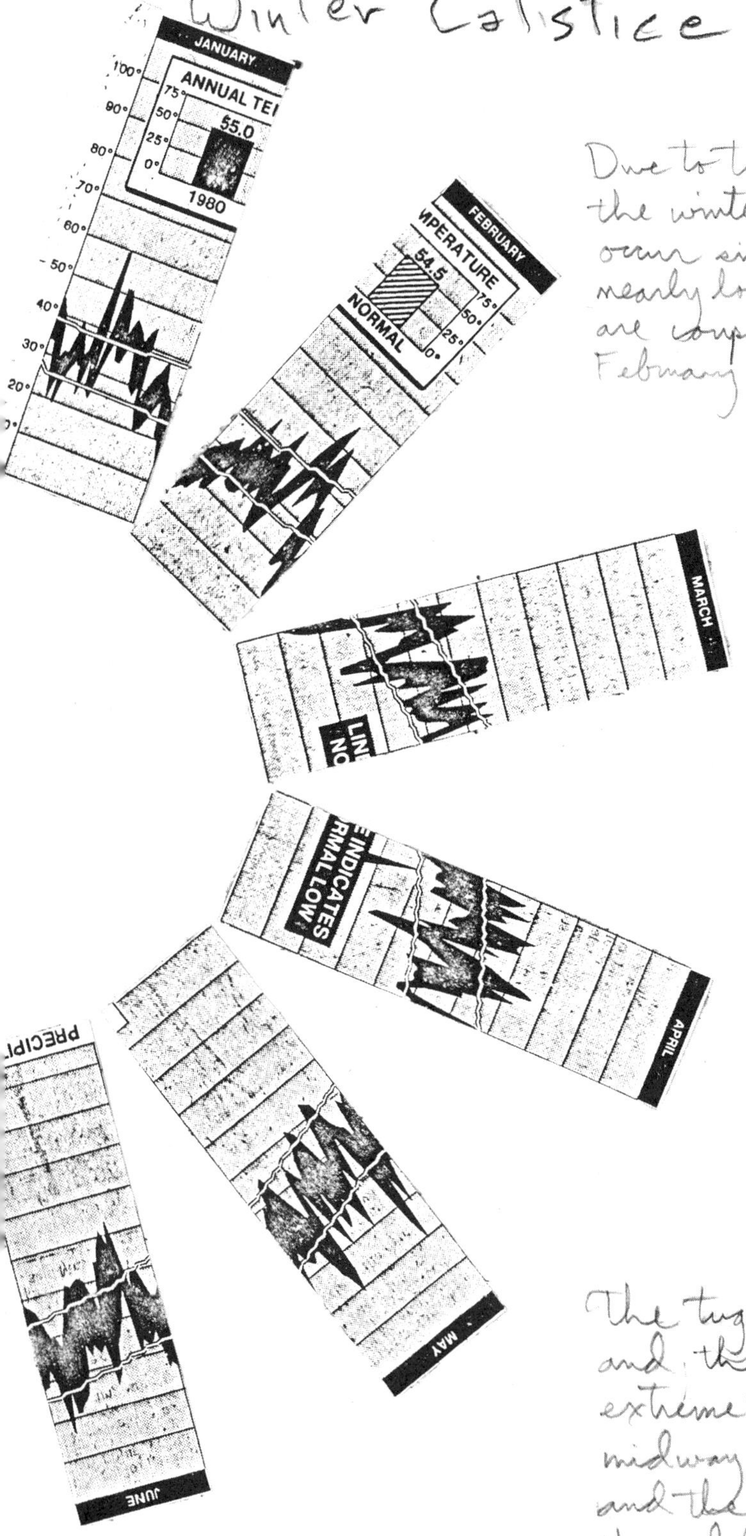

Due to the nature of heat flowing to cold, the winter calstice and winter calperi occur simultaneously when the nearly lowest angle and lowest heat storage are coupled in January. In the far north February is the winter calperi.

Vernal Equinox

Vernal Equical

The tug-of-war between the cold earth and the solar gain causes frequent extreme temperature changes exactly midway between the Vernal Equical and the Summer Solstice, and at a one-month thermal lag between Calstices.

Summer Solstice

50. Various climates are created at various sites because natures differ.

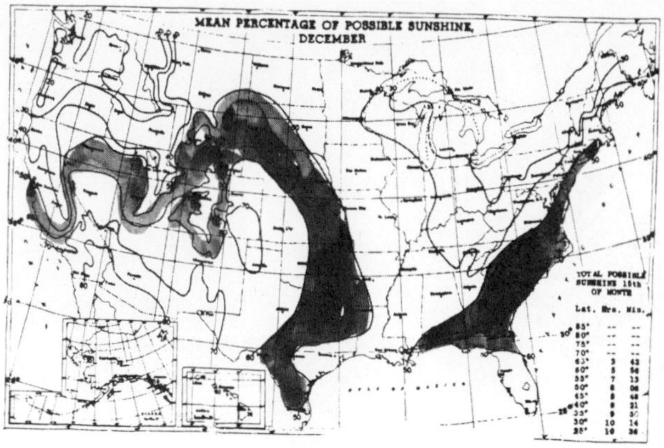

WINTER SOLSTICE

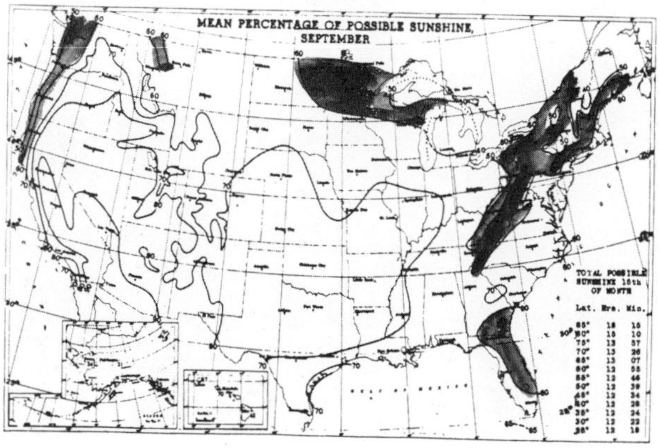

AUTUMNAL EQUINOX

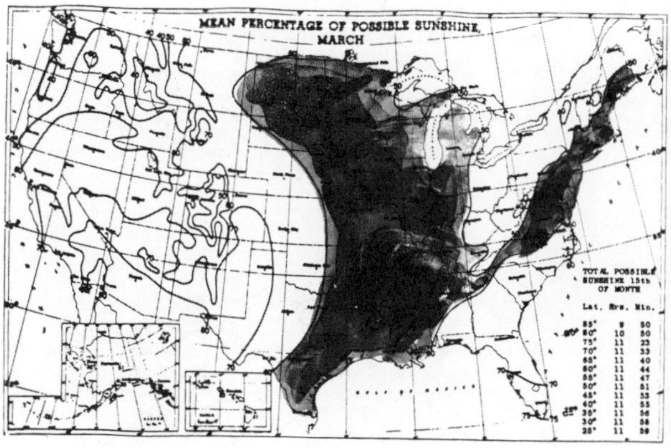

VERNAL EQUINOX

SUMMER SOLSTICE

51. The darkened proportions show the monthly movement of a temperature as the sun changes on the slopes of America.

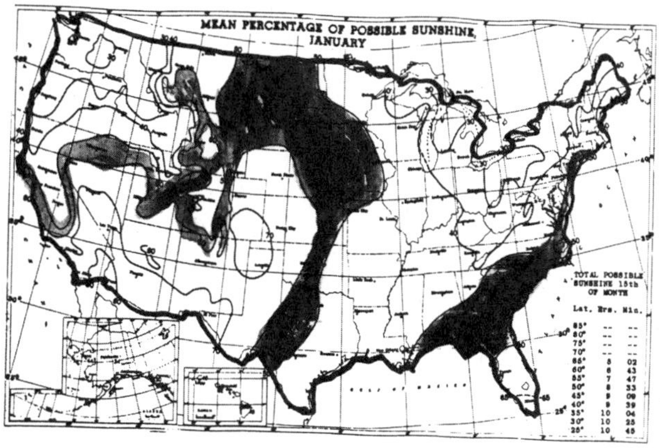

WINTER CALSTICE

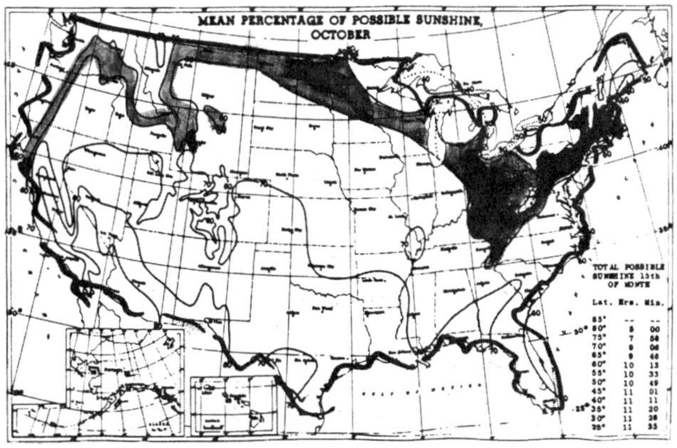

AUTUMNAL EQUICAL

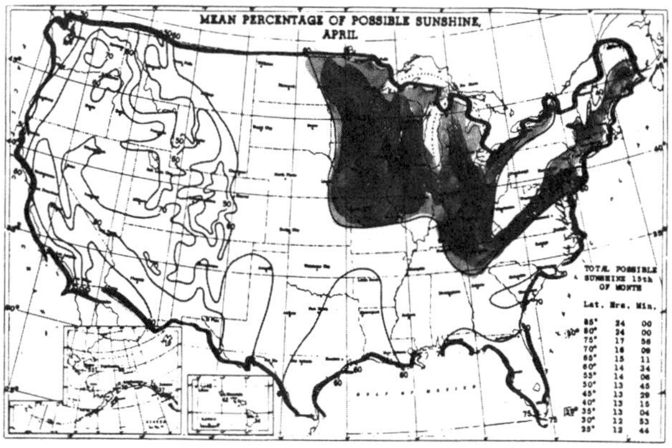

VERNAL EQUICAL

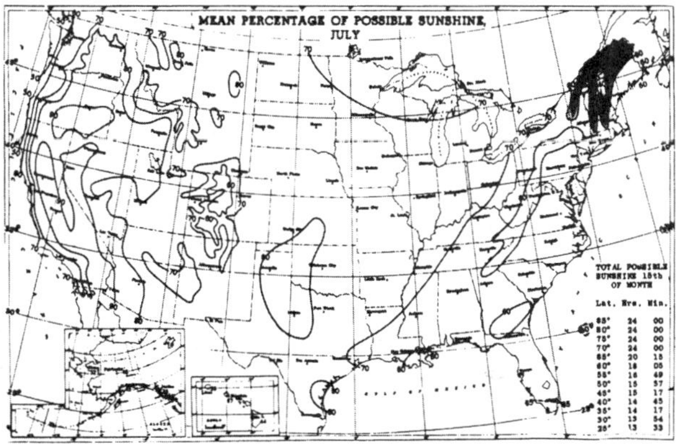

SUMMER CALSTICE

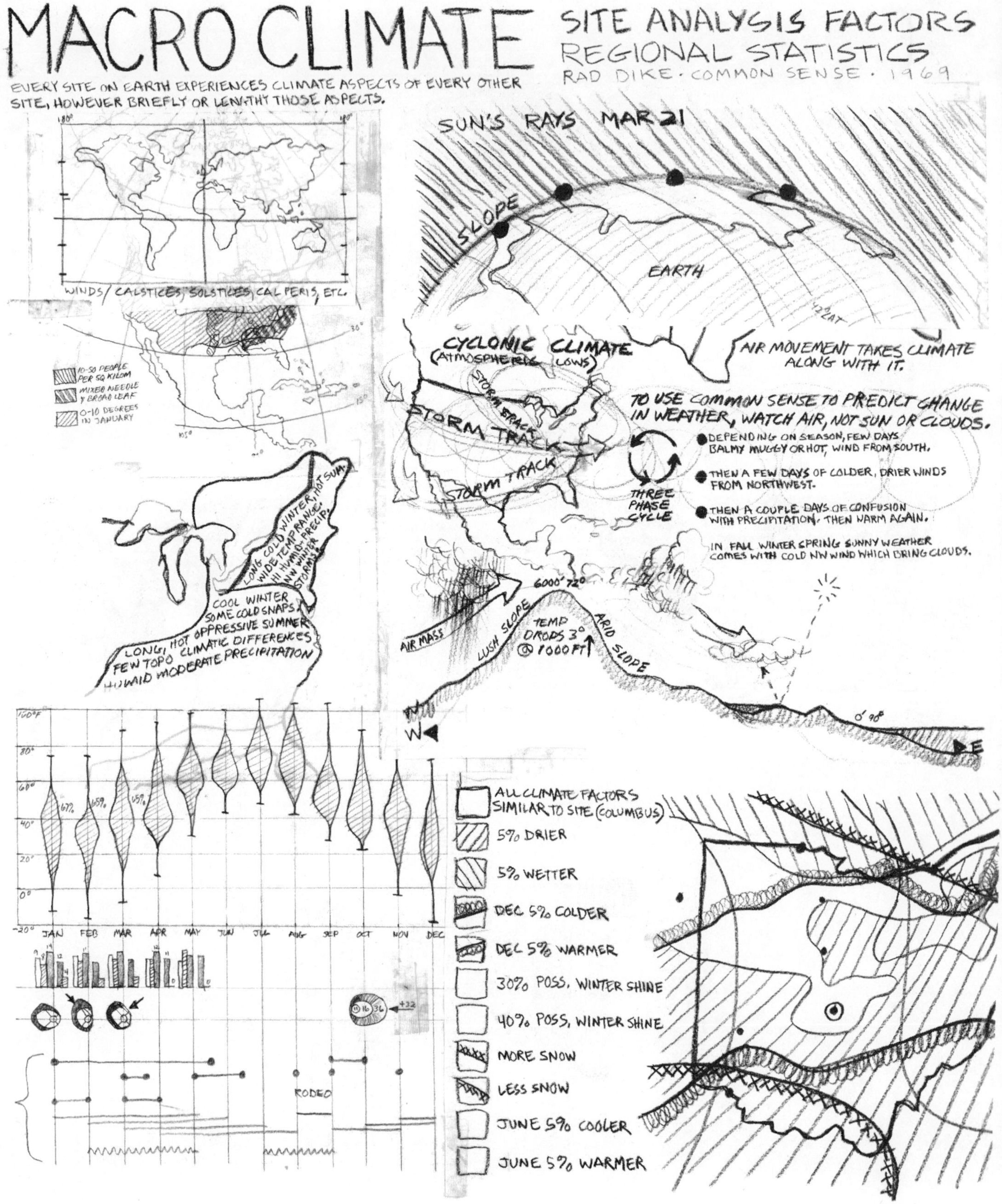

52. Sun and slope create large climates.

53. Sun and slope create small climates.

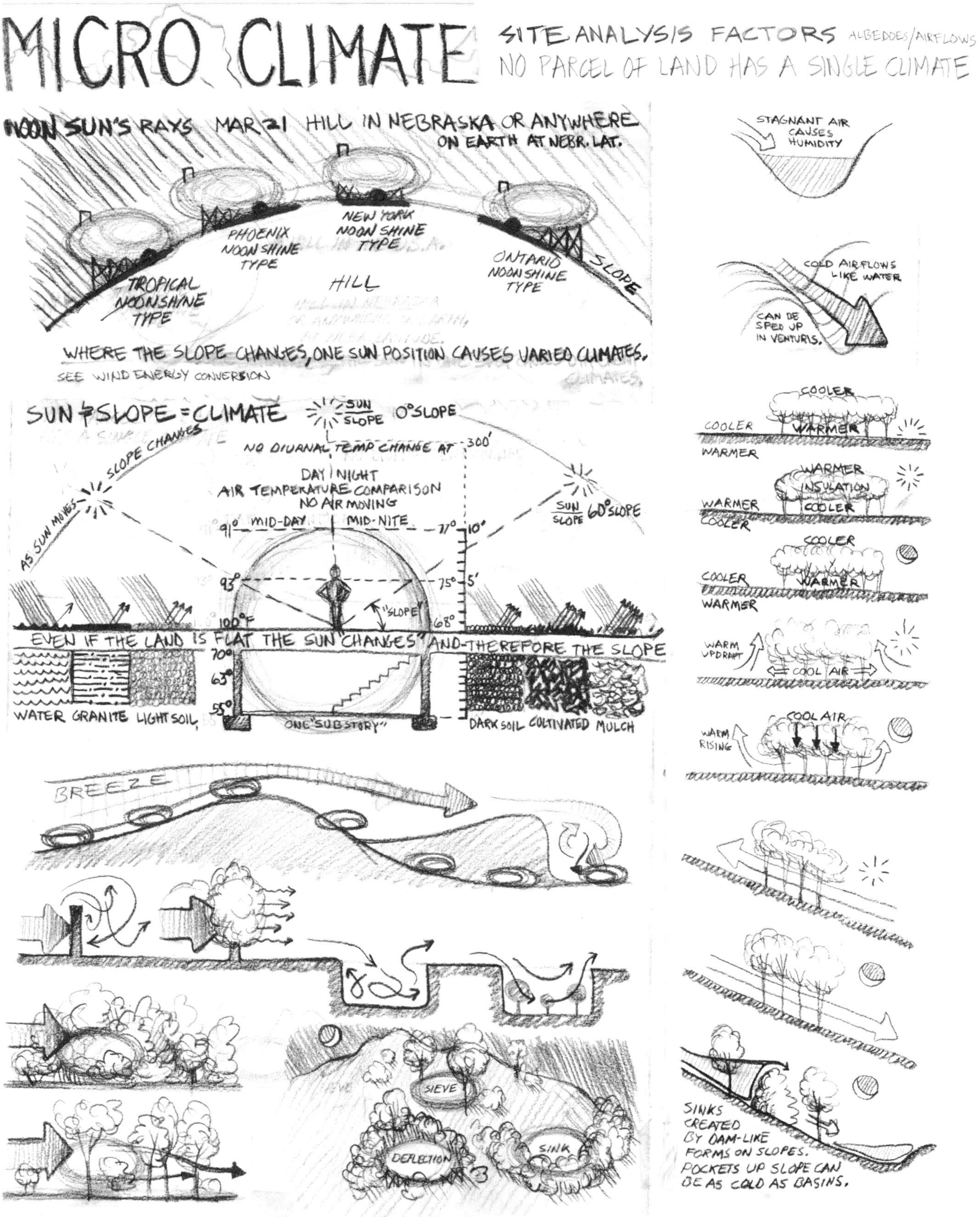

54. Sun and slope create wind. The air is moistened and heated by the sun's reaction with the slope and is made tangible as wind.

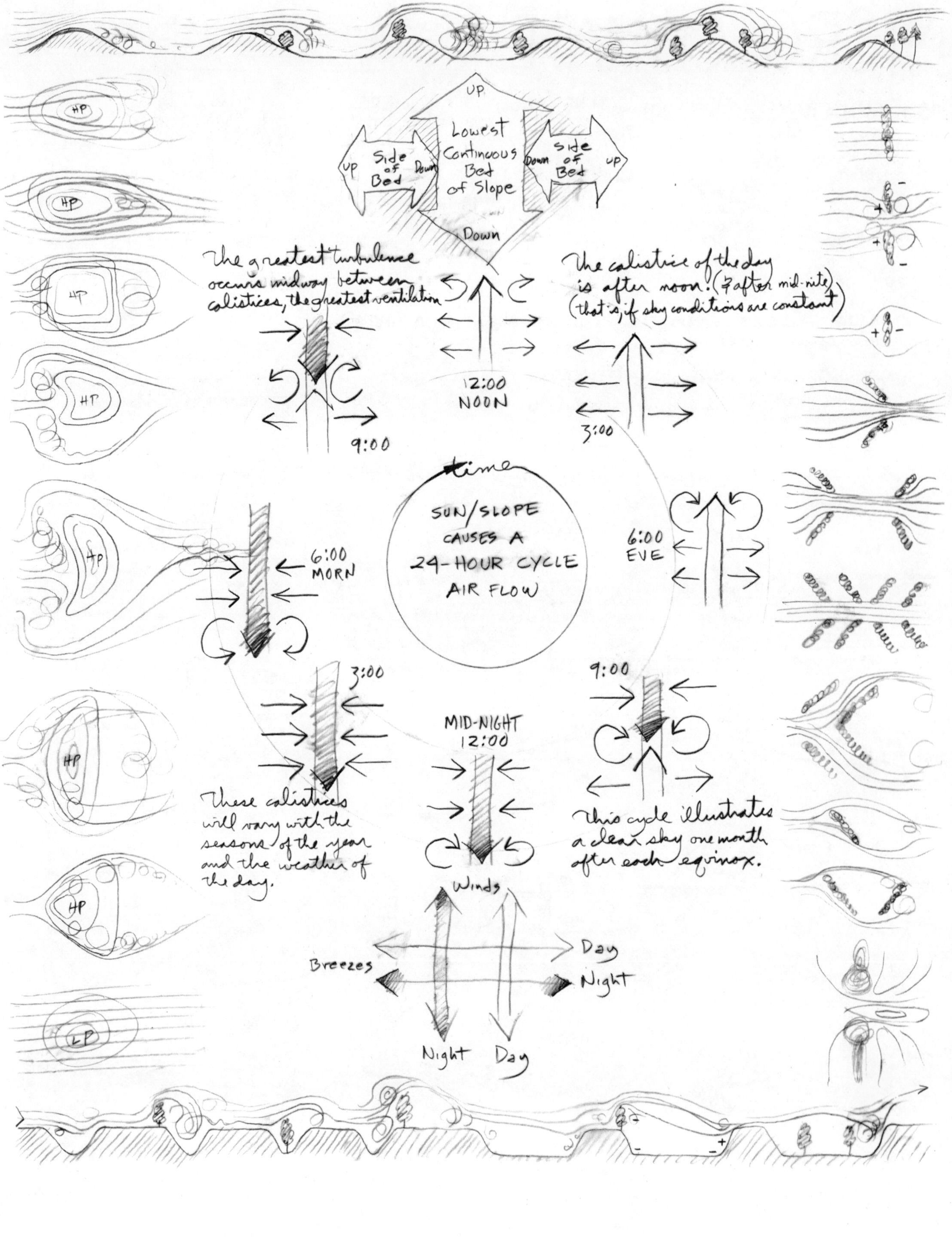

55. Wind wooes water, flower, and flame to bring about movement.

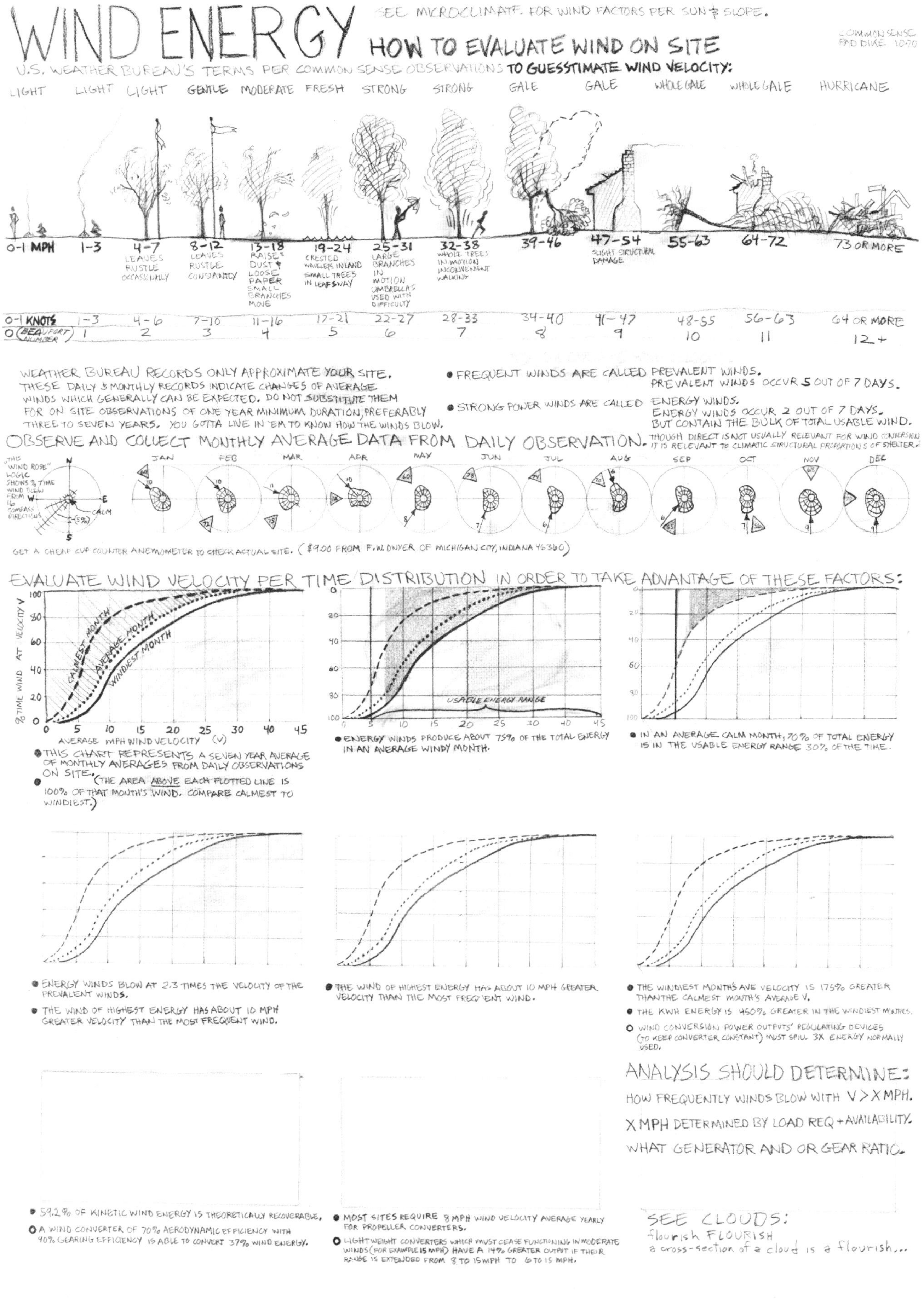

WIND ENERGY
SEE MICROCLIMATE FOR WIND FACTORS PER SUN & SLOPE.
HOW TO EVALUATE WIND ON SITE
COMMON SENSE
PAD DIKE 10-70
U.S. WEATHER BUREAU'S TERMS PER COMMON SENSE OBSERVATIONS TO GUESSTIMATE WIND VELOCITY:
LIGHT
LIGHT
LIGHT
GENTLE
MODERATE
FRESH
STRONG
STRONG
GALE
GALE
WHOLE GALE
WHOLE GALE
HURRICANE
0-1 MPH
1-3
4-7 LEAVES RUSTLE OCCASIONALLY
8-12 LEAVES RUSTLE CONSTANTLY
13-18 RAISES DUST & LOOSE PAPER SMALL BRANCHES MOVE
19-24 CRESTED WAVELETS INLAND SMALL TREES IN LEAF SWAY
25-31 LARGE BRANCHES IN MOTION UMBRELLAS USED WITH DIFFICULTY
32-38 WHOLE TREES IN MOTION INCONVENIENT WALKING
39-46
47-54 SLIGHT STRUCTURAL DAMAGE
55-63
64-72
73 OR MORE
0-1 KNOTS
1-3
4-6
7-10
11-16
17-21
22-27
28-33
34-40
41-47
48-55
56-63
64 OR MORE
0 (BEAUFORT NUMBER)
1
2
3
4
5
6
7
8
9
10
11
12+
WEATHER BUREAU RECORDS ONLY APPROXIMATE YOUR SITE. THESE DAILY & MONTHLY RECORDS INDICATE CHANGES OF AVERAGE WINDS WHICH GENERALLY CAN BE EXPECTED. DO NOT SUBSTITUTE THEM FOR ON SITE OBSERVATIONS OF ONE YEAR MINIMUM DURATION, PREFERABLY THREE TO SEVEN YEARS. YOU GOTTA LIVE IN 'EM TO KNOW HOW THE WINDS BLOW.
• FREQUENT WINDS ARE CALLED PREVALENT WINDS. PREVALENT WINDS OCCUR 5 OUT OF 7 DAYS.
• STRONG POWER WINDS ARE CALLED ENERGY WINDS. ENERGY WINDS OCCUR 2 OUT OF 7 DAYS. BUT CONTAIN THE BULK OF TOTAL USABLE WIND.
OBSERVE AND COLLECT MONTHLY AVERAGE DATA FROM DAILY OBSERVATION.
THOUGH DIRECT IS NOT USUALLY RELEVANT FOR WIND CONVERSION IT IS RELEVANT TO CLIMATIC STRUCTURAL PROPORTIONS OF SHELTER.
"THIS 'WIND ROSE' LOGIC SHOWS % TIME WIND BLEW FROM 16 COMPASS DIRECTIONS"
N
E
S
W
CALM
(5%)
JAN
FEB
MAR
APR
MAY
JUN
JUL
AUG
SEP
OCT
NOV
DEC
GET A CHEAP CUP COUNTER ANEMOMETER TO CHECK ACTUAL SITE. ($9.00 FROM F.W. DWYER OF MICHIGAN CITY, INDIANA 46360)
EVALUATE WIND VELOCITY PER TIME DISTRIBUTION IN ORDER TO TAKE ADVANTAGE OF THESE FACTORS:
% TIME WIND AT VELOCITY V
CALMEST MONTH
AVERAGE MONTH
WINDIEST MONTH
AVERAGE MPH WIND VELOCITY (V)
USABLE ENERGY RANGE
• THIS CHART REPRESENTS A SEVEN YEAR AVERAGE OF MONTHLY AVERAGES FROM DAILY OBSERVATIONS ON SITE.
• (THE AREA ABOVE EACH PLOTTED LINE IS 100% OF THAT MONTH'S WIND. COMPARE CALMEST TO WINDIEST.)
• ENERGY WINDS PRODUCE ABOUT 75% OF THE TOTAL ENERGY IN AN AVERAGE WINDY MONTH.
• IN AN AVERAGE CALM MONTH, 70% OF TOTAL ENERGY IS IN THE USABLE ENERGY RANGE 30% OF THE TIME.
• ENERGY WINDS BLOW AT 2.3 TIMES THE VELOCITY OF THE PREVALENT WINDS.
• THE WIND OF HIGHEST ENERGY HAS ABOUT 10 MPH GREATER VELOCITY THAN THE MOST FREQUENT WIND.
• THE WIND OF HIGHEST ENERGY HAS ABOUT 10 MPH GREATER VELOCITY THAN THE MOST FREQUENT WIND.
• THE WINDIEST MONTH'S AVE VELOCITY IS 175% GREATER THAN THE CALMEST MONTH'S AVERAGE V.
• THE KWH ENERGY IS 450% GREATER IN THE WINDIEST MONTHS.
o WIND CONVERSION POWER OUTPUTS' REGULATING DEVICES (TO KEEP CONVERTER CONSTANT) MUST SPILL 3X ENERGY NORMALLY USED.
ANALYSIS SHOULD DETERMINE:
HOW FREQUENTLY WINDS BLOW WITH V>X MPH.
X MPH DETERMINED BY LOAD REQ + AVAILABILITY.
WHAT GENERATOR AND OR GEAR RATIO.
• 59.2% OF KINETIC WIND ENERGY IS THEORETICALLY RECOVERABLE.
o A WIND CONVERTER OF 70% AERODYNAMIC EFFICIENCY WITH 90% GEARING EFFICIENCY IS ABLE TO CONVERT 37% WIND ENERGY.
• MOST SITES REQUIRE 8 MPH WIND VELOCITY AVERAGE YEARLY FOR PROPELLER CONVERTERS.
o LIGHTWEIGHT CONVERTERS WHICH MUST CEASE FUNCTIONING IN MODERATE WINDS (FOR EXAMPLE 15 MPH) HAVE A 14% GREATER OUTPUT IF THEIR RANGE IS EXTENDED FROM 8 TO 15 MPH TO 6 TO 15 MPH.
SEE CLOUDS:
flourish FLOURISH
a cross-section of a cloud is a flourish...

56. Wind not wound to hopes and expectations profits no one and is likely to be an annoyance if not a harm.

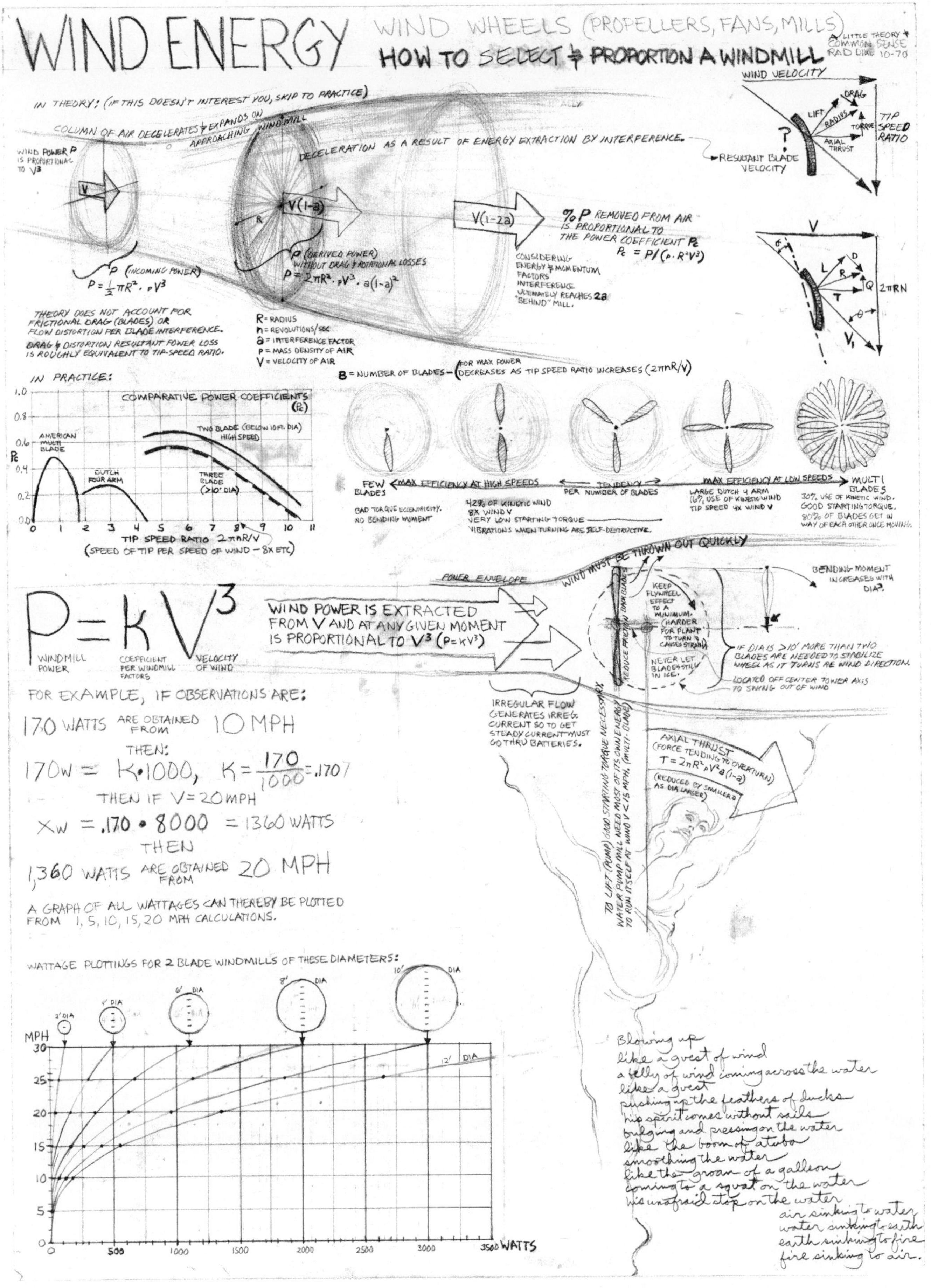
WIND ENERGY
WIND WHEELS (PROPELLERS, FANS, MILLS)
HOW TO SELECT & PROPORTION A WINDMILL
A LITTLE THEORY & COMMON SENSE
IN THEORY: (IF THIS DOESN'T INTEREST YOU, SKIP TO PRACTICE)
COLUMN OF AIR DECELERATES & EXPANDS ON APPROACHING WINDMILL
WIND POWER P IS PROPORTIONAL TO V³
DECELERATION AS A RESULT OF ENERGY EXTRACTION BY INTERFERENCE.
V(1-a)
V(1-2a)
%P REMOVED FROM AIR IS PROPORTIONAL TO THE POWER COEFFICIENT Pc
Pc = P/(ρ·R²V³)
P (DERIVED POWER) WITHOUT DRAG & ROTATIONAL LOSSES
P = 2πR²·ρV³·a(1-a)²
P (INCOMING POWER)
P = ½πR²·ρV³
CONSIDERING ENERGY & MOMENTUM FACTORS INTERFERENCE ULTIMATELY REACHES 2a "BEHIND" MILL.
WIND VELOCITY
DRAG
LIFT
RADIUS
TORQUE
AXIAL THRUST
TIP SPEED RATIO
RESULTANT BLADE VELOCITY
2πRN
THEORY DOES NOT ACCOUNT FOR FRICTIONAL DRAG (BLADES) OR FLOW DISTORTION PER BLADE INTERFERENCE.
DRAG & DISTORTION RESULTANT POWER LOSS IS ROUGHLY EQUIVALENT TO TIP-SPEED RATIO.
R = RADIUS
n = REVOLUTIONS/SEC
a = INTERFERENCE FACTOR
ρ = MASS DENSITY OF AIR
V = VELOCITY OF AIR
IN PRACTICE:
B = NUMBER OF BLADES — FOR MAX POWER DECREASES AS TIP SPEED RATIO INCREASES (2πnR/V)
COMPARATIVE POWER COEFFICIENTS (Pc)
AMERICAN MULTI BLADE
DUTCH FOUR ARM
TWO BLADE (BELOW 10FT. DIA) HIGH SPEED
THREE BLADE (>10' DIA)
TIP SPEED RATIO 2πnR/V
(SPEED OF TIP PER SPEED OF WIND — 8X ETC)
FEW BLADES
MAX EFFICIENCY AT HIGH SPEEDS
TENDENCY PER NUMBER OF BLADES
MAX EFFICIENCY AT LOW SPEEDS
MULTI BLADES
BAD TORQUE ECCENTRICITY. NO BENDING MOMENT
42% OF KINETIC WIND
8X WIND V
VERY LOW STARTING TORQUE
VIBRATIONS WHEN TURNING ARE SELF-DESTRUCTIVE.
LARGE DUTCH 4 ARM
16% USE OF KINETIC WIND
TIP SPEED 4X WIND V
30% USE OF KINETIC WIND.
GOOD STARTING TORQUE.
80% OF BLADES GET IN WAY OF EACH OTHER ONCE MOVING.
P = kV³
WINDMILL POWER
COEFFICIENT PER WINDMILL FACTORS
VELOCITY OF WIND
WIND POWER IS EXTRACTED FROM V AND AT ANY GIVEN MOMENT IS PROPORTIONAL TO V³ (P=kV³)
POWER ENVELOPE
WIND MUST BE THROWN OUT QUICKLY
KEEP FLYWHEEL EFFECT TO A MINIMUM. (HARDER FOR PLANT TO TURN & CAUSES STRAIN)
NEVER LET BLADES STILL IN ICE.
REDUCE FRICTION BACKLASH
BENDING MOMENT INCREASES WITH DIA.
IF DIA IS >10' MORE THAN TWO BLADES ARE NEEDED TO STABILIZE WHEEL AS IT TURNS RE WIND DIRECTION.
LOCATED OFF CENTER TOWER AXIS TO SWING OUT OF WIND
IRREGULAR FLOW GENERATES IRREG. CURRENT SO TO GET STEADY CURRENT MUST GO THRU BATTERIES.
TO LIFT (PUMP) GOOD STARTING TORQUE NECESSARY.
WATER PUMP MILL NEED MOST OF ITS OWN ENERGY TO RUN ITSELF AT WIND V < 15 MPH. (MULTI-BLADE)
AXIAL THRUST (FORCE TENDING TO OVERTURN)
T = 2πR²ρV³a(1-a)
(REDUCED BY SMALLER a AS DIA LARGER)
FOR EXAMPLE, IF OBSERVATIONS ARE:
170 WATTS ARE OBTAINED FROM 10 MPH
THEN:
170W = K·1000, K = 170/1000 = .170
THEN IF V = 20MPH
XW = .170 · 8000 = 1360 WATTS
THEN
1,360 WATTS ARE OBTAINED FROM 20 MPH
A GRAPH OF ALL WATTAGES CAN THEREBY BE PLOTTED FROM 1, 5, 10, 15, 20 MPH CALCULATIONS.
WATTAGE PLOTTINGS FOR 2 BLADE WINDMILLS OF THESE DIAMETERS:
2' DIA
4' DIA
6' DIA
8' DIA
10' DIA
12' DIA
MPH
WATTS
Blowing up
like a gust of wind
a belly of wind coming across the water
like a gust
pushing up the feathers of ducks
his spirit comes without sails
bulging and pressing on the water
like the boom of a tuba
smoothing the water
like the groan of a galleon
coming to a squat on the water
his unafraid step on the water
air sinking to water
water sinking to earth
earth sinking to fire
fire sinking to air.

57. Sails propel ships around the earth, as vanes propel the shaft.

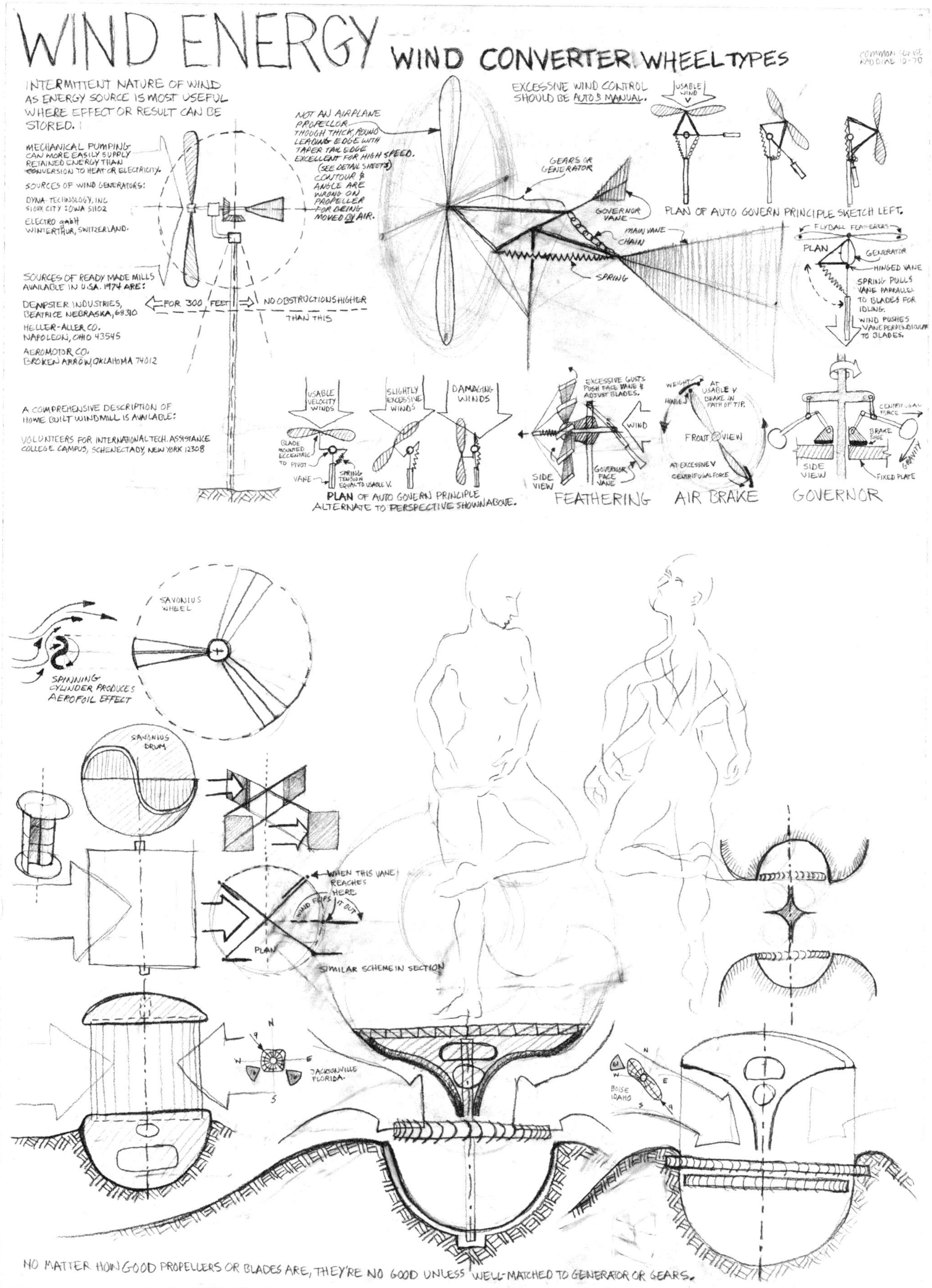

58. The primal yearn for water continues in the humidity ocean about us.

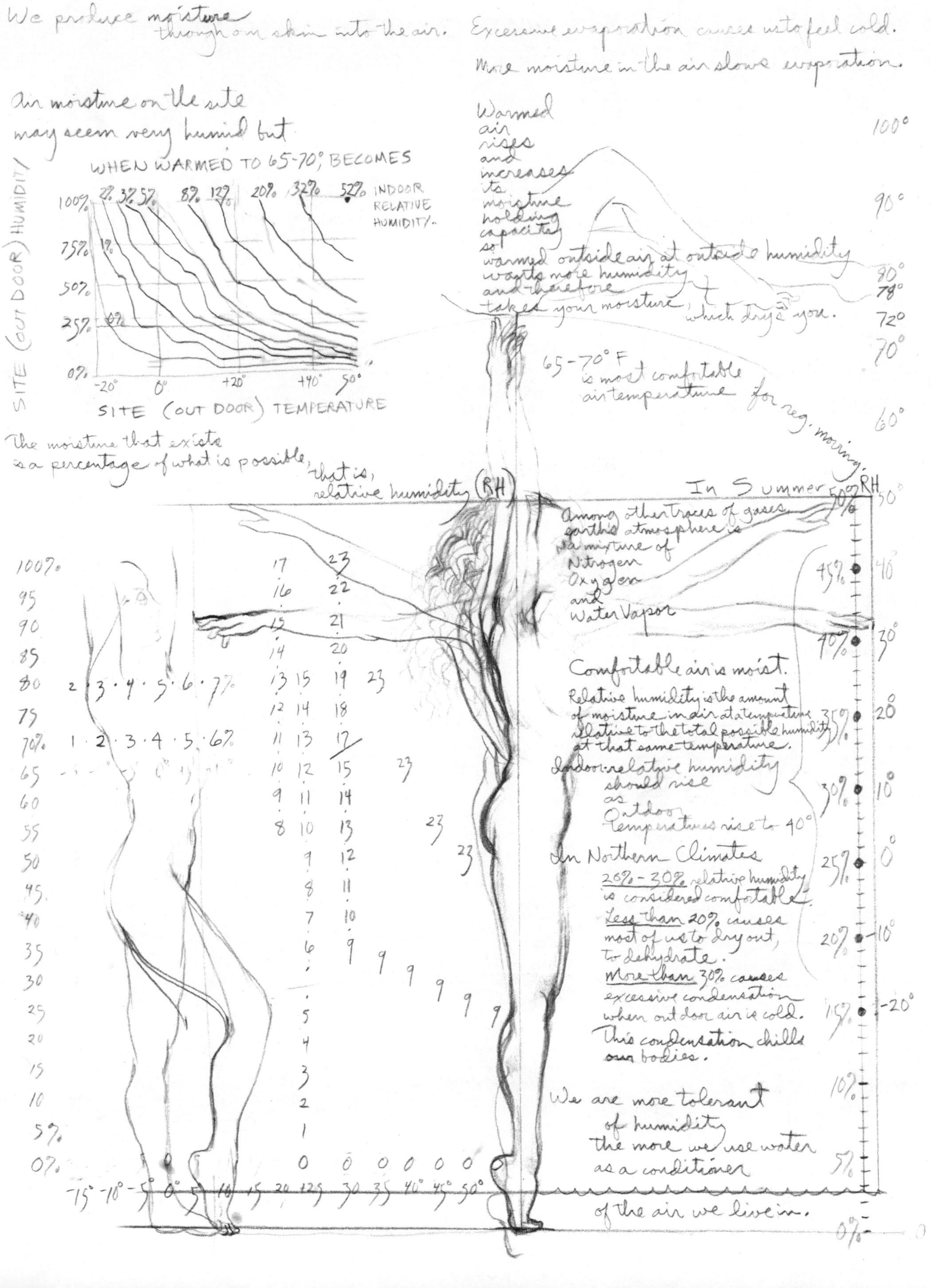
We produce moisture through our skin into the air.
Excessive evaporation causes us to feel cold.
More moisture in the air slows evaporation.
Air moisture on the site may seem very humid but
WHEN WARMED TO 65-70°, BECOMES
2% 3% 5% 8% 12% 20% 32% 52% INDOOR RELATIVE HUMIDITY.
SITE (OUT DOOR) HUMIDITY
100% 75% 50% 25% 0%
-20° 0° +20° +40° 50°
SITE (OUT DOOR) TEMPERATURE
The moisture that exists is a percentage of what is possible, that is, relative humidity (RH)
Warmed air rises and increases its moisture holding capacity so warmed outside air at outside humidity wants more humidity and therefore takes your moisture which drys you.
100° 90° 80° 78° 72° 70° 60°
65-70° F is most comfortable air temperature for reg. moving.
In Summer 50% RH
Among other traces of gases earth's atmosphere is a mixture of Nitrogen Oxygen and Water Vapor
Comfortable air is moist.
Relative humidity is the amount of moisture in air at a temperature relative to the total possible humidity at that same temperature.
Indoor relative humidity should rise as Outdoor Temperatures rise to 40°
In Northern Climates 20% - 30% relative humidity is considered comfortable.
Less than 20% causes most of us to dry out, to dehydrate.
More than 30% causes excessive condensation when outdoor air is cold.
This condensation chills our bodies.
We are more tolerant of humidity the more we use water as a conditioner of the air we live in.
50% 45% 40% 35% 30% 25% 20% 15% 10% 5% 0%
50° 40° 30° 20° 10° 0° -10° -20°
100% 95 90 85 80 75 70% 65 60 55 50 45 40 35 30 25 20 15 10 5% 0%
-15° -10° -5° 0° 5 10 15 20 25 30 35 40° 45° 50°

59. Sun powers wave motion, evaporates water to the clouds, and precipitates it to the ground.

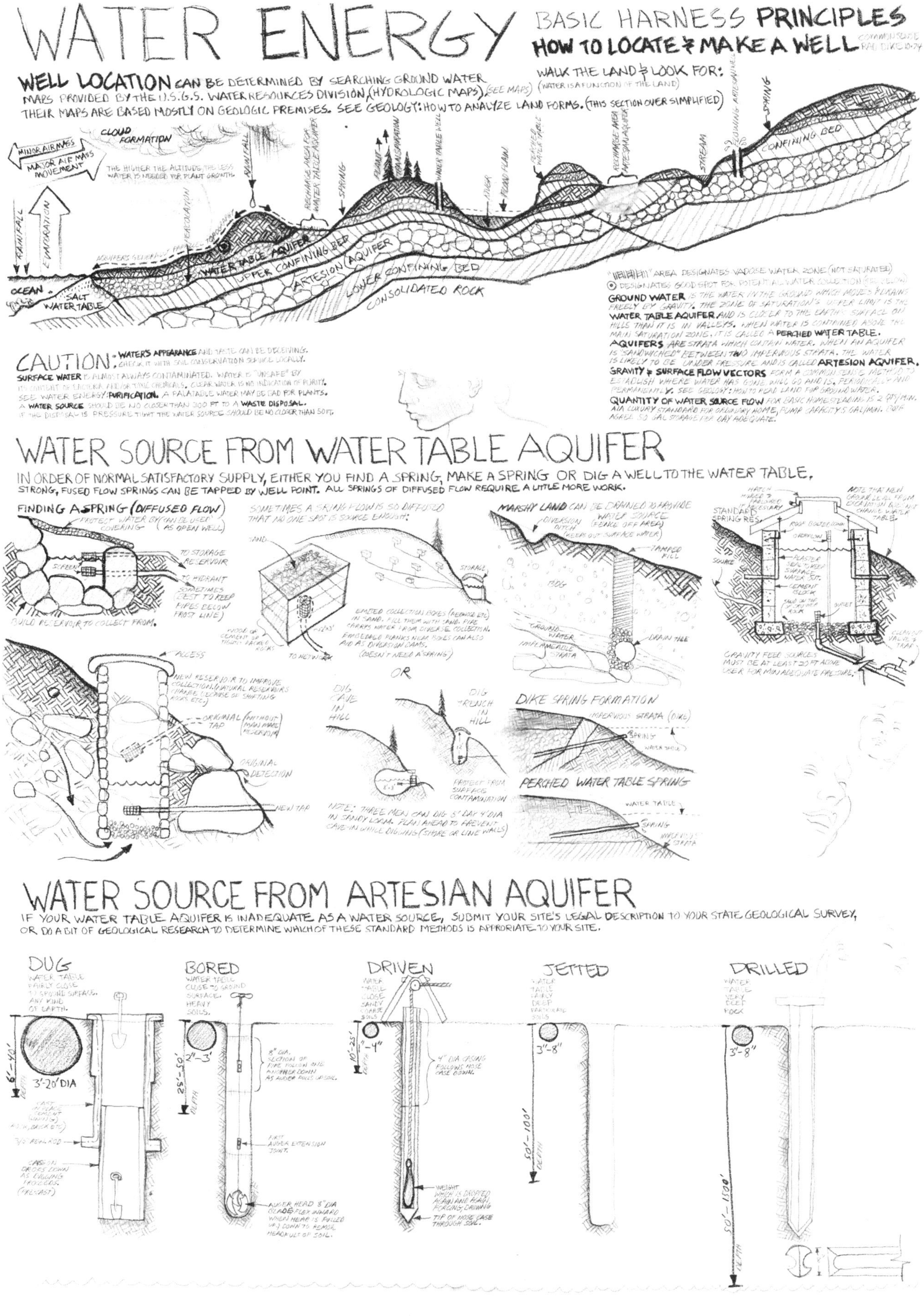

WATER ENERGY
BASIC HARNESS PRINCIPLES
HOW TO LOCATE & MAKE A WELL
WELL LOCATION CAN BE DETERMINED BY SEARCHING GROUND WATER MAPS PROVIDED BY THE U.S.G.S. WATER RESOURCES DIVISION (HYDROLOGIC MAPS) (SEE MAPS)
THEIR MAPS ARE BASED MOSTLY ON GEOLOGIC PREMISES. SEE GEOLOGY: HOW TO ANALYZE LAND FORMS. (THIS SECTION OVER SIMPLIFIED)
WALK THE LAND & LOOK FOR:
CLOUD FORMATION
MINOR AIR MASS
MAJOR AIR MASS MOVEMENT
RAINFALL
EVAPORATION
OCEAN
SALT WATER TABLE
WATER TABLE AQUIFER
UPPER CONFINING BED
ARTESIAN AQUIFER
LOWER CONFINING BED
CONSOLIDATED ROCK
CONFINING BED
STREAM
SPRING
CAUTION:
SURFACE WATER IS ALMOST ALWAYS CONTAMINATED.
GROUND WATER
WATER TABLE AQUIFER
AQUIFERS ARE STRATA WHICH CONTAIN WATER.
ARTESIAN AQUIFER.
GRAVITY & SURFACE FLOW VECTORS
QUANTITY OF WATER SOURCE FLOW
WATER SOURCE FROM WATER TABLE AQUIFER
IN ORDER OF NORMAL SATISFACTORY SUPPLY, EITHER YOU FIND A SPRING, MAKE A SPRING OR DIG A WELL TO THE WATER TABLE.
STRONG, FUSED FLOW SPRINGS CAN BE TAPPED BY WELL POINT. ALL SPRINGS OF DIFFUSED FLOW REQUIRE A LITTLE MORE WORK.
FINDING A SPRING (DIFFUSED FLOW)
MARSHY LAND
DIKE SPRING FORMATION
PERCHED WATER TABLE SPRING
OR
WATER SOURCE FROM ARTESIAN AQUIFER
IF YOUR WATER TABLE AQUIFER IS INADEQUATE AS A WATER SOURCE, SUBMIT YOUR SITE'S LEGAL DESCRIPTION TO YOUR STATE GEOLOGICAL SURVEY, OR DO A BIT OF GEOLOGICAL RESEARCH TO DETERMINE WHICH OF THESE STANDARD METHODS IS APPRORIATE TO YOUR SITE.
DUG
BORED
DRIVEN
JETTED
DRILLED
3'-20' DIA
2"-3'
3"-8"
3"-8"

 All the water that flows is detained in some way by something.

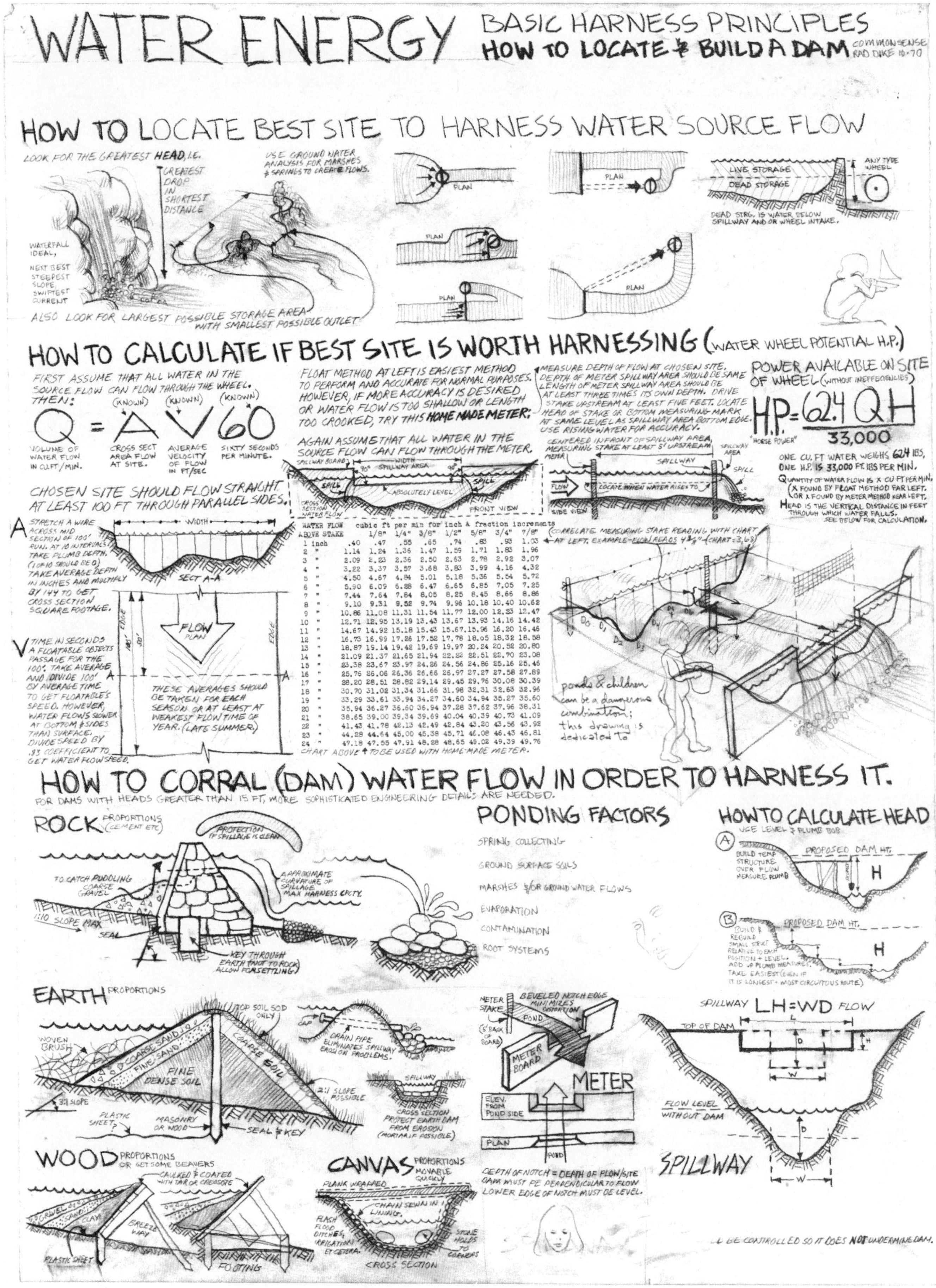
WATER ENERGY
BASIC HARNESS PRINCIPLES
HOW TO LOCATE & BUILD A DAM
COMMON SENSE
RAD DIKE 10-70
HOW TO LOCATE BEST SITE TO HARNESS WATER SOURCE FLOW
LOOK FOR THE GREATEST HEAD, I.E.
GREATEST DROP IN SHORTEST DISTANCE
USE GROUND WATER ANALYSIS FOR MARSHES & SPRINGS TO CREATE FLOWS.
WATERFALL IDEAL,
NEXT BEST STEEPEST SLOPE, SWIFTEST CURRENT
ALSO LOOK FOR LARGEST POSSIBLE STORAGE AREA WITH SMALLEST POSSIBLE OUTLET
PLAN
LIVE STORAGE
DEAD STORAGE
ANY TYPE WHEEL
DEAD STRG. IS WATER BELOW SPILLWAY AND OR WHEEL INTAKE.
HOW TO CALCULATE IF BEST SITE IS WORTH HARNESSING (WATER WHEEL POTENTIAL H.P.)
FIRST ASSUME THAT ALL WATER IN THE SOURCE FLOW CAN FLOW THROUGH THE WHEEL. THEN:
Q = A V 60
(KNOWN) (KNOWN) (KNOWN)
VOLUME OF WATER FLOW IN CU.FT/MIN.
CROSS SECT AREA FLOW AT SITE.
AVERAGE VELOCITY OF FLOW IN FT/SEC
SIXTY SECONDS PER MINUTE.
FLOAT METHOD AT LEFT IS EASIEST METHOD TO PERFORM AND ACCURATE FOR NORMAL PURPOSES. HOWEVER, IF MORE ACCURACY IS DESIRED OR WATER FLOW IS TOO SHALLOW OR LENGTH TOO CROOKED, TRY THIS HOME MADE METER:
MEASURE DEPTH OF FLOW AT CHOSEN SITE. DEPTH OF METER SPILLWAY AREA SHOULD BE SAME LENGTH OF METER SPILLWAY AREA SHOULD BE AT LEAST THREE TIMES ITS OWN DEPTH. DRIVE STAKE UPSTREAM AT LEAST FIVE FEET. LOCATE HEAD OF STAKE OR BOTTOM MEASURING MARK AT SAME LEVEL AS SPILLWAY AREA BOTTOM EDGE. USE RISING WATER FOR ACCURACY.
POWER AVAILABLE ON SITE OF WHEEL (WITHOUT INEFFECIENCIES)
H.P. = 62.4 QH / 33,000
"HORSE POWER"
ONE CU. FT WATER WEIGHS 62.4 LBS, ONE H.P. IS 33,000 FT. LBS PER MIN.
QUANTITY OF WATER FLOW IS X CU FT PER MIN. (X FOUND BY FLOAT METHOD FAR LEFT. OR X FOUND BY METER METHOD NEAR LEFT.
HEAD IS THE VERTICAL DISTANCE IN FEET THROUGH WHICH WATER FALLS. SEE BELOW FOR CALCULATION.
AGAIN ASSUME THAT ALL WATER IN THE SOURCE FLOW CAN FLOW THROUGH THE METER.
SPILLWAY BOARD
WIDTH
SPILLWAY AREA
90°
SPILL
ABSOLUTELY LEVEL
CROSS SECTION WATER FLOW
FRONT VIEW
CENTERED IN FRONT OF SPILLWAY AREA, MEASURING STAKE AT LEAST 5' UPSTREAM
METER
SPILLWAY
SPILLWAY AREA
SPILL
FLOW
LOCATE WHEN WATER RISES TO
SIDE VIEW
CHOSEN SITE SHOULD FLOW STRAIGHT AT LEAST 100 FT THROUGH PARALLEL SIDES.
A
STRETCH A WIRE ACROSS MID SECTION OF 100' RUN. AT 10 INTERVALS TAKE PLUMB DEPTH. (1 OF 10 SHOULD BE 0) TAKE AVERAGE DEPTH IN INCHES AND MULTIPLY BY 144 TO GET CROSS SECTION SQUARE FOOTAGE.
WIDTH
SECT A-A
FLOW
PLAN
EDGE
V
TIME IN SECONDS A FLOATABLE OBJECTS PASSAGE FOR THE 100'. TAKE AVERAGE AND DIVIDE 100' BY AVERAGE TIME TO GET FLOATABLE'S SPEED. HOWEVER, WATER FLOWS SLOWER AT BOTTOM & SIDES THAN SURFACE. DIVIDE SPEED BY .93 COEFFICIENT TO GET WATER FLOW SPEED.
THESE AVERAGES SHOULD BE TAKEN FOR EACH SEASON, OR AT LEAST AT WEAKEST FLOW TIME OF YEAR. (LATE SUMMER)
CORRELATE MEASURING STAKE READING WITH CHART AT LEFT.
ponds & children can be a dangerous combination; this drawing is dedicated to
HOW TO CORRAL (DAM) WATER FLOW IN ORDER TO HARNESS IT.
FOR DAMS WITH HEADS GREATER THAN 15 FT, MORE SOPHISTICATED ENGINEERING DETAILS ARE NEEDED.
ROCK PROPORTIONS (CEMENT ETC)
PROTECTION IF SPILLAGE IS CLEAN
TO CATCH PUDDLING COARSE GRAVEL
APPROXIMATE CURVATURE OF SPILLAGE MAX HARNESS EFCTY.
1:10 SLOPE MAX
SEAL
KEY THROUGH EARTH (NOT TO ROCK) ALLOW FOR SETTLING
PONDING FACTORS
SPRING COLLECTING
GROUND SURFACE SOILS
MARSHES &/OR GROUND WATER FLOWS
EVAPORATION
CONTAMINATION
ROOT SYSTEMS
HOW TO CALCULATE HEAD
USE LEVEL & PLUMB BOB
A
PROPOSED DAM HT.
BUILD TEMP STRUCTURE OVER FLOW MEASURE PLUMB
H
B
PROPOSED DAM HT.
TAKE EASIEST (EVEN IF IT IS LONGEST & MOST CIRCUITOUS ROUTE)
EARTH PROPORTIONS
TOP SOIL SOD ONLY
WOVEN BRUSH
COARSE SAND
FINE SAND
FINE DENSE SOIL
COARSE SOIL
3:1 SLOPE
2:1 SLOPE POSSIBLE
PLASTIC SHEET
MASONRY OR WOOD
SEAL & KEY
DRAIN PIPE ELIMINATES SPILLWAY EROSION PROBLEMS.
CROSS SECTION PROTECT EARTH DAM FROM EROSION (MORTAR IF POSSIBLE)
METER STAKE
BEVELED NOTCH EDGE MINIMIZES DISTORTION
POND
METER BOARD
METER
ELEV. FROM POND SIDE
PLAN
DEPTH OF NOTCH = DEPTH OF FLOW/SITE DAM MUST BE PERPENDICULAR TO FLOW LOWER EDGE OF NOTCH MUST BE LEVEL.
SPILLWAY
LH=WD
FLOW
TOP OF DAM
FLOW LEVEL WITHOUT DAM
SPILLWAY
WOOD PROPORTIONS OR GET SOME BEAVERS
CAULKED & COATED WITH TAR OR CREOSOTE
GRAVEL
SAND
CLAY
BREEZE WAY
PLASTIC SHEET
FOOTING
CANVAS PROPORTIONS MOVABLE QUICKLY
PLANK WRAPPED
CHAIN SEWN IN LINING
FLASH FLOOD, DITCHES, IRRIGATION ET CETERA.
STONE HOLDS TO CORNERS
CROSS SECTION
BE CONTROLLED SO IT DOES NOT UNDERMINE DAM.

WATER FLOW ABOVE STAKE — cubic ft per min for inch & fraction increments

WATER FLOW ABOVE STAKE		1/8"	1/4"	3/8"	1/2"	5/8"	3/4"	7/8"
1 inch	.40	.47	.55	.65	.74	.83	.93	1.03
2 "	1.14	1.24	1.36	1.47	1.59	1.71	1.83	1.96
3 "	2.09	2.23	2.36	2.50	2.63	2.78	2.92	3.07
4 "	3.22	3.37	3.57	3.68	3.83	3.99	4.16	4.32
5 "	4.50	4.67	4.84	5.01	5.18	5.36	5.54	5.72
6 "	5.90	6.09	6.28	6.47	6.65	6.85	7.05	7.25
7 "	7.44	7.64	7.84	8.05	8.25	8.45	8.66	8.86
8 "	9.10	9.31	9.52	9.74	9.96	10.18	10.40	10.62
9 "	10.86	11.08	11.31	11.54	11.77	12.00	12.23	12.47
10 "	12.71	12.95	13.19	13.43	13.67	13.93	14.16	14.42
11 "	14.67	14.92	15.18	15.43	15.67	15.96	16.20	16.45
12 "	16.73	16.99	17.26	17.52	17.78	18.05	18.32	18.58
13 "	18.87	19.14	19.42	19.69	19.97	20.24	20.52	20.80
14 "	21.09	21.37	21.65	21.94	22.22	22.51	22.70	23.08
15 "	23.38	23.67	23.97	24.26	24.56	24.86	25.16	25.46
16 "	25.76	26.06	26.36	26.66	26.97	27.27	27.58	27.89
17 "	28.20	28.51	28.82	29.14	29.45	29.76	30.08	30.39
18 "	30.70	31.02	31.34	31.66	31.98	32.31	32.63	32.96
19 "	33.29	33.61	33.94	34.27	34.60	34.94	35.27	35.60
20 "	35.94	36.27	36.60	36.94	37.28	37.62	37.96	38.31
21 "	38.65	39.00	39.34	39.69	40.04	40.39	40.73	41.09
22 "	41.43	41.78	42.13	42.49	42.84	43.20	43.56	43.92
23 "	44.28	44.64	45.00	45.38	45.71	46.08	46.43	46.81
24 "	47.18	47.55	47.91	48.28	48.65	49.02	49.39	49.76

CHART ABOVE ↑ TO BE USED WITH HOME-MADE METER.

61. The lowering of water can raise a wheel and the lowering of a wheel can raise water.

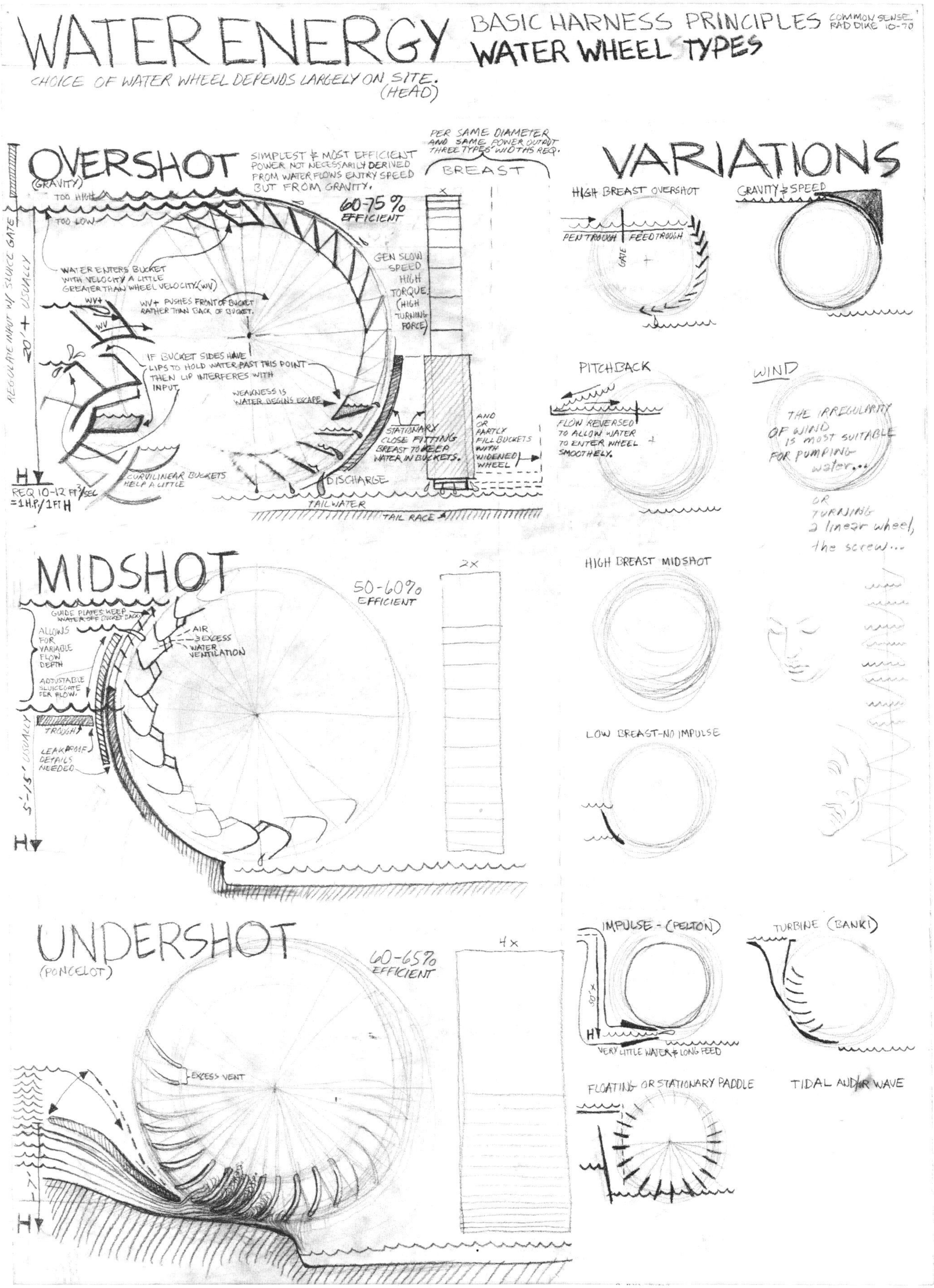

62. Your right and left hands hold enough soil to understand the slope you stand on.

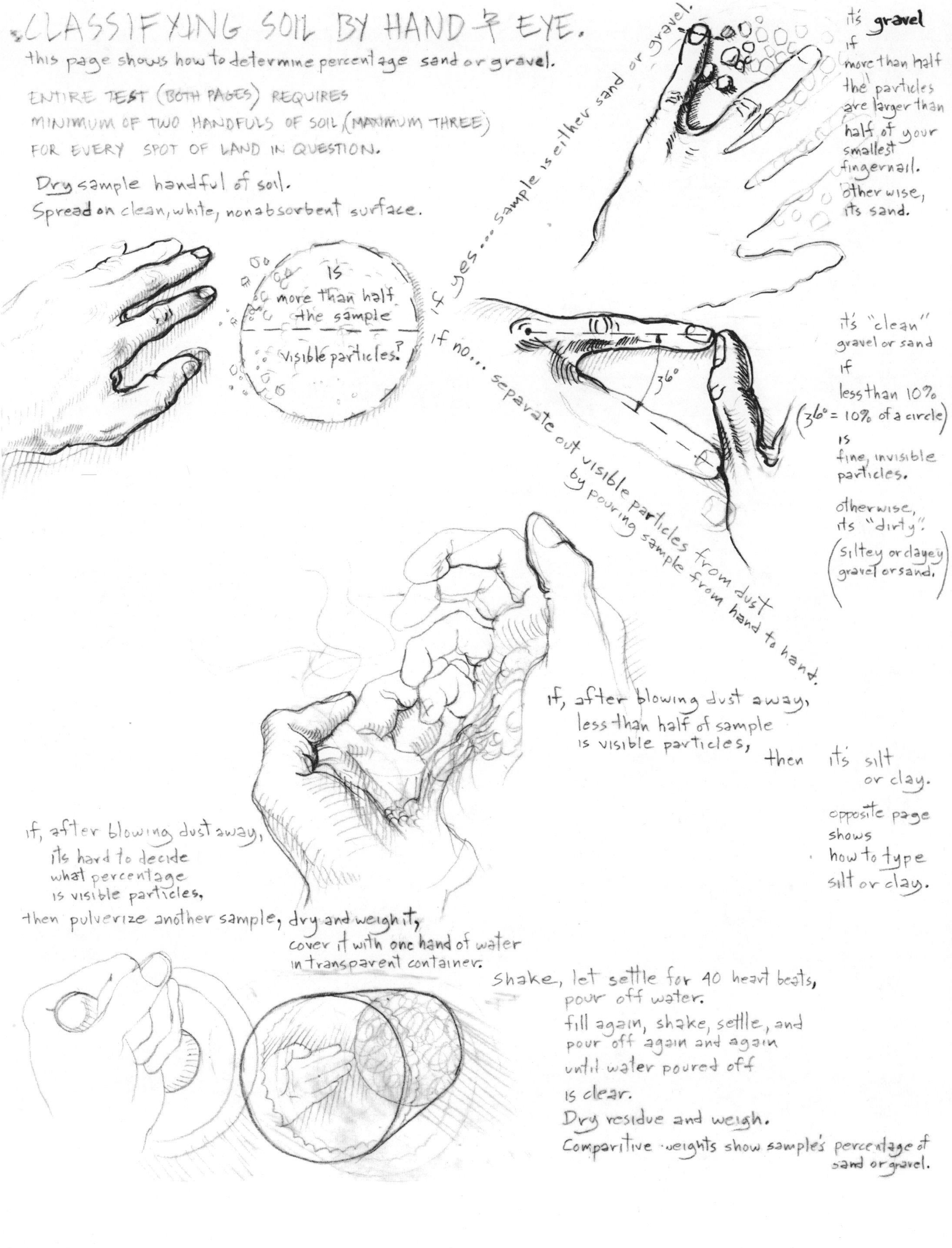
CLASSIFYING SOIL BY HAND & EYE.
this page shows how to determine percentage sand or gravel.
ENTIRE TEST (BOTH PAGES) REQUIRES
MINIMUM OF TWO HANDFULS OF SOIL (MAXIMUM THREE)
FOR EVERY SPOT OF LAND IN QUESTION.
Dry sample handful of soil.
Spread on clean, white, nonabsorbent surface.
is more than half the sample visible particles?
if yes... sample is either sand or gravel.
it's gravel if more than half the particles are larger than half of your smallest fingernail. otherwise, its sand.
if no... separate out visible particles from dust by pouring sample from hand to hand.
36°
it's "clean" gravel or sand if less than 10% (36° = 10% of a circle) is fine, invisible particles.
otherwise, its "dirty". (siltey or clayey gravel or sand.)
if, after blowing dust away, less than half of sample is visible particles, then it's silt or clay.
opposite page shows how to type silt or clay.
if, after blowing dust away, its hard to decide what percentage is visible particles,
then pulverize another sample, dry and weigh it,
cover it with one hand of water in transparent container.
Shake, let settle for 40 heart beats, pour off water.
fill again, shake, settle, and pour off again and again until water poured off is clear.
Dry residue and weigh.
Comparitive weights show sample's percentage of sand or gravel.

63. Distinguish gravel from sand, visually; clay from silt, tensilely; plastic from organic, sensorially.

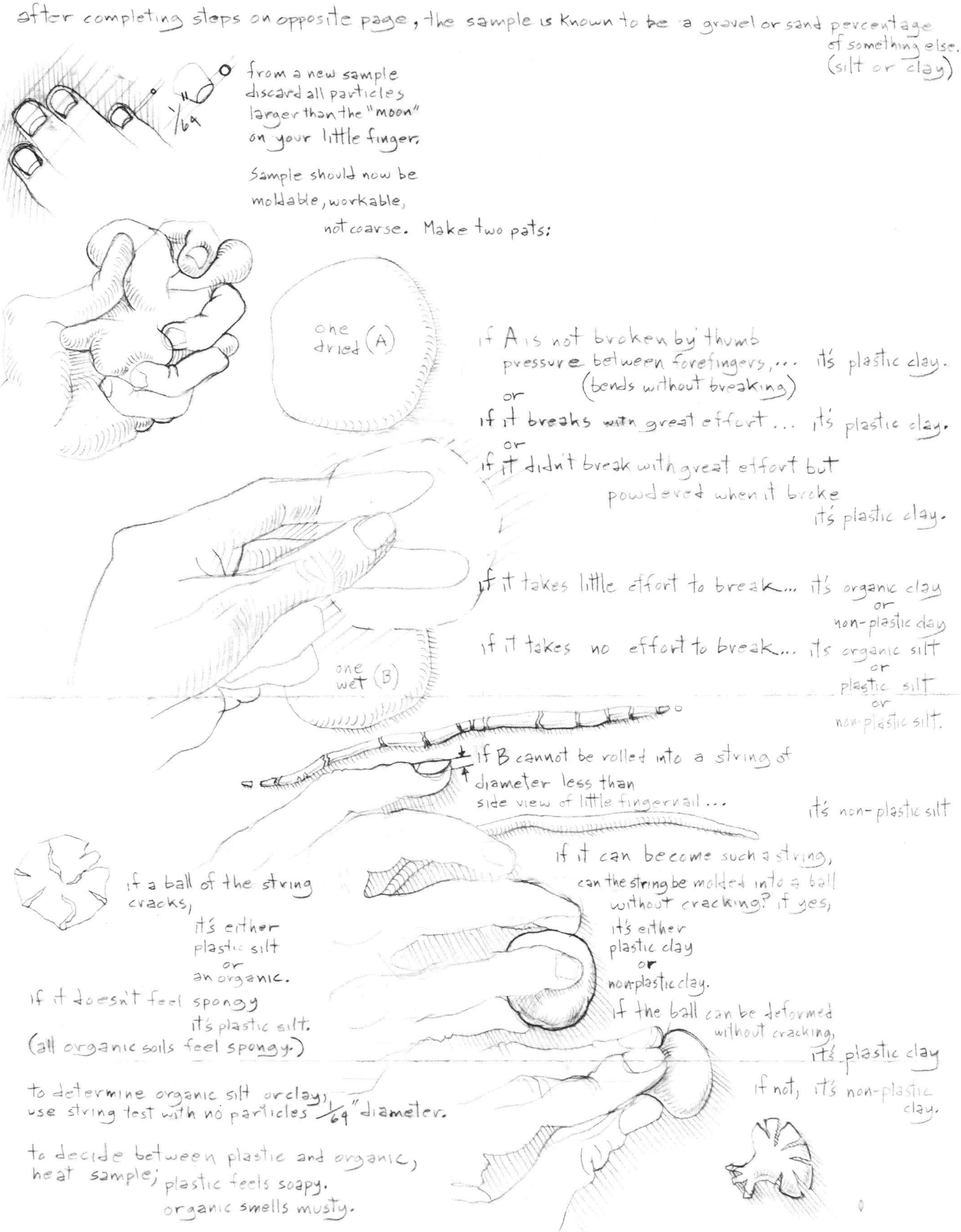
after completing steps on opposite page, the sample is known to be a gravel or sand percentage of something else. (silt or clay)
from a new sample discard all particles larger than the "moon" on your little finger.
1/64"
Sample should now be moldable, workable, not coarse. Make two pats:
one dried (A)
one wet (B)
if A is not broken by thumb pressure between forefingers,... it's plastic clay. (bends without breaking)
or
if it breaks with great effort ... it's plastic clay.
or
if it didn't break with great effort but powdered when it broke it's plastic clay.
if it takes little effort to break... it's organic clay or non-plastic clay
if it takes no effort to break... it's organic silt or plastic silt or non-plastic silt.
if B cannot be rolled into a string of diameter less than side view of little fingernail ... it's non-plastic silt
if it can become such a string, can the string be molded into a ball without cracking? if yes, it's either plastic clay or nonplastic clay.
if the ball can be deformed without cracking, it's plastic clay
if not, it's non-plastic clay.
if a ball of the string cracks, it's either plastic silt or an organic.
if it doesn't feel spongy it's plastic silt.
(all organic soils feel spongy.)
to determine organic silt or clay, use string test with no particles 1/64" diameter.
to decide between plastic and organic, heat sample; plastic feels soapy. organic smells musty.

64. Dirt plus sun, water, wind, and life equals soil.

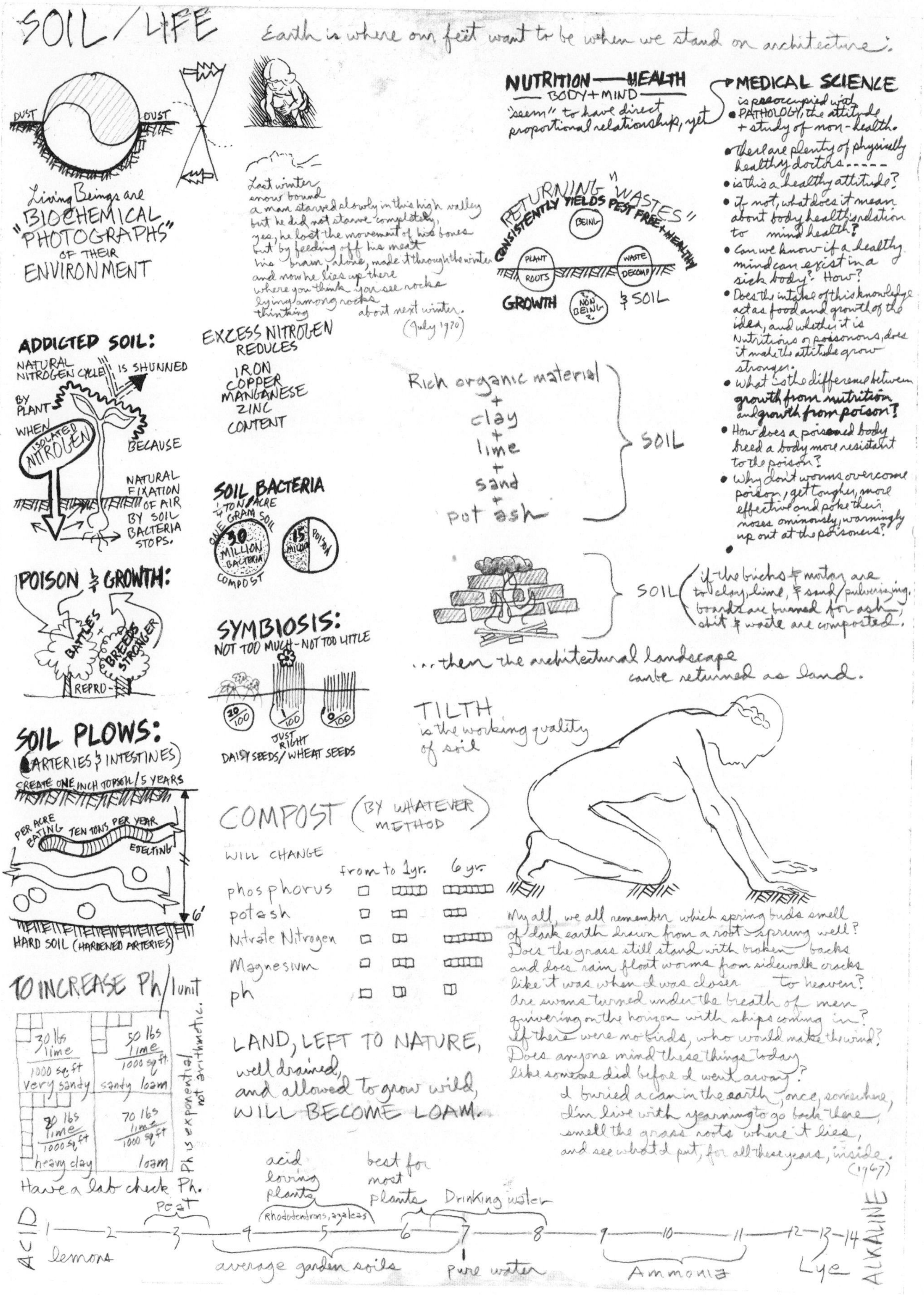
SOIL / LIFE
Earth is where our feet want to be when we stand on architecture.
DUST
DUST
Living Beings are "BIOCHEMICAL PHOTOGRAPHS" OF THEIR ENVIRONMENT
Last winter
snow bound
a man starved slowly in this high valley
but he did not starve completely,
yes, he lost the movement of his bones
but by feeding off his meat
his brain alone, made it through the winter
and now he lies up there
where you think you see rocks
lying among rocks
thinking about next winter.
(July 1970)
NUTRITION — HEALTH
BODY + MIND
"seem" to have direct proportional relationship, yet
MEDICAL SCIENCE
is preoccupied with
• PATHOLOGY, the attitude + study of non-health.
• there are plenty of physically healthy doctors - - - -
• is this a healthy attitude?
• if not, what does it mean about body health's relation to mind health?
• Can we know if a healthy mind can exist in a sick body? How?
• Does the intake of this knowledge act as food and growth of the idea, and whether it is nutritious or poisonous, does it make the attitude grow stronger.
• What is the difference between growth from nutrition and growth from poison?
• How does a poisoned body breed a body more resistant to the poison?
• Why don't worms overcome poison, get tougher, more effective and poke their noses ominously, warningly up out at the poisoners?
RETURNING "WASTES" CONSISTENTLY YIELDS PEST FREE + HEALTHY
BEING
PLANT
ROOTS
WASTE
DECOMP
GROWTH
NON BEING
& SOIL
ADDICTED SOIL:
NATURAL NITROGEN CYCLE IS SHUNNED BY PLANT WHEN
ISOLATED NITROGEN
BECAUSE
NATURAL FIXATION OF AIR BY SOIL BACTERIA STOPS.
EXCESS NITROGEN REDUCES
IRON
COPPER
MANGANESE
ZINC
CONTENT
Rich organic material + clay + lime + sand + pot ash
SOIL
SOIL BACTERIA
¼ TON/ACRE
ONE GRAM SOIL
30 MILLION BACTERIA
COMPOST
15 MILLION
POISON
SOIL
if the bricks & mortar are to clay, lime, & sand pulverizing, boards are burned for ash, shit & waste are composted.
...then the architectural landscape can be returned as land.
POISON & GROWTH:
BATTLES + BREEDS STRONGER
REPRO-
SYMBIOSIS:
NOT TOO MUCH - NOT TOO LITTLE
20/100
1/100
0/100
JUST RIGHT
DAISY SEEDS / WHEAT SEEDS
TILTH
is the working quality of soil
SOIL PLOWS:
(ARTERIES & INTESTINES)
CREATE ONE INCH TOPSOIL / 5 YEARS
PER ACRE EATING TEN TONS PER YEAR EJECTING
6'
HARD SOIL (HARDENED ARTERIES)
COMPOST (BY WHATEVER METHOD)
WILL CHANGE
from to 1 yr. 6 yr.
phosphorus
potash
Nitrate Nitrogen
Magnesium
ph
LAND, LEFT TO NATURE,
well drained,
and allowed to grow wild,
WILL BECOME LOAM.
My all, we all remember which spring buds smell
of dark earth drawn from a root spring well?
Does the grass still stand with broken backs
and does rain float worms from sidewalk cracks
like it was when I was closer to heaven?
Are swans turned under the breath of men
quivering on the horizon with ships coming in?
If there were no birds, who would make the wind?
Does anyone mind these things today
like someone did before I went away?
I buried a can in the earth, once, somewhere,
I'm live with yearning to go back there,
smell the grass roots where it lies,
and see what I put, for all these years, inside.
(1967)
TO INCREASE Ph / 1 unit
30 lbs lime / 1000 sq ft very sandy
50 lbs lime / 1000 sq ft sandy loam
30 lbs lime / 1000 sq ft heavy clay
70 lbs lime / 1000 sq ft loam
Ph is exponential not arithmetic.
Have a lab check Ph.
acid loving plants
(Rhododendrons, azaleas)
best for most plants
Drinking water
Peat
ACID 1 — 2 — 3 — 4 — 5 — 6 — 7 — 8 — 9 — 10 — 11 — 12 — 13 — 14 ALKALINE
lemons
average garden soils
pure water
Ammonia
Lye

65. As soil is the obvious living material, so the material beneath the soil is the obvious building material.

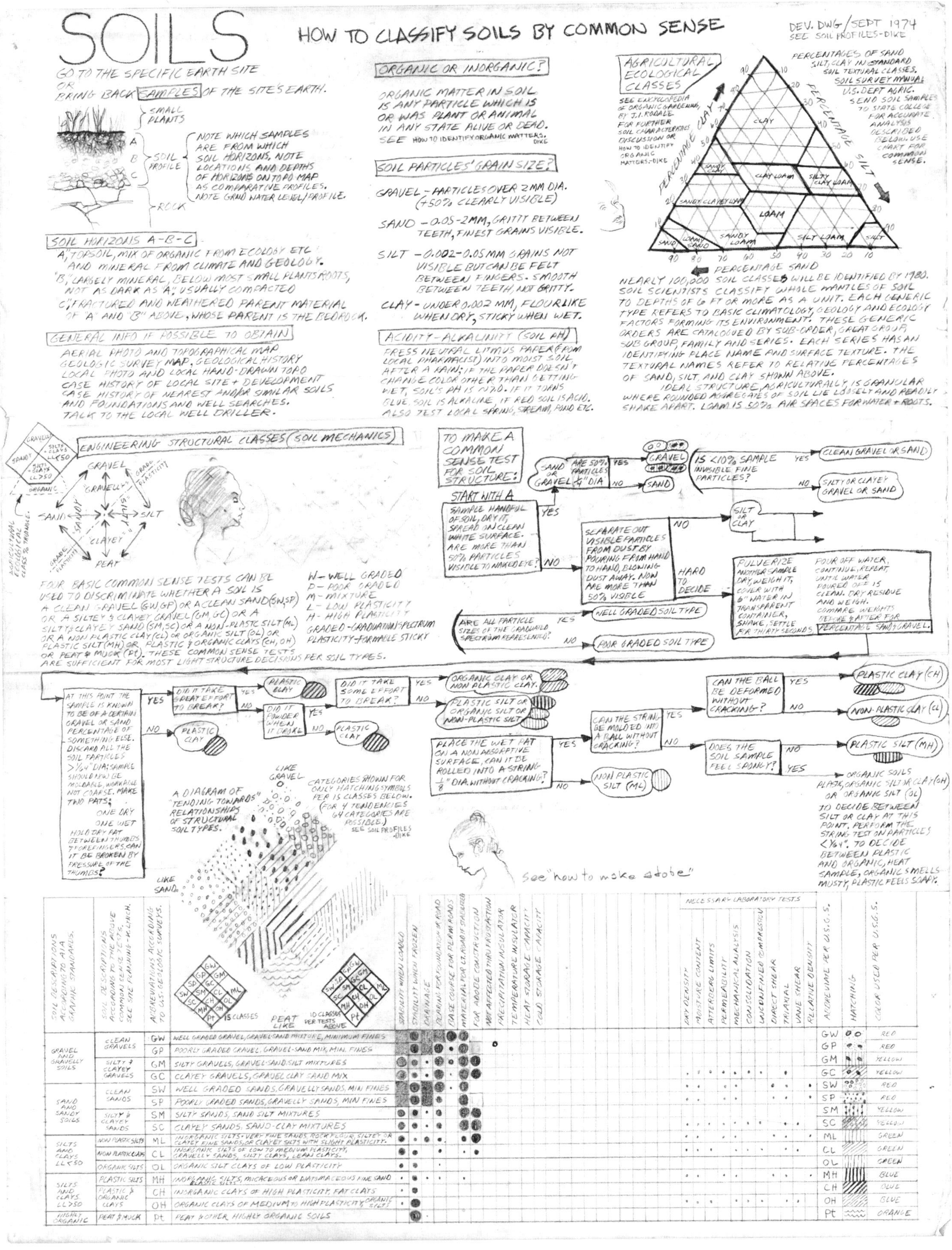

SOILS
HOW TO CLASSIFY SOILS BY COMMON SENSE
DEV. DWG/SEPT 1974
SEE SOIL PROFILES-DIKE
GO TO THE SPECIFIC EARTH SITE OR BRING BACK SAMPLES OF THE SITE'S EARTH.
SMALL PLANTS
A
B
C
SOIL PROFILE
ROCK
NOTE WHICH SAMPLES ARE FROM WHICH SOIL HORIZONS. NOTE LOCATIONS AND DEPTHS OF HORIZONS ON TOPO MAP AS COMPARATIVE PROFILES. NOTE GRND WATER LEVEL/PROFILE.
SOIL HORIZONS A-B-C
"A," TOPSOIL, MIX OF ORGANIC FROM ECOLOGY ETC AND MINERAL FROM CLIMATE AND GEOLOGY.
"B," LARGELY MINERAL, BELOW MOST SMALL PLANTS ROOTS, NOT AS DARK AS "A," USUALLY COMPACTED
"C," FRACTURED AND WEATHERED PARENT MATERIAL OF "A" AND "B" ABOVE, WHOSE PARENT IS THE BEDROCK.
GENERAL INFO IF POSSIBLE TO OBTAIN
AERIAL PHOTO AND TOPOGRAPHICAL MAP
GEOLOGIC SURVEY MAP, GEOLOGICAL HISTORY
LOCAL PHOTO AND LOCAL HAND-DRAWN TOPO
CASE HISTORY OF LOCAL SITE + DEVELOPMENT
CASE HISTORY OF NEAREST AND/OR SIMILAR SOILS AND FOUNDATIONS AND WELL SEARCHES.
TALK TO THE LOCAL WELL DRILLER.
ORGANIC OR INORGANIC?
ORGANIC MATTER IN SOIL IS ANY PARTICLE WHICH IS OR WAS PLANT OR ANIMAL IN ANY STATE ALIVE OR DEAD.
SEE HOW TO IDENTIFY ORGANIC MATTERS. DIKE
SOIL PARTICLES' GRAIN SIZE?
GRAVEL - PARTICLES OVER 2MM DIA. (+50% CLEARLY VISIBLE)
SAND - 0.05-2MM, GRITTY BETWEEN TEETH, FINEST GRAINS VISIBLE.
SILT - 0.002-0.05MM GRAINS NOT VISIBLE BUT CAN BE FELT BETWEEN FINGERS. SMOOTH BETWEEN TEETH, NOT GRITTY.
CLAY - UNDER 0.002 MM, FLOURLIKE WHEN DRY, STICKY WHEN WET.
ACIDITY-ALKALINITY (SOIL PH)
PRESS NEUTRAL LITMUS PAPER (FROM LOCAL PHARMACIST) INTO MOIST SOIL AFTER A RAIN; IF THE PAPER DOESN'T CHANGE COLOR OTHER THAN GETTING WET, SOIL'S PH IS 7.0. IF IT TURNS BLUE, SOIL IS ALKALINE, IF RED SOIL IS ACID. ALSO TEST LOCAL SPRING, STREAM, POND ETC.
AGRICULTURAL ECOLOGICAL CLASSES
SEE ENCYCLOPEDIA OF ORGANIC GARDENING, BY J.I. RODALE FOR FURTHER SOIL CHARACTERISTICS DISCUSSION OR HOW TO IDENTIFY ORGANIC MATTERS-DIKE
PERCENTAGE CLAY
PERCENTAGE SILT
PERCENTAGE SAND
CLAY
SANDY CLAY
CLAY LOAM
SILTY CLAY LOAM
SANDY CLAY LOAM
LOAM
SANDY LOAM
SILT LOAM
LOAMY SAND
SAND
SILT
PERCENTAGES OF SAND SILT, CLAY IN STANDARD SOIL TEXTURAL CLASSES. SOIL SURVEY MANUAL U.S. DEPT AGRIC. SEND SOIL SAMPLES TO STATE COLLEGE FOR ACCURATE ANALYSIS DESCRIBED BELOW. USE CHART FOR COMMON SENSE.
NEARLY 100,000 SOIL CLASSES WILL BE IDENTIFIED BY 1980. SOIL SCIENTISTS CLASSIFY WHOLE MANTLES OF SOIL TO DEPTHS OF 6 FT OR MORE AS A UNIT. EACH GENERIC TYPE REFERS TO BASIC CLIMATOLOGY, GEOLOGY AND ECOLOGY FACTORS FORMING ITS ENVIRONMENT. THESE GENERIC ORDERS ARE CATALOGUED BY SUB-ORDER, GREAT GROUP, SUB GROUP, FAMILY AND SERIES. EACH SERIES HAS AN IDENTIFYING PLACE NAME AND SURFACE TEXTURE. THE TEXTURAL NAMES REFER TO RELATIVE PERCENTAGES OF SAND, SILT, AND CLAY SHOWN ABOVE.
IDEAL STRUCTURE, AGRICULTURALLY, IS GRANULAR WHERE ROUNDED AGGREGATES OF SOIL LIE LOOSELY AND READILY SHAKE APART. LOAM IS 50% AIR SPACES FOR WATER + ROOTS.
ENGINEERING STRUCTURAL CLASSES (SOIL MECHANICS)
GRAVEL
SANDY
SILTS CLAYS LL<50
LL>50
ORGANIC
SAND
SILT
"GRAVELLY"
"SANDY"
"SILTY"
"CLAYEY"
PEAT
GRADE PLASTICITY
AGRICULTURAL ECOLOGICAL CLASS TO TRIANGLE.
FOUR BASIC COMMON SENSE TESTS CAN BE USED TO DISCRIMINATE WHETHER A SOIL IS A CLEAN GRAVEL (GW, GP) OR A CLEAN SAND (SW, SP) OR A SILTY & CLAYEY GRAVEL (GM, GC) OR A SILTY & CLAYEY SAND (SM, SC) OR A NON-PLASTIC SILT (ML) OR A NON PLASTIC CLAY (CL) OR ORGANIC SILT (OL) OR PLASTIC SILT (MH) OR PLASTIC & ORGANIC CLAYS (CH, OH) OR PEAT & MUCK (Pt). THESE COMMON SENSE TESTS ARE SUFFICIENT FOR MOST LIGHT STRUCTURE DECISIONS PER SOIL TYPES.
W - WELL GRADED
P - POOR GRADED
M - MIXTURE
L - LOW PLASTICITY
H - HIGH PLASTICITY
GRADED - GRADUATION SPECTRUM
PLASTICITY - FORMABLE STICKY
TO MAKE A COMMON SENSE TEST FOR SOIL STRUCTURE:
START WITH A SAMPLE HANDFUL OF SOIL, DRY IT, SPREAD ON CLEAN WHITE SURFACE. ARE MORE THAN 50% PARTICLES VISIBLE TO NAKED EYE?
YES
NO
SAND OR GRAVEL
ARE 50% PARTICLES 1/4" DIA
GRAVEL
SAND
IS <10% SAMPLE INVISIBLE FINE PARTICLES?
CLEAN GRAVEL OR SAND
SILTY OR CLAYEY GRAVEL OR SAND
SEPARATE OUT VISIBLE PARTICLES FROM DUST BY POURING FROM HAND TO HAND, BLOWING DUST AWAY. NOW ARE MORE THAN 50% VISIBLE
SILT OR CLAY
HARD TO DECIDE
PULVERIZE ANOTHER SAMPLE DRY, WEIGH IT, COVER WITH 6" WATER IN TRANSPARENT CONTAINER, SHAKE, SETTLE FOR THIRTY SECONDS
POUR OFF WATER, CONTINUE, REPEAT, UNTIL WATER POURED OFF IS CLEAN. DRY RESIDUE AND WEIGH. COMPARE WEIGHTS BEFORE & AFTER FOR PERCENTAGE SAND & GRAVEL.
ARE ALL PARTICLE SIZES OF THE GRADUATED SPECTRUM REPRESENTED?
WELL GRADED SOIL TYPE
POOR GRADED SOIL TYPE
AT THIS POINT THE SAMPLE IS KNOWN TO BE OF A CERTAIN GRAVEL OR SAND PERCENTAGE OF SOMETHING ELSE. DISCARD ALL THE SOIL PARTICLES >1/64" DIA; SAMPLE SHOULD NOW BE MOLDABLE, WORKABLE NOT COARSE. MAKE TWO PATS: ONE DRY ONE WET
HOLD DRY PAT BETWEEN THUMBS & FOREFINGERS, CAN IT BE BROKEN BY PRESSURE OF THE THUMBS?
DID IT TAKE GREAT EFFORT TO BREAK?
PLASTIC CLAY
DID IT POWDER WHEN IT BROKE
PLASTIC CLAY
DID IT TAKE SOME EFFORT TO BREAK?
ORGANIC CLAY OR NON PLASTIC CLAY.
PLASTIC SILT OR ORGANIC SILT OR NON-PLASTIC SILT
PLACE THE WET PAT ON A NON ABSORPTIVE SURFACE, CAN IT BE ROLLED INTO A STRING 1/8" DIA. WITHOUT CRACKING?
NON PLASTIC SILT (ML)
CAN THE STRING BE MOLDED INTO A BALL WITHOUT CRACKING?
CAN THE BALL BE DEFORMED WITHOUT CRACKING?
PLASTIC CLAY (CH)
NON-PLASTIC CLAY (CL)
DOES THE SOIL SAMPLE FEEL SPONGY?
PLASTIC SILT (MH)
ORGANIC SOILS PLASTIC, ORGANIC SILT OR CLAY (OH) OR ORGANIC SILT (OL)
TO DECIDE BETWEEN SILT OR CLAY AT THIS POINT, PERFORM THE STRING TEST ON PARTICLES <1/64". TO DECIDE BETWEEN PLASTIC AND ORGANIC, HEAT SAMPLE, ORGANIC SMELLS MUSTY, PLASTIC FEELS SOAPY.
A DIAGRAM OF "TENDING TOWARDS" RELATIONSHIPS OF STRUCTURAL SOIL TYPES.
LIKE GRAVEL
LIKE SAND
PEAT LIKE
CATEGORIES SHOWN FOR ONLY HATCHING SYMBOLS PER 15 CLASSES BELOW. (FOR 4 TENDENCIES 64 CATEGORIES ARE POSSIBLE.) SEE SOIL PROFILES -DIKE
15 CLASSES
10 CLASSES PER TESTS ABOVE
see "how to make adobe"
SOIL DESCRIPTIONS ACCORDING TO AIA GRAPHIC STANDARDS.
SOIL DESCRIPTIONS ACCORDING TO THE ABOVE COMMON SENSE TESTS. SEE SITE PLANNING - K. LYNCH.
ABBREVIATIONS ACCORDING TO US GEOLOGIC SURVEYS.
STABILITY WHEN LOADED
STABILITY WHEN FROZEN
DRAINAGE
PLAIN: FOR FOUNDATION OR ROAD
BASE COURSE FOR PLAIN ROADS
MATERIAL FOR LT. ROAD IF SURFACED
FOR ADOBE CONSTRUCTION
NOT AFFECTED THRU PRECIPITATION
PRECIPITATION INSULATOR
TEMPERATURE INSULATOR
HEAT STORAGE CAPACITY
COLD STORAGE CAPACITY
NECESSARY LABORATORY TESTS
DRY DENSITY
MOISTURE CONTENT
ATTERBERG LIMITS
PERMEABILITY
MECHANICAL ANALYSIS
CONSOLIDATION
UNCONFINED COMPRESSION
DIRECT SHEAR
TRIAXIAL
VANE SHEAR
RELATIVE DENSITY
ABBREVIATE PER U.S.G.S.
HATCHING
COLOR USED PER U.S.G.S.
GRAVEL AND GRAVELLY SOILS
CLEAN GRAVELS
GW WELL GRADED GRAVEL, GRAVEL-SAND MIXTURE, MINIMUM FINES — GW — RED
GP POORLY GRADED GRAVEL, GRAVEL-SAND MIX, MIN. FINES — GP — RED
SILTY & CLAYEY GRAVELS
GM SILTY GRAVELS, GRAVEL-SAND-SILT MIXTURES — GM — YELLOW
GC CLAYEY GRAVELS, GRAVEL-CLAY-SAND MIX — GC — YELLOW
SAND AND SANDY SOILS
CLEAN SANDS
SW WELL GRADED SANDS, GRAVELLY SANDS, MIN FINES — SW — RED
SP POORLY GRADED SANDS, GRAVELLY SANDS, MIN FINES — SP — RED
SILTY & CLAYEY SANDS
SM SILTY SANDS, SAND SILT MIXTURES — SM — YELLOW
SC CLAYEY SANDS, SAND-CLAY MIXTURES — SC — YELLOW
SILTS AND CLAYS LL<50
NON PLASTIC SILTS — ML INORGANIC SILTS-VERY FINE SANDS, ROCK FLOUR, SILTY OR CLAYEY FINE SANDS, OR CLAYEY SILTS WITH SLIGHT PLASTICITY. — ML — GREEN
NON PLASTIC CLAYS — CL INORGANIC CLAYS OF LOW TO MEDIUM PLASTICITY, GRAVELLY SANDS, SILTY CLAYS, LEAN CLAYS. — CL — GREEN
ORGANIC SILTS — OL ORGANIC SILT CLAYS OF LOW PLASTICITY — OL — GREEN
SILTS AND CLAYS LL>50
PLASTIC SILTS — MH INORGANIC SILTS, MICACEOUS OR DIATOMACEOUS FINE SAND ELASTIC SILTS — MH — BLUE
PLASTIC & ORGANIC CLAYS — CH INORGANIC CLAYS OF HIGH PLASTICITY, FAT CLAYS — CH — BLUE
OH ORGANIC CLAYS OF MEDIUM TO HIGH PLASTICITY, ORGANIC SILTS — OH — BLUE
HIGHLY ORGANIC
PEAT & MUCK — Pt PEAT & OTHER HIGHLY ORGANIC SOILS — Pt — ORANGE

66. A root is the seed's downward thrust from the sun.

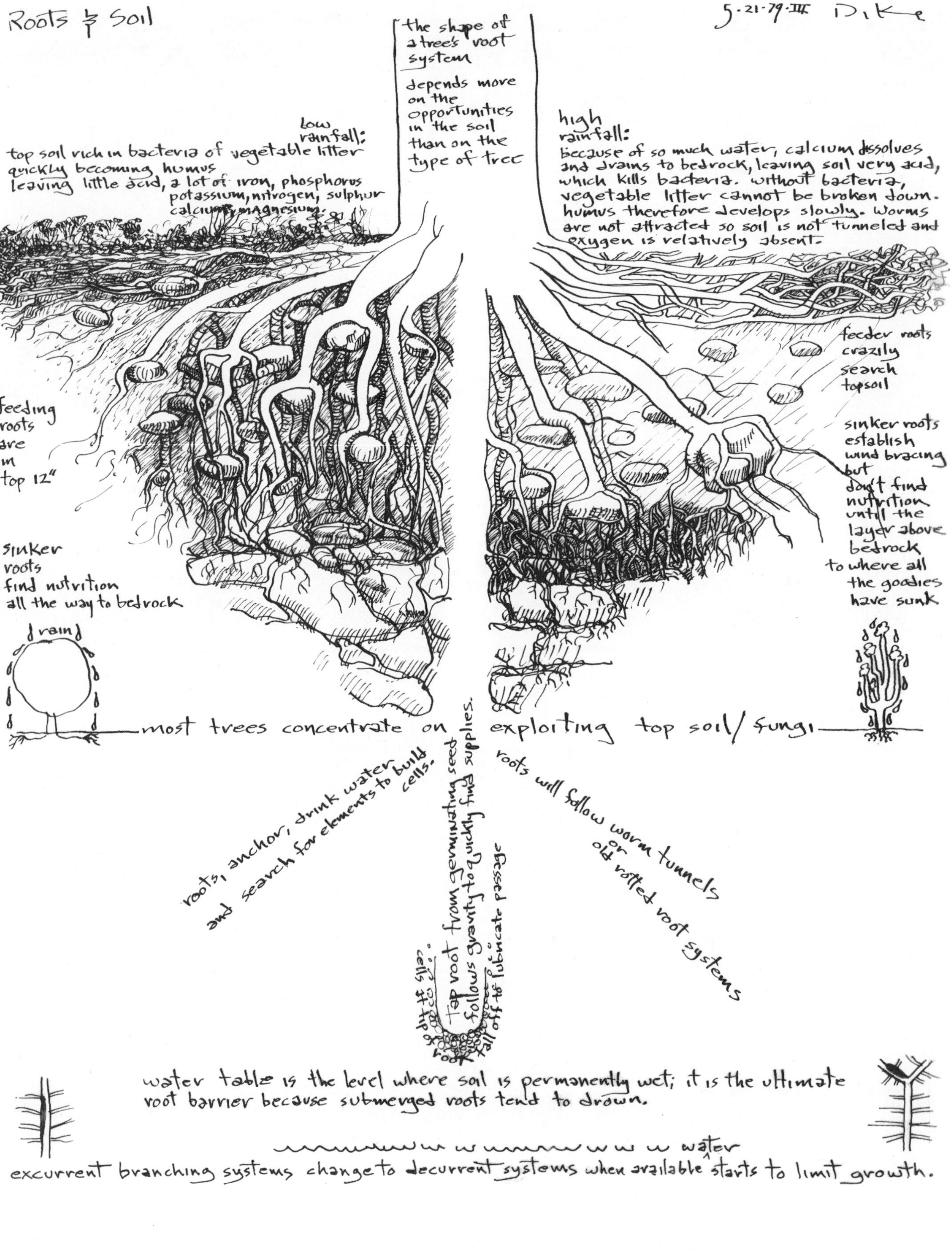
Roots & Soil
5.21.79.III Dike
the shape of a tree's root system
depends more on the opportunities in the soil than on the type of tree
low rainfall:
top soil rich in bacteria of vegetable litter quickly becoming humus leaving little acid, a lot of iron, phosphorus potassium, nitrogen, sulphur calcium, magnesium.
high rainfall:
because of so much water, calcium dissolves and drains to bedrock, leaving soil very acid, which kills bacteria. without bacteria, vegetable litter cannot be broken down. humus therefore develops slowly. Worms are not attracted so soil is not tunneled and oxygen is relatively absent.
feeder roots crazily search topsoil
feeding roots are in top 12"
sinker roots establish wind bracing but don't find nutrition until the layer above bedrock to where all the goodies have sunk
sinker roots find nutrition all the way to bedrock
rain
most trees concentrate on exploiting top soil/fungi
roots, anchor, drink water and search for elements to build cells.
tap root from germinating seed follows gravity to quickly find supplies.
roots will follow worm tunnels or old rotted root systems
cells at tip of root fall off to lubricate passage
water table is the level where soil is permanently wet; it is the ultimate root barrier because submerged roots tend to drown.
water
excurrent branching systems change to decurrent systems when available water starts to limit growth.

67. A branch is the seed's upward thrust from the slope to the sun.

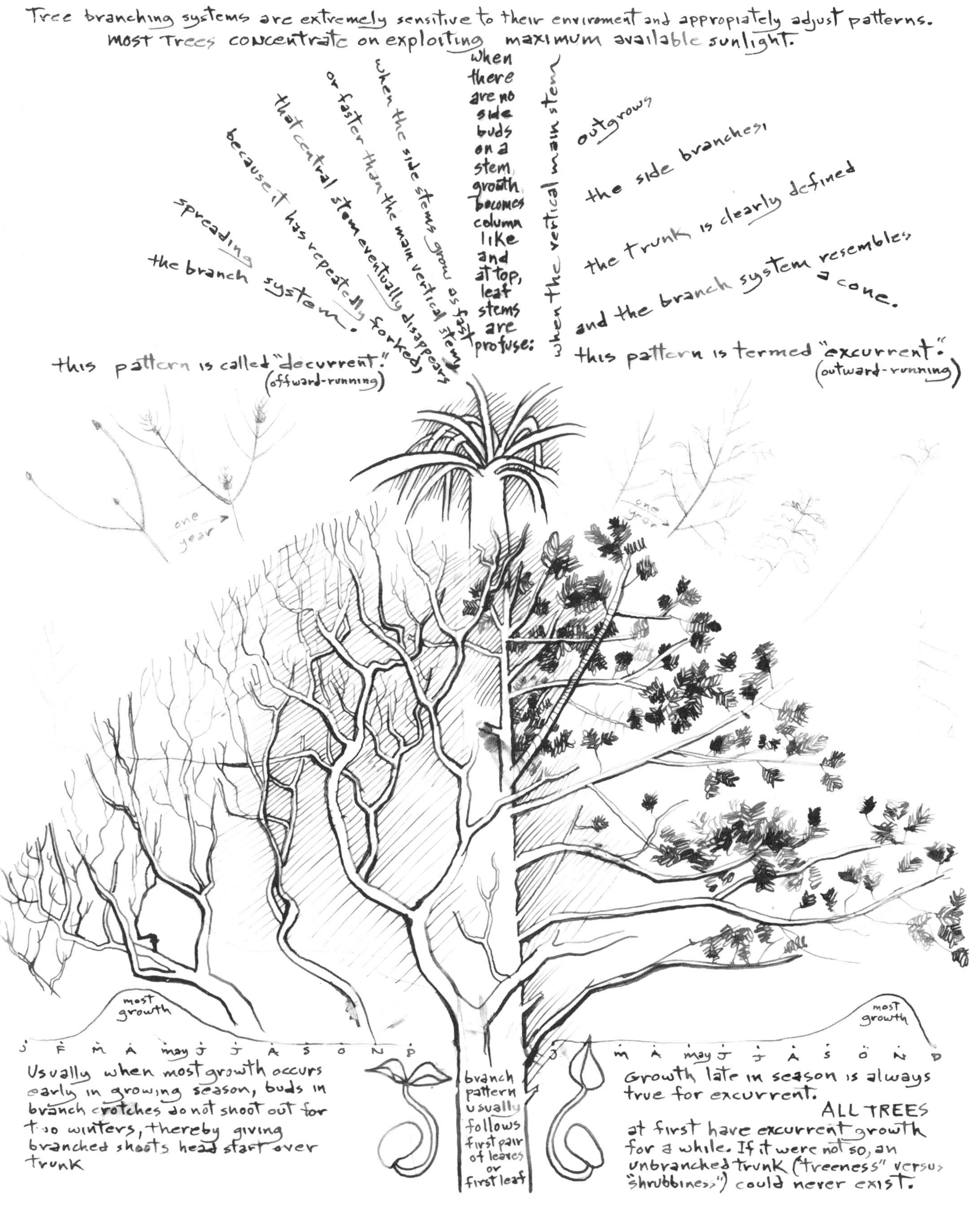
Tree branching systems are extremely sensitive to their environment and appropiately adjust patterns.
most trees concentrate on exploiting maximum available sunlight.
spreading the branch system.
because it has repeatedly forked)
that central stem eventually disappears
or faster than the main vertical stem)
when the side stems grow as fast
this pattern is called "decurrent." (offward-running)
when there are no side buds on a stem, growth becomes column like and at top, leaf stems are profuse:
when the vertical main stem
outgrows
the side branches,
the trunk is clearly defined
the trunk resembles
and the branch system resembles a cone.
this pattern is termed "excurrent." (outward-running)
one year
one year
most growth
J F M A may J J A S O N D
most growth
J M A may J J A S O N D
Usually when most growth occurs early in growing season, buds in branch crotches do not shoot out for two winters, thereby giving branched shoots head start over trunk
branch pattern usually follows first pair of leaves or first leaf
Growth late in season is always true for excurrent.
ALL TREES at first have excurrent growth for a while. If it were not so, an unbranched trunk ("treeness" versus "shrubbiness") could never exist.

68. Speculations of electromagnetic and genetic control may someday architecture growth.

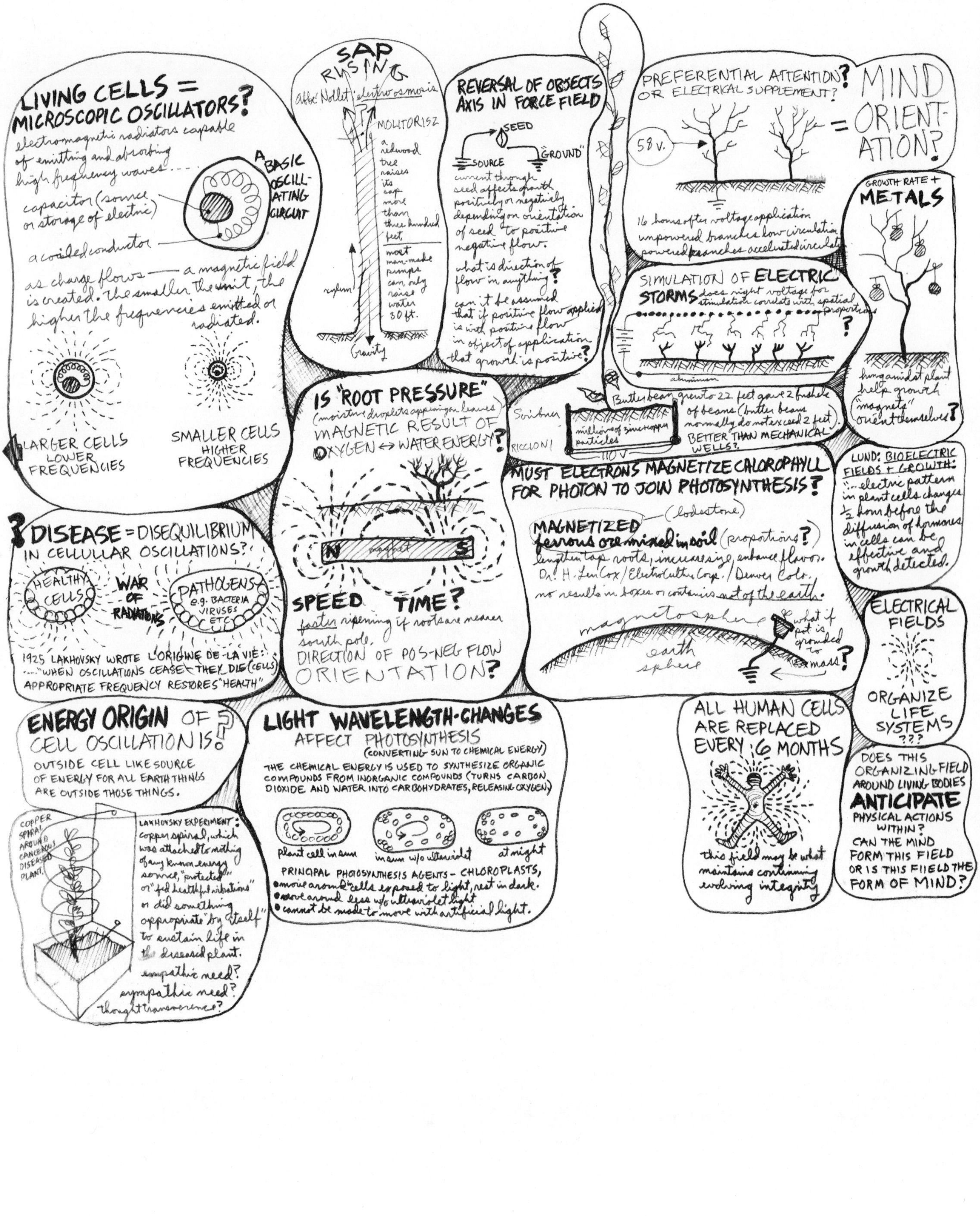

LIVING CELLS = MICROSCOPIC OSCILLATORS?
electromagnetic radiators capable of emitting and absorbing high frequency waves
A BASIC OSCILLATING CIRCUIT
capacitor (source or storage of electric)
a coiled conductor
as charge flows — a magnetic field is created. The smaller the unit, the higher the frequencies emitted or radiated.
LARGER CELLS LOWER FREQUENCIES
SMALLER CELLS HIGHER FREQUENCIES
SAP RISING
Abbé Nollet / electro-osmosis
MOLITORIS 32
a redwood tree raises its sap more than three hundred feet
most man-made pumps can only raise water 30 ft.
xylem
Gravity
REVERSAL OF OBJECTS AXIS IN FORCE FIELD
SEED
SOURCE
"GROUND"
current through seed affects growth positively or negatively depending on orientation of seed to positive negative flow.
what is direction of flow in anything?
can it be assumed that if positive flow applied is with positive flow in object of application that growth is positive?
PREFERENTIAL ATTENTION? OR ELECTRICAL SUPPLEMENT?
58 v.
= MIND ORIENTATION?
16 hours after voltage application unpowered branches low circulation powered branches accelerated circulation
GROWTH RATE + METALS
hung amidst plant help growth "magnets" orient themselves?
SIMULATION OF ELECTRIC STORMS
does night voltage for stimulation correlate with spatial proportions?
aluminum
Butter bean grew to 22 feet gave 2 bushels of beans (butter beans normally do not exceed 2 feet). BETTER THAN MECHANICAL WELLS?
millions of zinc-copper particles
110 V
RICCIONI
IS "ROOT PRESSURE" (moisture droplets appearing on leaves) MAGNETIC RESULT OF OXYGEN ↔ WATER ENERGY?
N
magnet
S
SPEED TIME?
faster ripening if roots are nearer south pole.
DIRECTION OF POS-NEG FLOW ORIENTATION?
MUST ELECTRONS MAGNETIZE CALOROPHYLL FOR PHOTON TO JOIN PHOTOSYNTHESIS?
(lodestone)
MAGNETIZED ferrous ore mixed in soil (proportions?)
lengthen tap roots, increasing enhance flavor.
Dr. H. LenCox / Electroculture Corp. / Denver Colo.
no results in boxes or containers out of the earth.
magnetosphere
earth sphere
what if pot is grounded to mass?
LUND: BIOELECTRIC FIELDS + GROWTH:
electric pattern in plant cells changes ½ hour before the diffusion of hormones in cells can be effective and growth detected.
ELECTRICAL FIELDS
ORGANIZE LIFE SYSTEMS ???
DISEASE = DISEQUILIBRIUM IN CELLULAR OSCILLATIONS?
HEALTHY CELLS
WAR OF RADIATIONS
PATHOGENS e.g. BACTERIA VIRUSES ETC.
1925 LAKHOVSKY WROTE L'ORIGINE DE LA VIE: "WHEN OSCILLATIONS CEASE THEY DIE (CELLS)" APPROPRIATE FREQUENCY RESTORES "HEALTH"
ENERGY ORIGIN OF CELL OSCILLATION IS?
OUTSIDE CELL LIKE SOURCE OF ENERGY FOR ALL EARTH THINGS ARE OUTSIDE THOSE THINGS.
COPPER SPIRAL AROUND CANCEROUS DISEASED PLANT.
LAKHOVSKY EXPERIMENT: copper spiral, which was attached to nothing of any known energy source, "protected" or "fed healthful vibrations" or did something appropriate "by itself" to sustain life in the diseased plant.
empathic need?
sympathic need?
thought transference?
LIGHT WAVELENGTH-CHANGES AFFECT PHOTOSYNTHESIS
(CONVERTING SUN TO CHEMICAL ENERGY)
THE CHEMICAL ENERGY IS USED TO SYNTHESIZE ORGANIC COMPOUNDS FROM INORGANIC COMPOUNDS (TURNS CARBON DIOXIDE AND WATER INTO CARBOHYDRATES, RELEASING OXYGEN)
plant cell in sun
in sun w/o ultraviolet
at night
PRINCIPAL PHOTOSYNTHESIS AGENTS - CHLOROPLASTS,
move around cells exposed to light, rest in dark.
move around less w/o ultraviolet light
cannot be made to move with artificial light.
ALL HUMAN CELLS ARE REPLACED EVERY 6 MONTHS
this field may be what maintains continuing evolving integrity
DOES THIS ORGANIZING FIELD AROUND LIVING BODIES ANTICIPATE PHYSICAL ACTIONS WITHIN?
CAN THE MIND FORM THIS FIELD OR IS THIS FIELD THE FORM OF MIND?

69. But fundamental facts of climate must be respected if we are the selectors of what grows where.

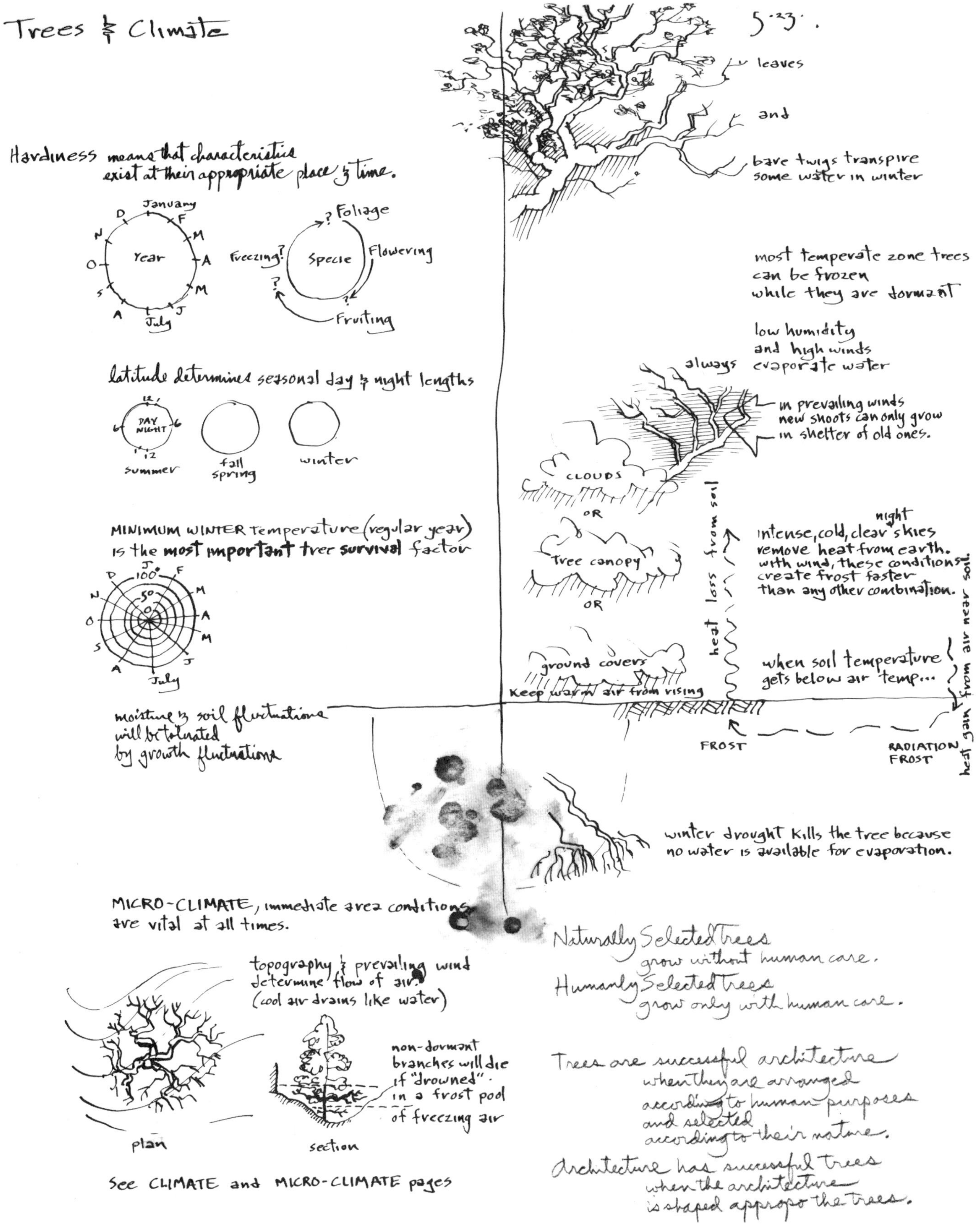
Trees & Climate
5·23·
leaves
and
bare twigs transpire
some water in winter
Hardiness means that characteristics
exist at their appropriate place & time.
January
F
M
A
M
J
July
A
S
O
N
D
Year
Foliage
Flowering
Fruiting
Freezing?
Specie
latitude determines seasonal day & night lengths
DAY
NIGHT
12
6
summer
fall
spring
winter
most temperate zone trees
can be frozen
while they are dormant
low humidity
and high winds
always evaporate water
in prevailing winds
new shoots can only grow
in shelter of old ones.
CLOUDS
OR
tree canopy
OR
ground covers
Keep warm air from rising
heat loss from soil
night
intense, cold, clear skies
remove heat from earth.
with wind, these conditions
create frost faster
than any other combination.
when soil temperature
gets below air temp...
heat gain from air near soil
FROST
RADIATION
FROST
MINIMUM WINTER temperature (regular year)
is the most important tree survival factor
100
50
0
moisture & soil fluctuations
will be tolerated
by growth fluctuations
winter drought kills the tree because
no water is available for evaporation.
MICRO-CLIMATE, immediate area conditions
are vital at all times.
Naturally Selected trees
grow without human care.
Humanly Selected trees
grow only with human care.
topography & prevailing wind
determine flow of air.
(cool air drains like water)
non-dormant
branches will die
if "drowned"
in a frost pool
of freezing air
plan
section
Trees are successful architecture
when they are arranged
according to human purposes
and selected
according to their nature.
Architecture has successful trees
when the architecture
is shaped appropo the trees.
See CLIMATE and MICRO-CLIMATE pages

70. Both branches and leaves need protection and are protection.

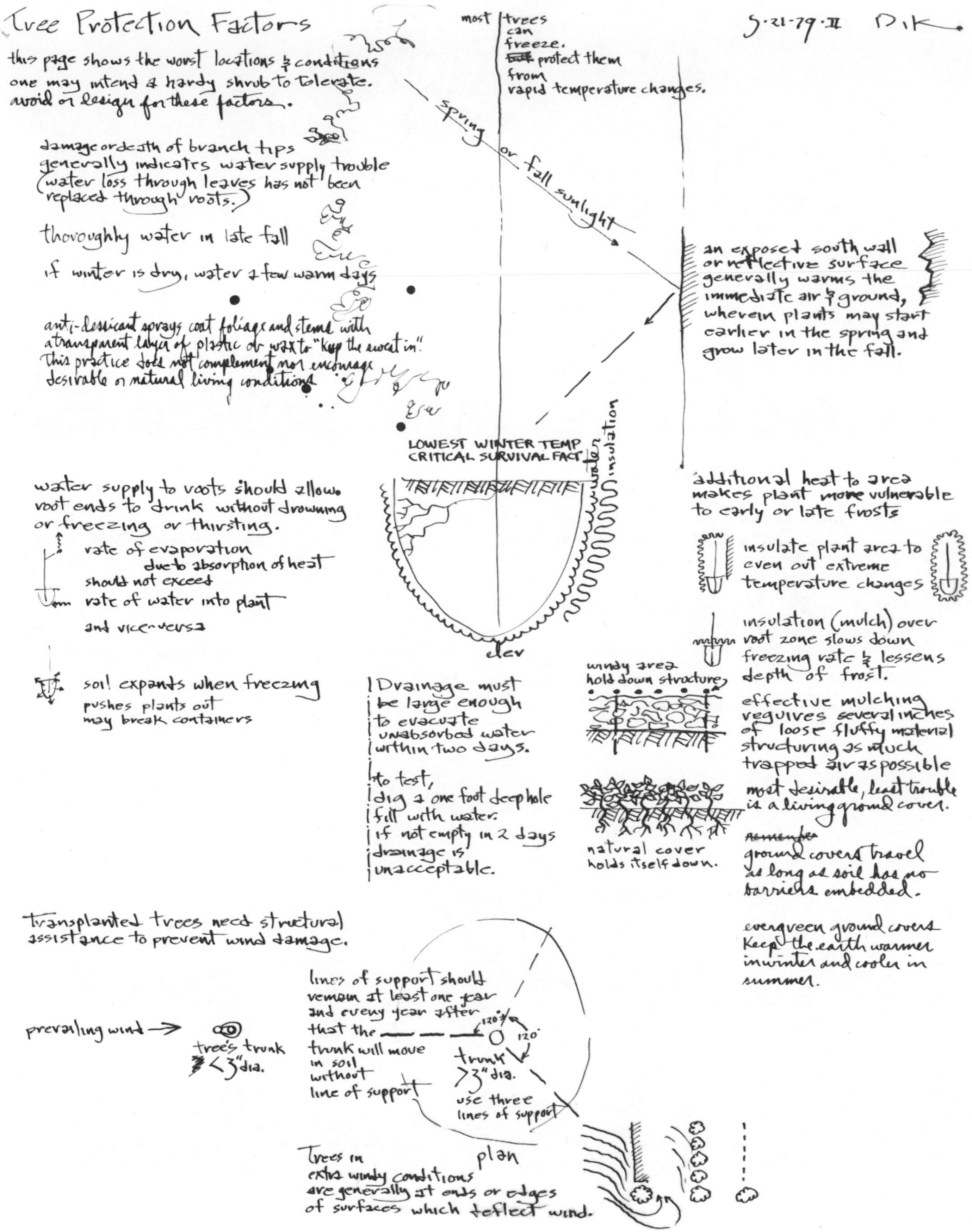
Tree Protection Factors
5·21·79·II DIK.
this page shows the worst locations & conditions one may intend a hardy shrub to tolerate. avoid or design for these factors.
most trees can freeze. protect them from rapid temperature changes.
spring or fall sunlight
damage or death of branch tips generally indicates water supply trouble (water loss through leaves has not been replaced through roots.)
thoroughly water in late fall
if winter is dry, water a few warm days
anti-dessicant sprays coat foliage and stems with a transparent layer of plastic or wax to "keep the sweat in". This practice does not complement nor encourage desirable or natural living conditions.
an exposed south wall or reflective surface generally warms the immediate air & ground, wherein plants may start earlier in the spring and grow later in the fall.
LOWEST WINTER TEMP CRITICAL SURVIVAL FACT
water
insulation
elev
additional heat to area makes plant more vulnerable to early or late frosts
water supply to roots should allow root ends to drink without drowning or freezing or thirsting.
rate of evaporation due to absorption of heat should not exceed rate of water into plant and vice-versa
insulate plant area to even out extreme temperature changes
insulation (mulch) over root zone slows down freezing rate & lessens depth of frost.
soil expands when freezing pushes plants out may break containers
Drainage must be large enough to evacuate unabsorbed water within two days.
to test, dig a one foot deep hole fill with water. if not empty in 2 days drainage is unacceptable.
windy area hold down structure
natural cover holds itself down.
effective mulching requires several inches of loose fluffy material structuring as much trapped air as possible
most desirable, least trouble is a living ground cover.
ground covers travel as long as soil has no barriers embedded.
evergreen ground covers keep the earth warmer in winter and cooler in summer.
transplanted trees need structural assistance to prevent wind damage.
prevailing wind
tree's trunk <3" dia.
lines of support should remain at least one year and every year after that the trunk will move in soil without line of support
120°
120°
trunk >3" dia.
use three lines of support
plan
Trees in extra windy conditions are generally at ends or edges of surfaces which deflect wind.

71. Protected roots are contained, but must be protected from their containment.

within the root resides a will, a dull nudge with cunning strategy...
patiently follows up the rising civilization-hill,
then surely brings it down, beyond mere tragedy.

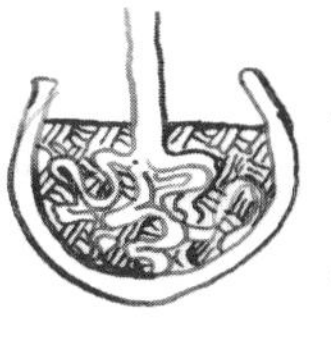

roots growing in contained earth will push the soil upwards (outwards)

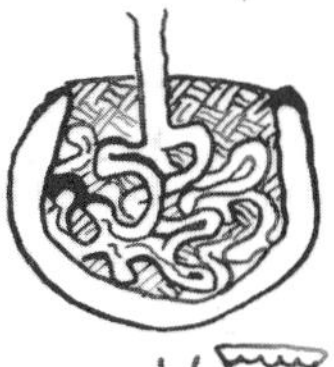

when earth fills water catch-basin, water runs off and roots dry.

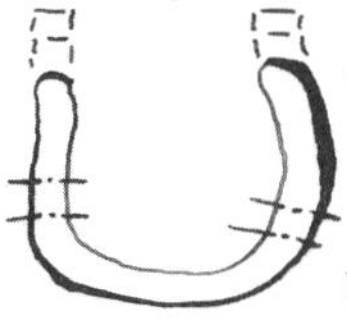

therefore design for container expansion, access for pruning roots,

and possible water source below roots.

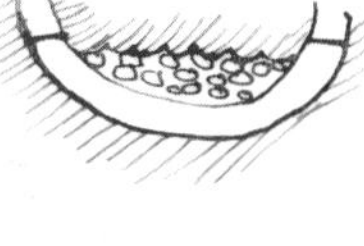

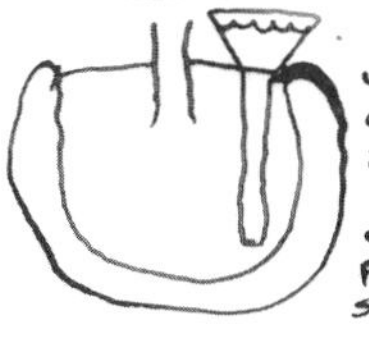

when this has occurred, a funnel can reservoir the water while a pipe lets it slowly seep into the soil.

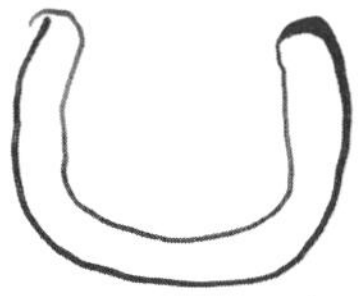

Masonry containers, though less movable than Wooden containers
are more durable
more insulative for winter summer extremes
more expensive, immediately

(5-8 years average life span)
(must water every day at peak of summer, unless larger than frost line depth, will freeze)
(more expensive in the long run after rebuilding 3-4 times in 20 years.

therefore, when designing:

brick is the most adaptable
(easy to move to building site in loads appropriate to your capacity.
(easy to construct irregular walls
but
difficult to build a brick floor without an under floor.

stone is heavy, bulky and if not quarried, unwieldy and visually demands large space around it. Otherwise it has brick's adaptability.

concrete must either be poured at the site or transported as pre-cast slabs, both of which require form-work and transport labor, large expenses. reinforced concrete's advantage over brick and stone as a container is that it is self supporting in a horizontal plane. therefore circulation and drainage feel freest with concrete.

modular concrete, lighter man made stones

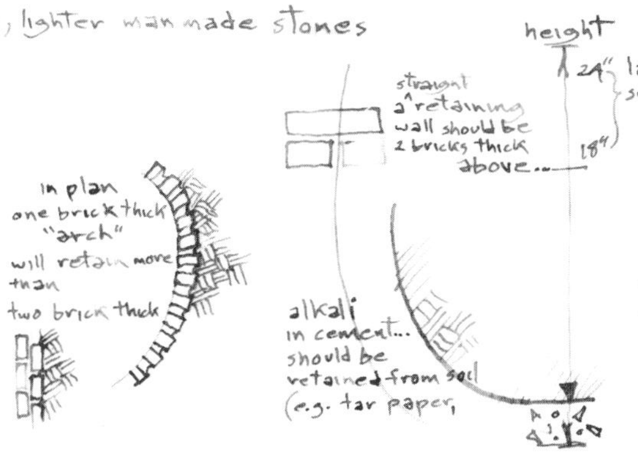
in plan
one brick thick "arch"
will retain more than
two brick thick
height
24"
18"
large shrubs
small trees
a straight retaining wall should be 2 bricks thick above...
alkali in cement... should be retained from soil (e.g. tar paper,

visual relationship of wall thickness may ask for one brick thick wall...

if "aesthetic decision" builds such then reinforce ...

reinforcement is not visible in present construction techniques therefore such a wall does not totally "say" what it is doing and may "look" like it is too thin to retain the mass of earth behind it.

wood containing or retaining soil wants to become soil
(rot)
because moisture and fungi in soil spread through porousness and find themselves moving against any containing surface.

woods most resistant to rotting are california redwood and red cedar especially the heartwood.

less resistant but especially strong is Douglas fir.

high grades of pine are durable and less expensive.

cypress resists rot well but it is not strong.

preservatives applied to wood do help it resist rot longer.
creosote is dangerous to plants.
tar paint is cheap, messy and lasts max. 2 yrs.
liquid plastics & liquid rubbers are air-tight seals

joints & drainage holes must be preserved as well as surfaces.

72. Stem begets stem, a system of sunlight exploitation.

shrubs and trees do not seem to be definable by difference in growth rate, for I have observed shrubs growing faster and larger in early years than trees which, of course, in later years overtook them. (a few years)

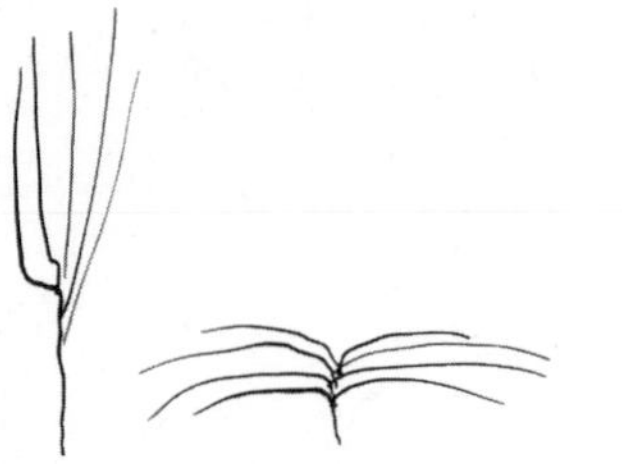

It appears that, when an upright stem bends away from vertical (what would cause this?) gravity no longer restrains the transport of food at the end of the shoot in the same way. travelling horizontally and continuing the habit from budding branch to budding branch, ~~creates~~ this sprawling pattern ~~of x~~ creates "shrubbiness" rather than "tree-ness".

It seems probable that duct ~~size~~ diameter, either to carry the nourishment upward or downward, limits the quantity of nourishment and therefore quantity of length and number of stems.

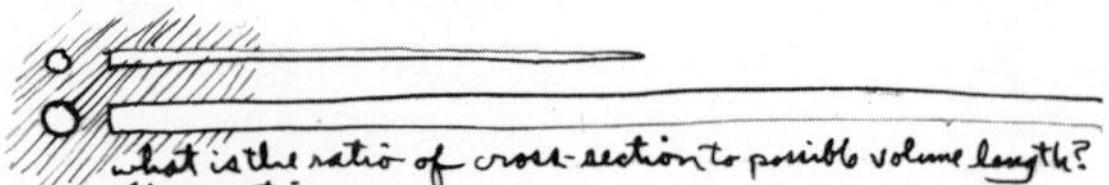

Certainly the structural nature of the duct and multi-duct wall determines whether the stem wants to be a column or a cantilever.

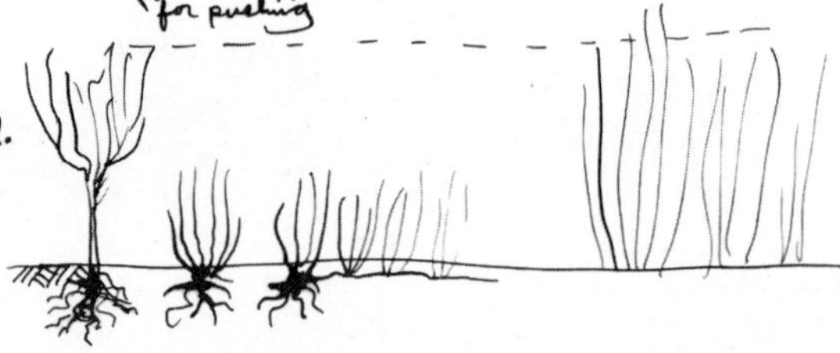

Assuming that genetics determines these patterns, regularity as well as irregularity may be the potential. If "regular plant growth is from one stem begetting many stems, then the irregularity of many stems starting from root level may be called "shrubbiness" and the former "tree-ness"

However, I have seen medium tree-size multi-stems rising from the ground.

That which is normally a tree will become a shrub if winter is so cold that its survival potential is greater under a snow drift, which is warmer than standing in a fierce wind. This was one of my first observations about trees after many visits to timber-line on Trail-Ridge in the Rockies.

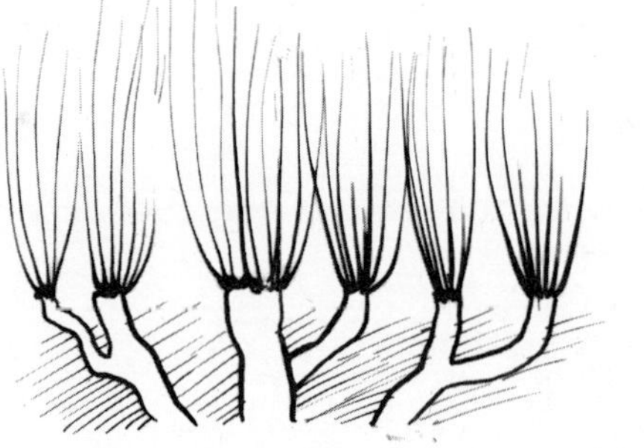

Chemical and genetic facts I have not observed, and perhaps that is where the definition of "shrubbiness" and "treeness" lies. For me, the terms and definitions don't matter. However, the visible facts above and the structural deductions are useful both in detailing a site and dreaming about swaying buildings!

Brutally pruned older trees causes simultaneous release of buds, each initially acting like a tree.

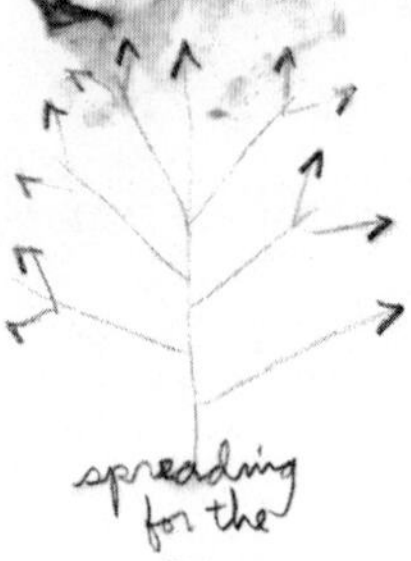

spreading for the sun

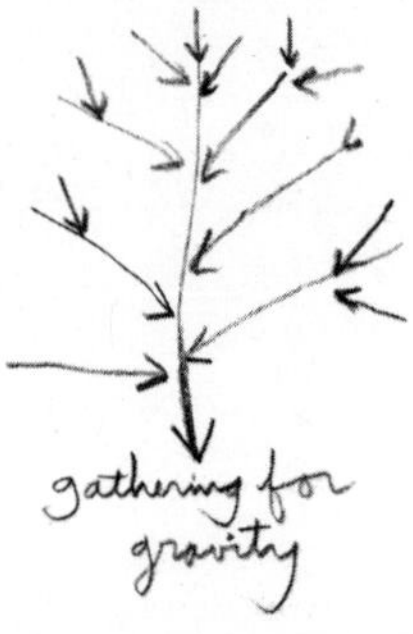

gathering for gravity

73. All trees follow the same branching laws to define different patterns.

Tree branches act like people trying peacefully to co-exist while striving towards the top.
(All routes make the fascination of a tree)

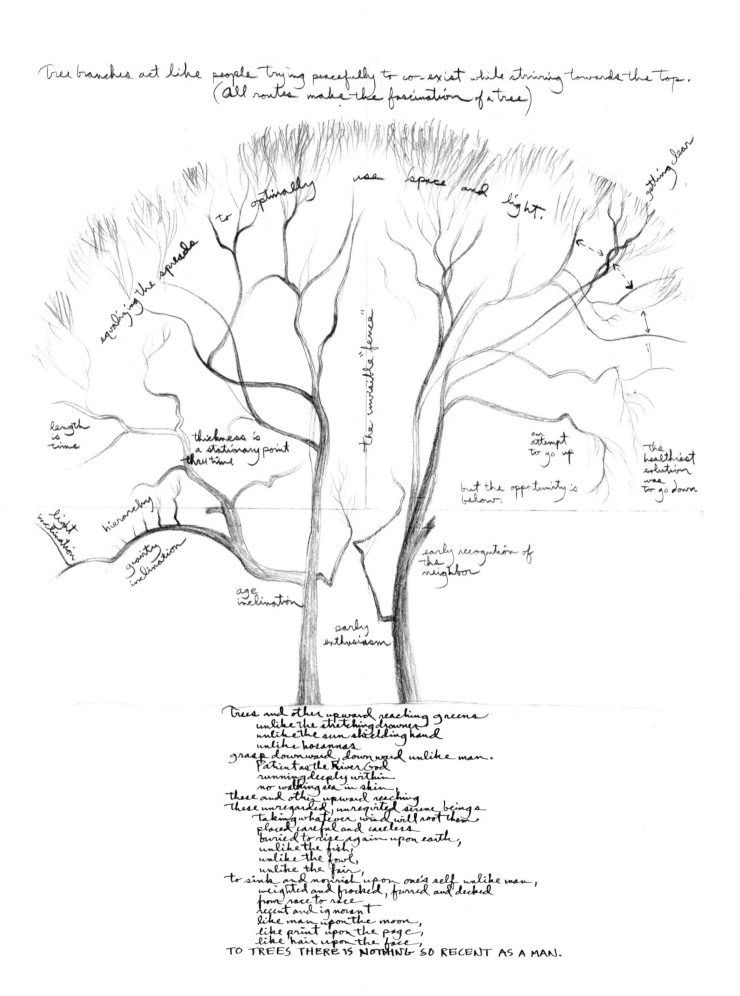

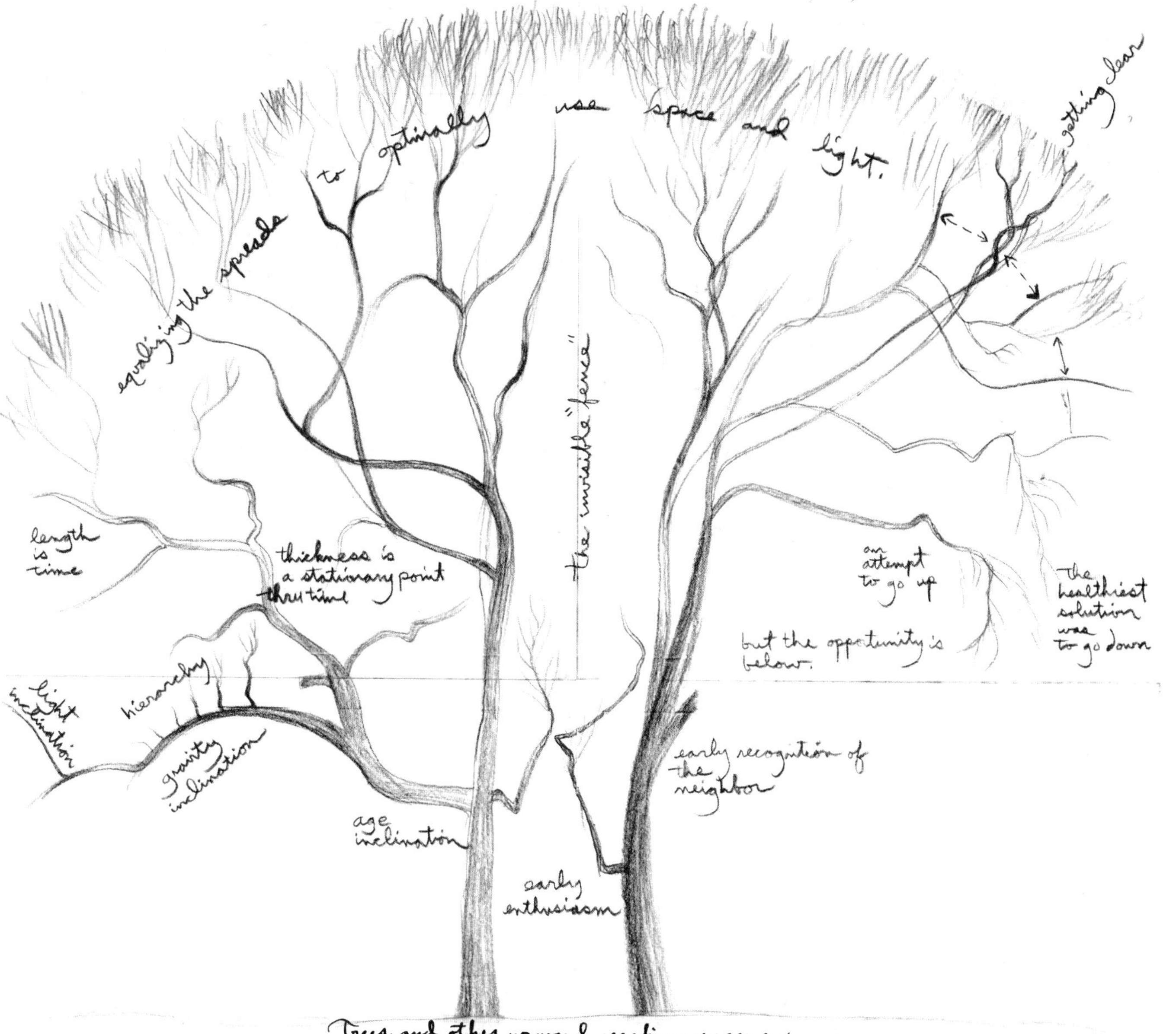

Trees and other upward reaching greens
unlike the stretching drowner
unlike the sun shielding hand
unlike hosannas
grasp downward, downward unlike man.
Patient as the River God
running deeply within
no walking sea in skin,
these and other upward reaching
these unregarded, unrequited serene beings
taking whatever wind will root them
placed careful and careless
buried to rise again upon earth,
unlike the fish,
unlike the fowl,
unlike the fair,
to sink and nourish upon one's self unlike man,
weighted and frocked, furred and decked
from race to race
recent and ignorant
like man upon the moon,
like print upon the page,
like hair upon the face,
TO TREES THERE IS NOTHING SO RECENT AS A MAN.

74. Shrubs can be arranged to move color throughout a space as the seasons change.

75. Fruiting trees are ornamental with usefulness. All trees shown here are well known for their hardiness.

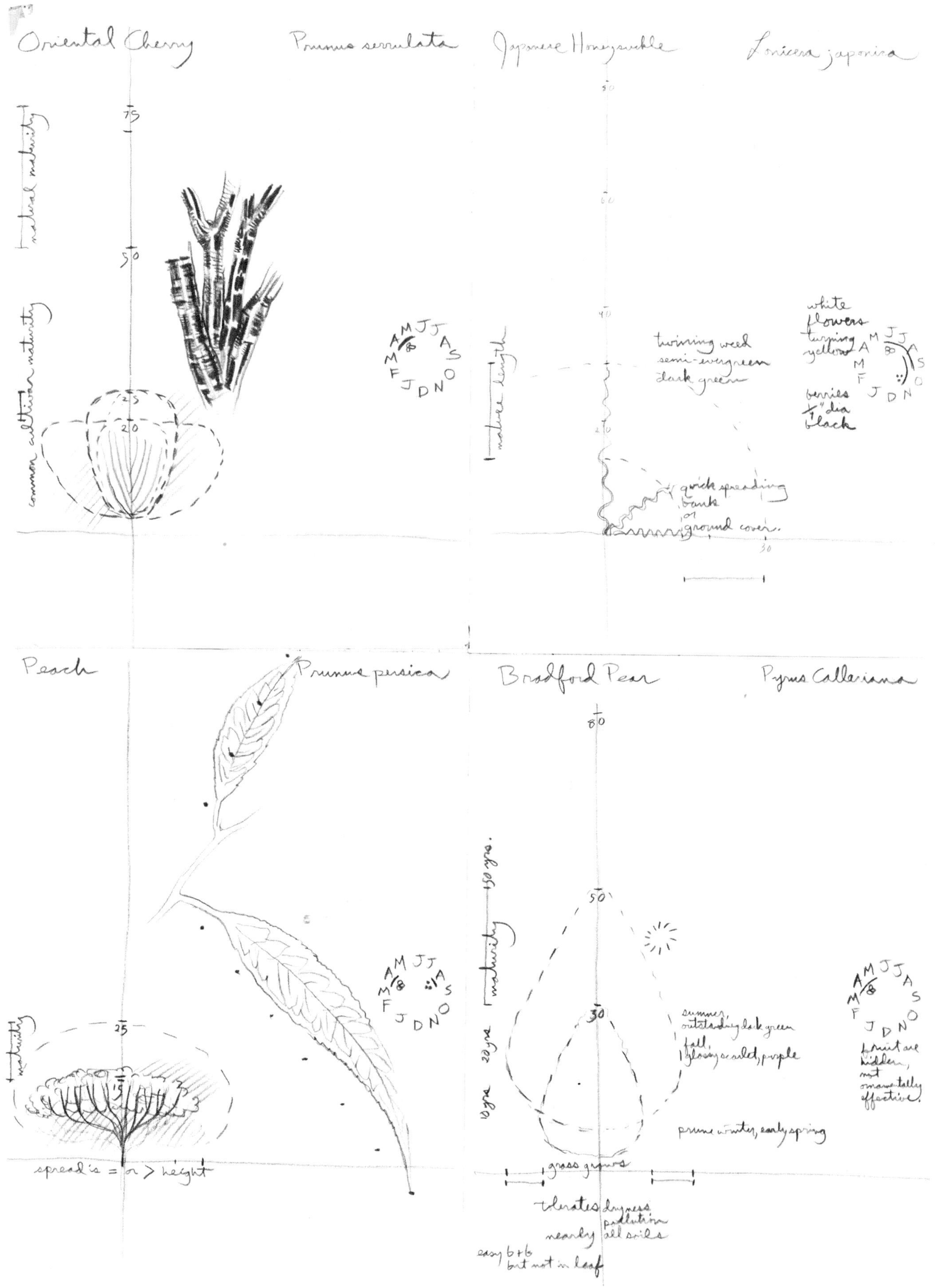

76. Small deciduous trees are reasonably pruned and offer transitional scale between shrubs and giants.

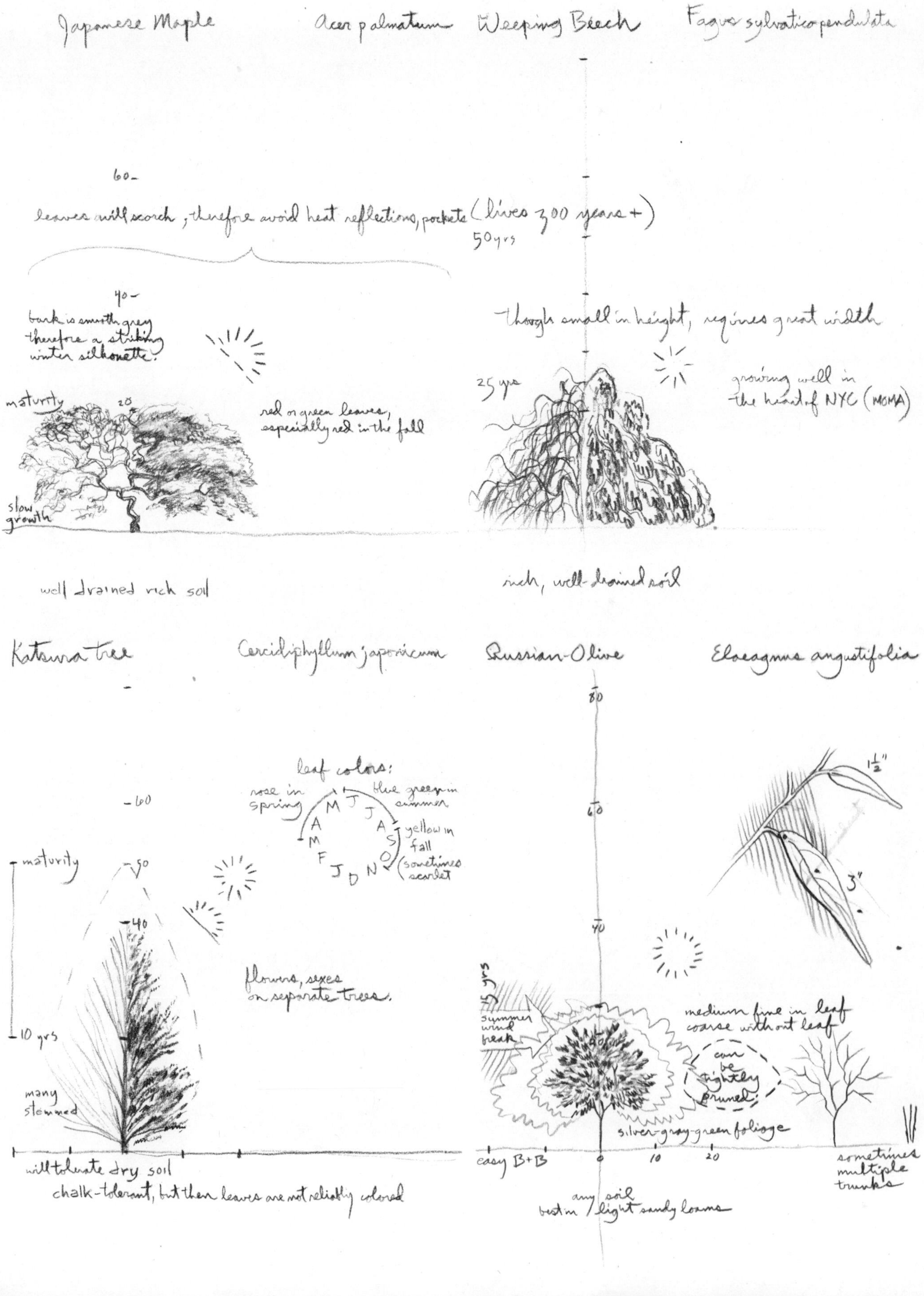

77. Large deciduous trees speckle light on the ground or densely shade, causing great visual commotion or color.

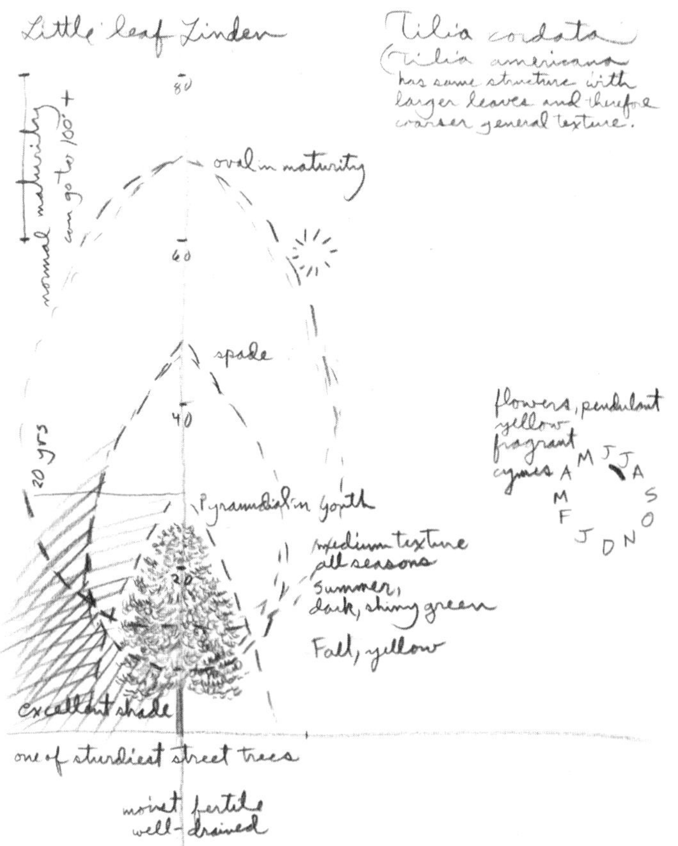

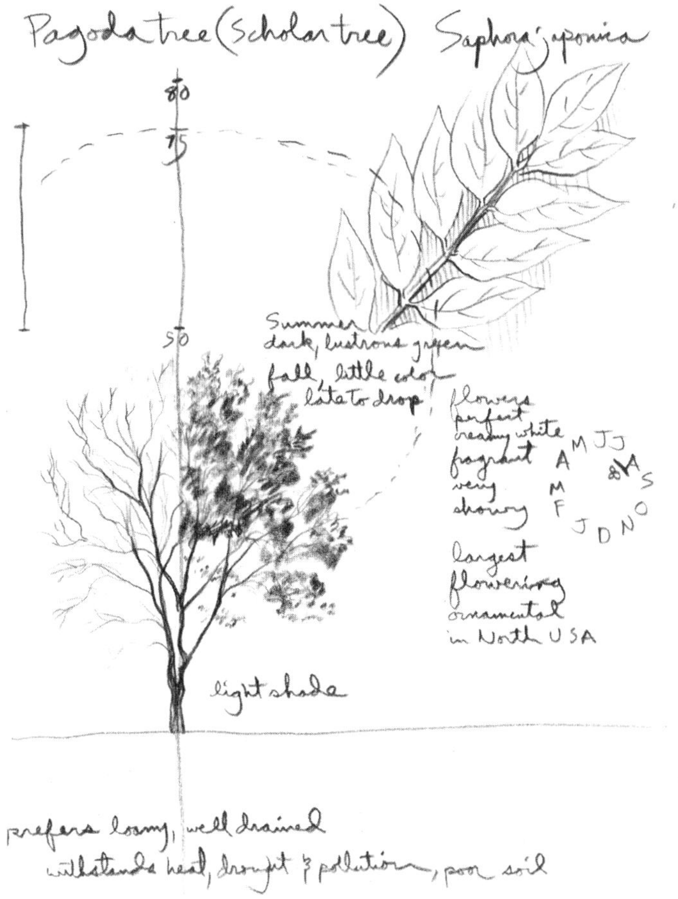

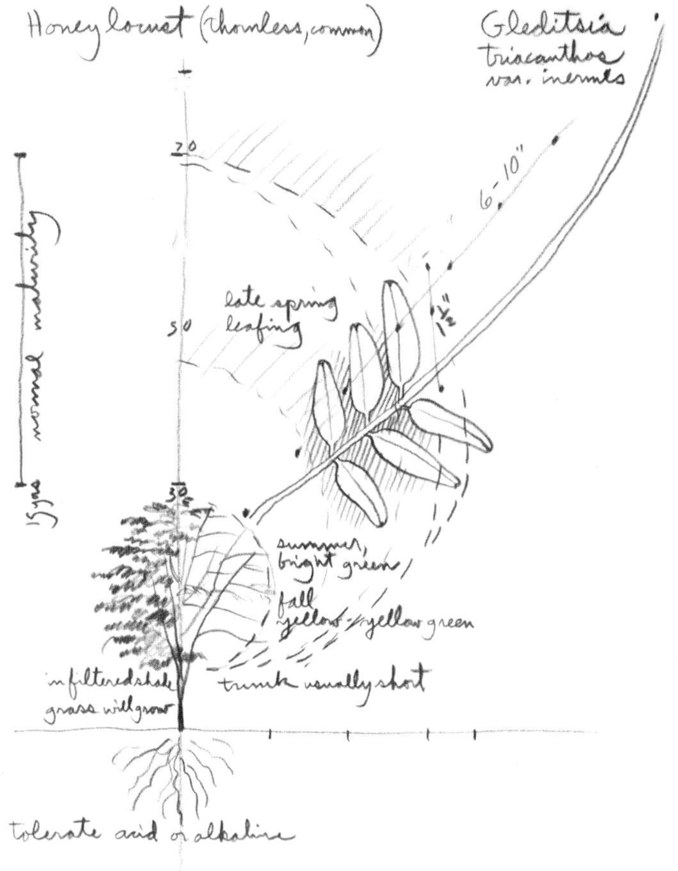

78. Small evergreen trees can form platforms, ledges, and walls.

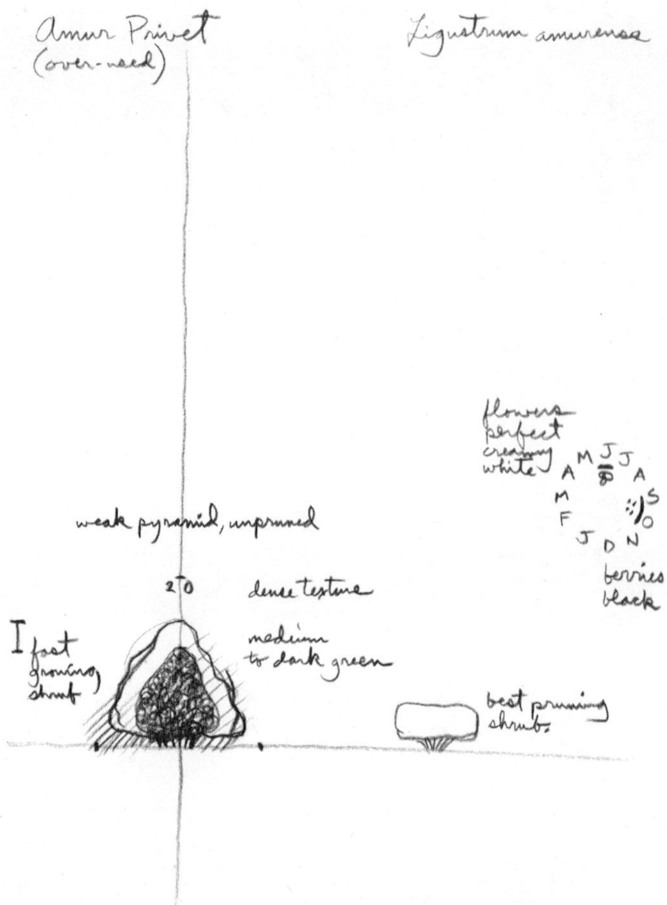

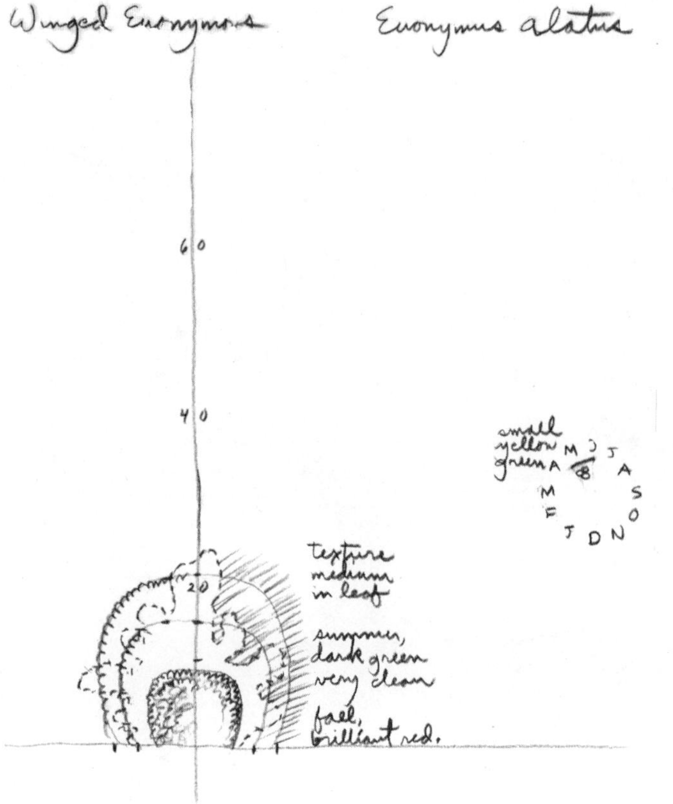

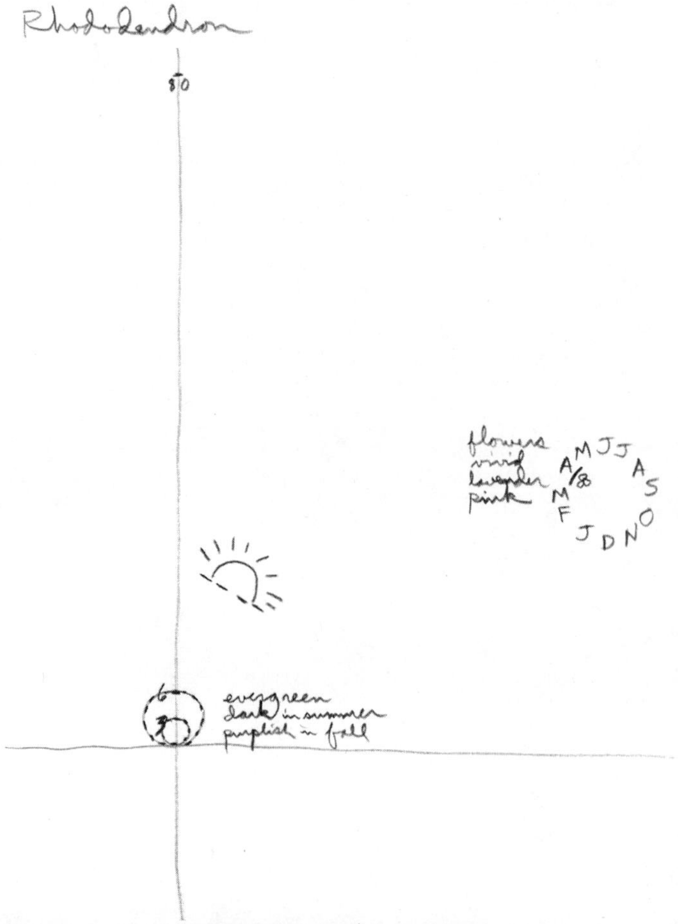

79. Large evergreen trees can form columns, walls, windows, ceilings, and skylights.

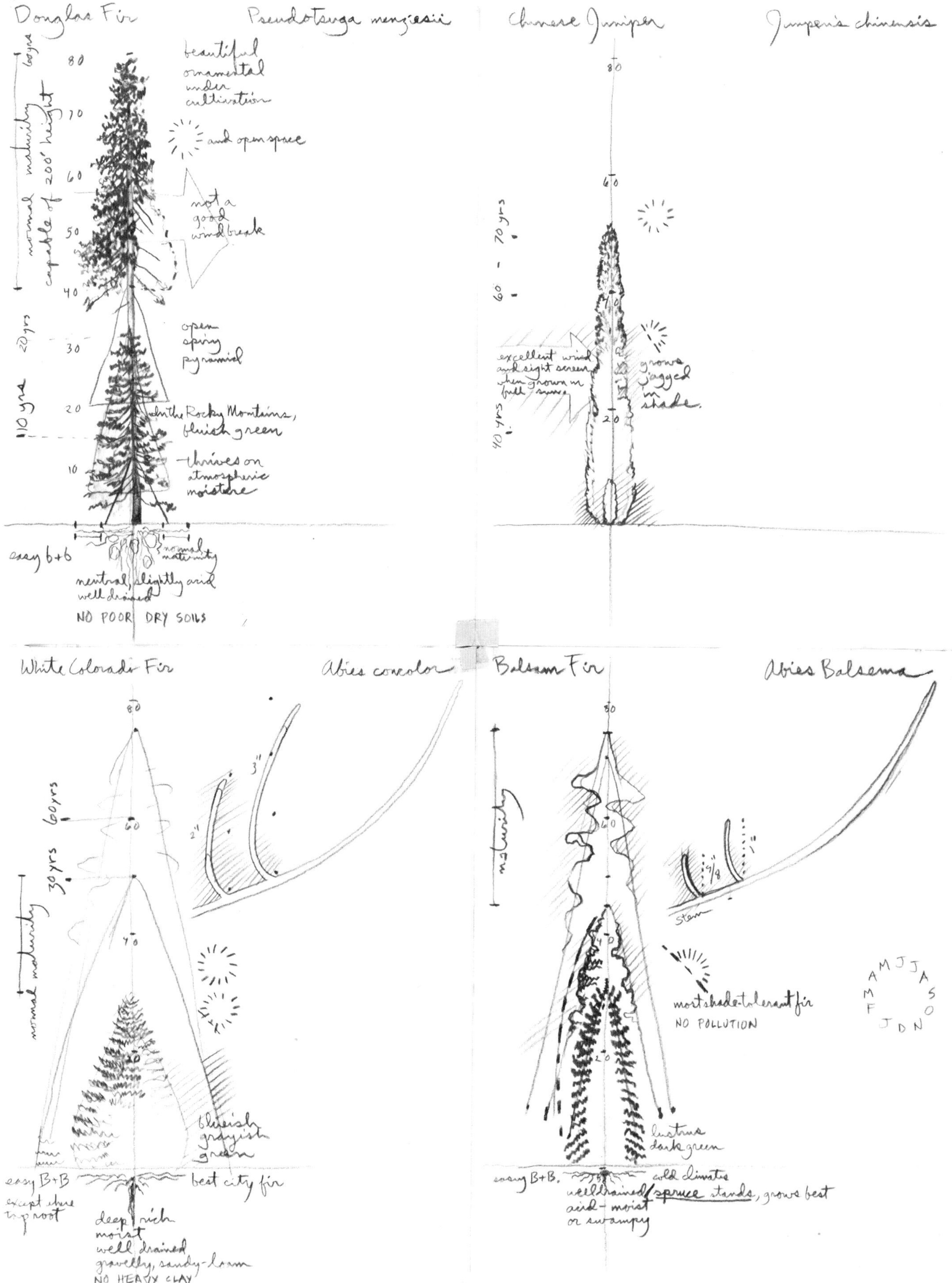

80. A natural architectural order can be developed for tree ceilings and tree walls.

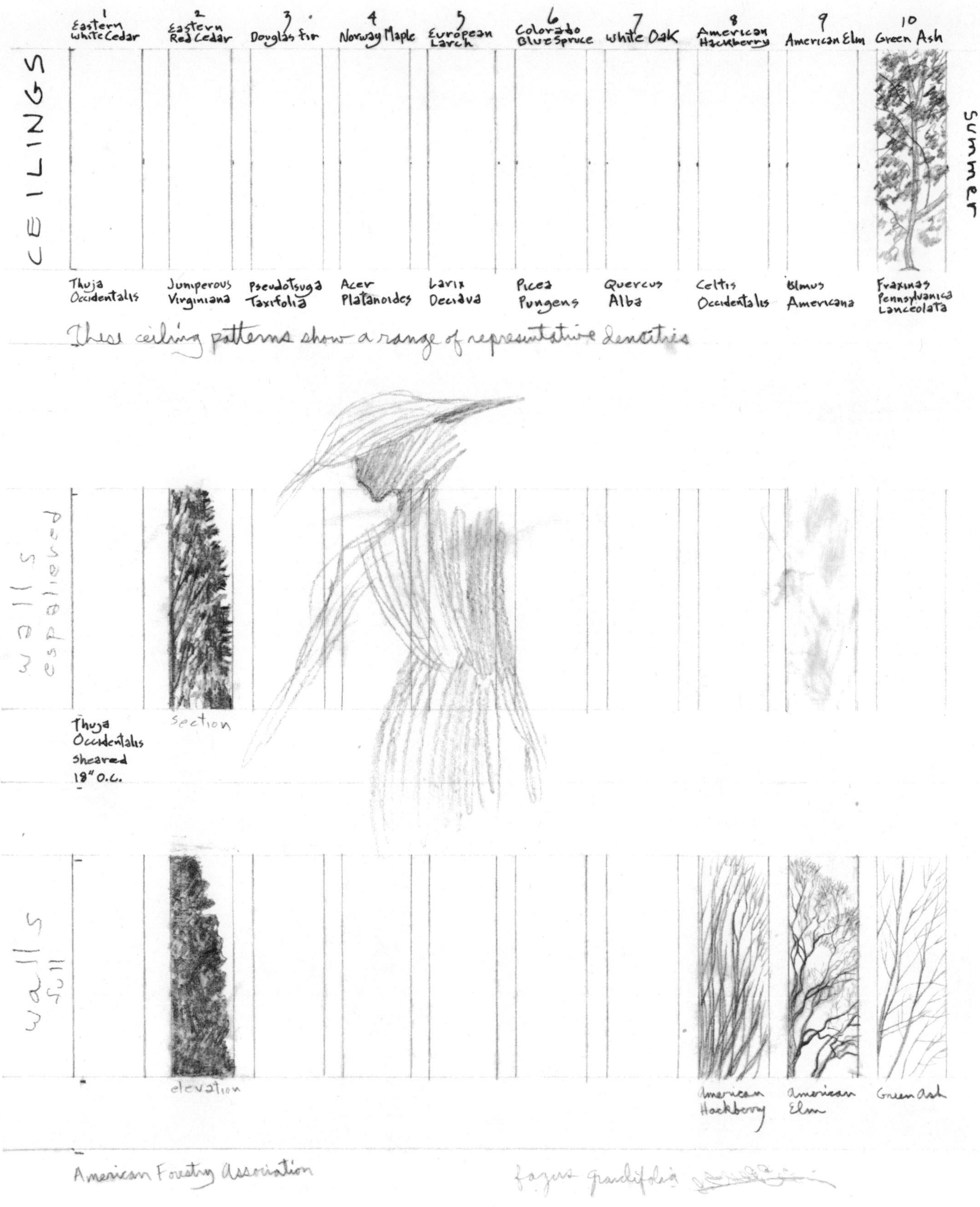

81. Weave and density can be arranged as living "wallpaper."

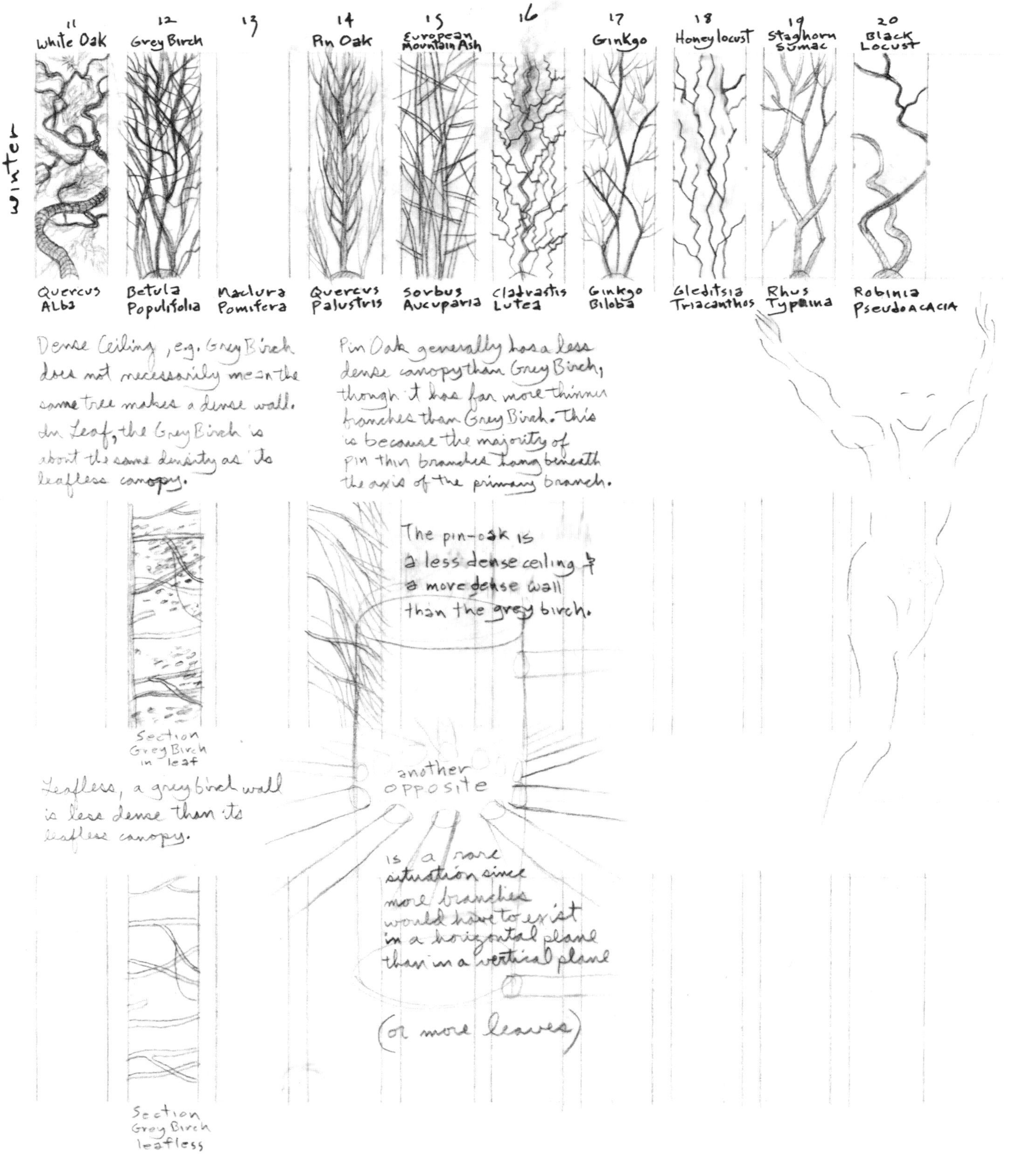
winter
11 White Oak
12 Grey Birch
13
14 Pin Oak
15 European Mountain Ash
16
17 Ginkgo
18 Honeylocust
19 Staghorn Sumac
20 Black Locust
Quercus Alba
Betula Populifolia
Maclura Pomifera
Quercus Palustris
Sorbus Aucuparia
Cladrastis Lutea
Ginkgo Biloba
Gleditsia Triacanthos
Rhus Typhina
Robinia Pseudoacacia
Dense Ceiling, e.g. Grey Birch does not necessarily mean the same tree makes a dense wall. In Leaf, the Grey Birch is about the same density as its leafless canopy.
Pin Oak generally has a less dense canopy than Grey Birch, though it has far more thinner branches than Grey Birch. This is because the majority of pin thin branches hang beneath the axis of the primary branch.
The pin-oak is a less dense ceiling & a more dense wall than the grey birch.
Section Grey Birch in leaf
Leafless, a grey birch wall is less dense than its leafless canopy.
another opposite
is a rare situation since more branches would have to exist in a horizontal plane than in a vertical plane
(or more leaves)
Section Grey Birch leafless

82. Unrelated species are related by similarity of canopy textures.

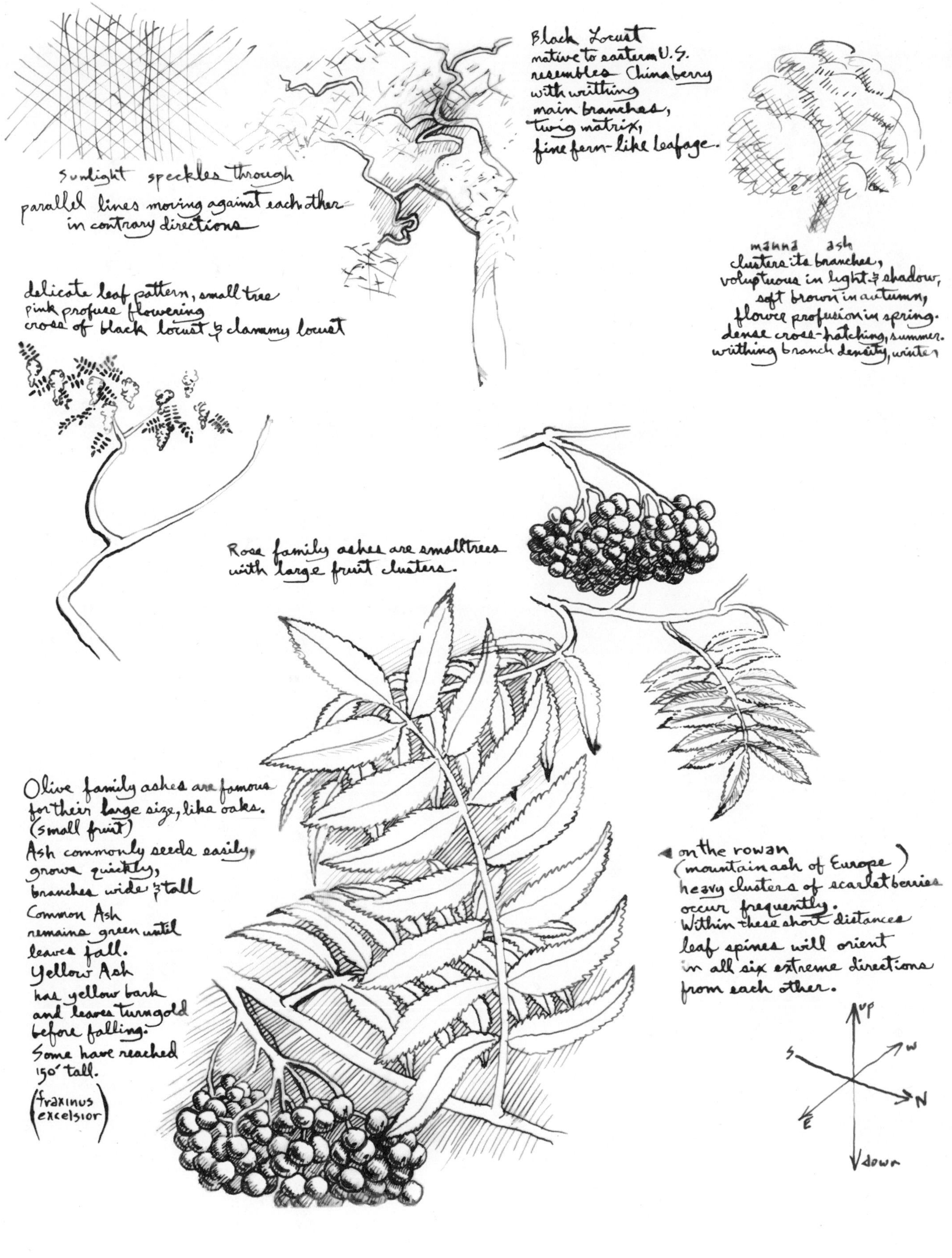
Sunlight speckles through
parallel lines moving against each other
in contrary directions
Black Locust
native to eastern U.S.
resembles Chinaberry
with writhing
main branches,
twig matrix,
fine fern-like leafage.
manna ash
clusters its branches,
voluptuous in light & shadow,
soft brown in autumn,
flower profusion in spring.
dense cross-hatching, summer.
writhing branch density, winter
delicate leaf pattern, small tree
pink profuse flowering
cross of black locust & clammy locust
Rose family ashes are small trees
with large fruit clusters.
Olive family ashes are famous
for their large size, like oaks.
(small fruit)
Ash commonly seeds easily,
grows quickly,
branches wide & tall
Common Ash
remains green until
leaves fall.
Yellow Ash
has yellow bark
and leaves turn gold
before falling.
Some have reached
150' tall.
(fraxinus
excelsior)
on the rowan
(mountain ash of Europe)
heavy clusters of scarlet berries
occur frequently.
Within these short distances
leaf spines will orient
in all six extreme directions
from each other.
up
W
S
N
E
down

83. For example, Manna Ash, Black Locust, Olive, Rowan, China Berry, Ailanthus, and Chinese Cedar are a group of canopies of different species consistently similar in overall texture.

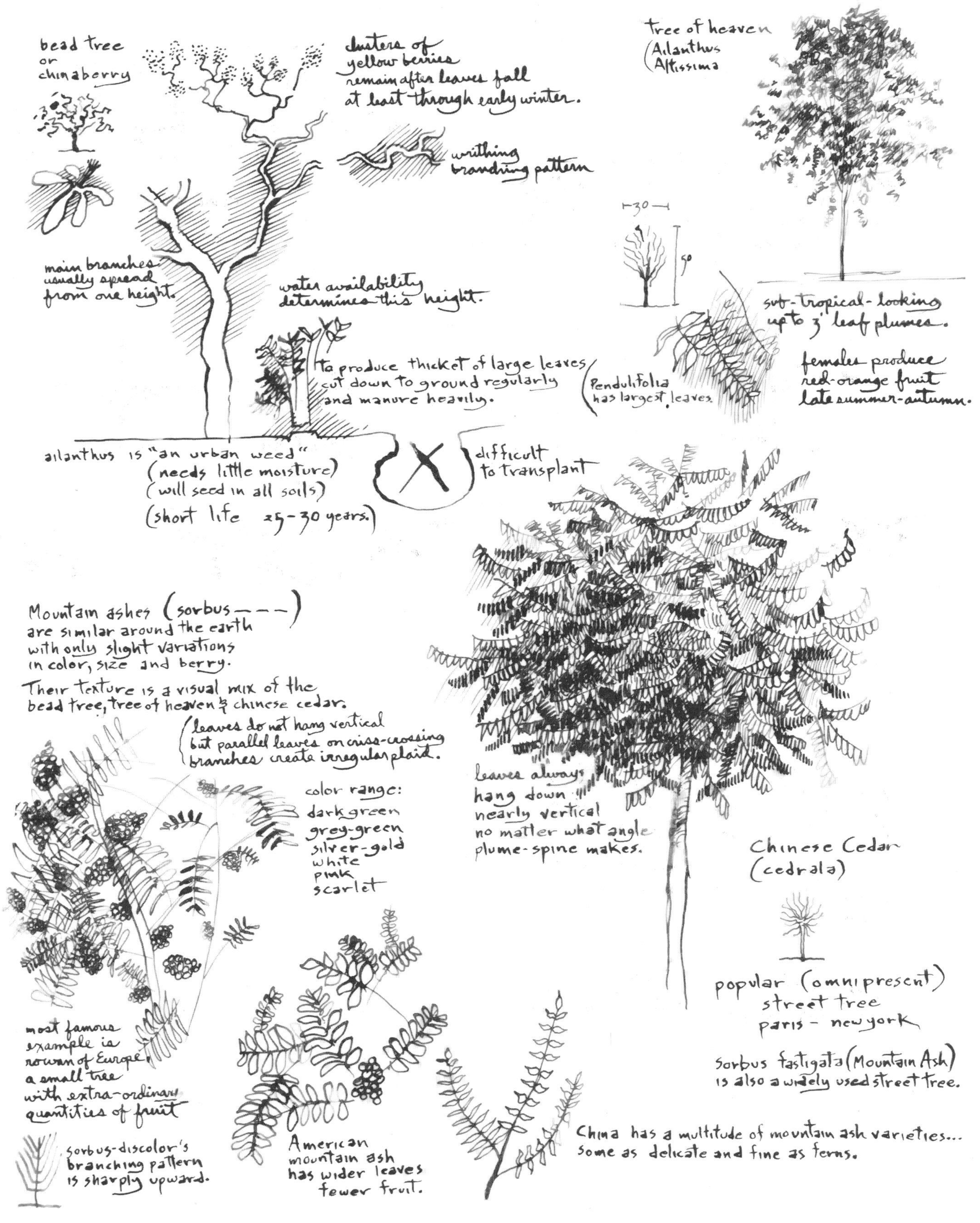
bead tree or chinaberry
clusters of yellow berries remain after leaves fall at least through early winter.
writhing branching pattern
main branches usually spread from one height.
water availability determines this height.
to produce thicket of large leaves cut down to ground regularly and manure heavily.
tree of heaven (Ailanthus Altissima
30
50
sub-tropical-looking up to 3' leaf plumes.
Pendulifolia has largest leaves.
females produce red-orange fruit late summer-autumn.
ailanthus is "an urban weed" (needs little moisture) (will seed in all soils) (short life 25-30 years.)
difficult to transplant
Mountain ashes (sorbus---) are similar around the earth with only slight variations in color, size and berry.
Their texture is a visual mix of the bead tree, tree of heaven & chinese cedar.
(leaves do not hang vertical but parallel leaves on criss-crossing branches create irregular plaid.
color range: dark green grey-green silver-gold white pink scarlet
leaves always hang down nearly vertical no matter what angle plume-spine makes.
Chinese Cedar (cedrala)
popular (omnipresent) street tree paris - new york
Sorbus fastigata (Mountain Ash) is also a widely used street tree.
most famous example is rowan of Europe a small tree with extra-ordinary quantities of fruit
sorbus-discolor's branching pattern is sharply upward.
American mountain ash has wider leaves fewer fruit.
China has a multitude of mountain ash varieties... some as delicate and fine as ferns.

84. Weave and density tree classes will be detailed just as are masonry, wood, and steel.

plants & trees will eventually be as much a detail of building design & working drawings as now are masonry, glass, wood & steel.

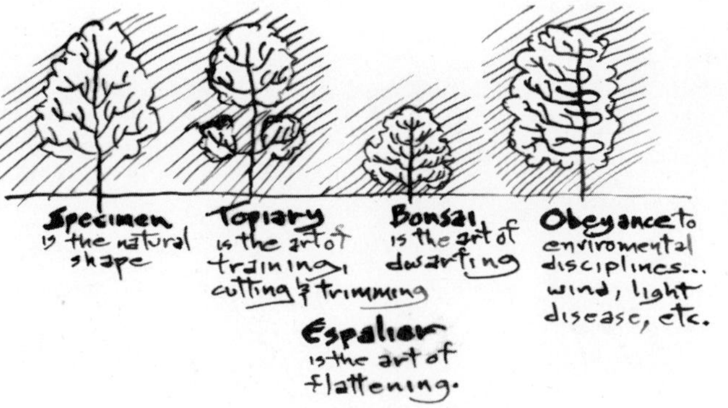
Specimen
is the natural shape
Topiary
is the art of training, cutting & trimming
Bonsai
is the art of dwarfing
Espalier
is the art of flattening.
Obeyance to
environmental disciplines... wind, light disease, etc.

Until now, the prime architectural use of foliage has been aesthetics. To a similar extent that architecture can be sculpture, trees will be sculpted as part of the architecture rather than just being added or left-over or landscaped.

Historically Useful Principles:

silhouette
calligraphy line
sculpture
decoration
pattern

background
foreground
framing
walling

unifying
separating
enhancing
distracting
softening
not-softening

that is...
defined relationships depend on definitions of relative characteristics.

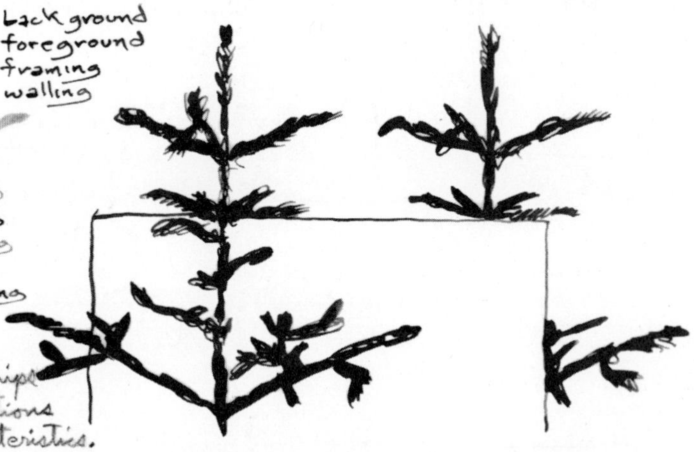

Lines of a tree are generally most striking in relation to strong direct light, which frequently causes the tree to appear as flat as its shadow, also potentially striking.

Volumes of trees are also most pronounced by strong light when the observer sees both lit and shadowed sides, simultaneously.

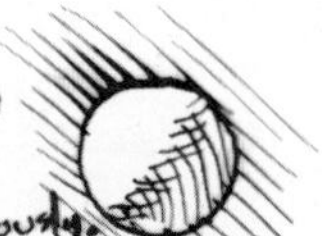

Haze, smoke, fog, rain & snow tend to cause both lines & volumes to appear as planes. (Contrast is less when light is diffuse.)

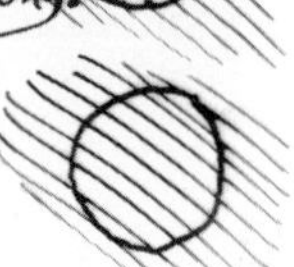

Suggestion & Simplicity

expand the possibilities of perception.

some minds (eyes) discover beauty by completing the incomplete.

Other minds find beauty where complexity provides more than is completable.

Dictation & Complexity limit . . . if attention is kept to them.

Most texts today describe plants & trees as the element which softens stark, hard modern architecture.

Naturalness,
Apparently undisciplined,
movement, colors, texture, change,
Birth, growth, decay, death,

Trees & plants by their own nature do not need master planners to unify, synthesize and organize the visual chaotic ugliness of most man-made environments.

We claim to be able to use trees & plants to emphasize, de-emphasize, attract and detract as if we could control people with plants & trees. I doubt it.

Instead of trying to understand abstract symbolic diagrams of landscaping... Look at lots of examples; conclude what you want to do; find appropriate trees & plants. DESIGN WITH THEM.
THE SITE WILL SUGGEST SOLUTIONS

85. Branches will become the mullions in our views and leaves the frames. To see is to move and movement is ordered by windows.

Trees & plants can be architecturally composed for viewing. (Wood, masonry, etc)

The key problem to planning for a view is that both the eyes and the viewers move. Therefore, encouraging their movement in order to see pre-planned, valued scenes is the art of spatial orchestration.

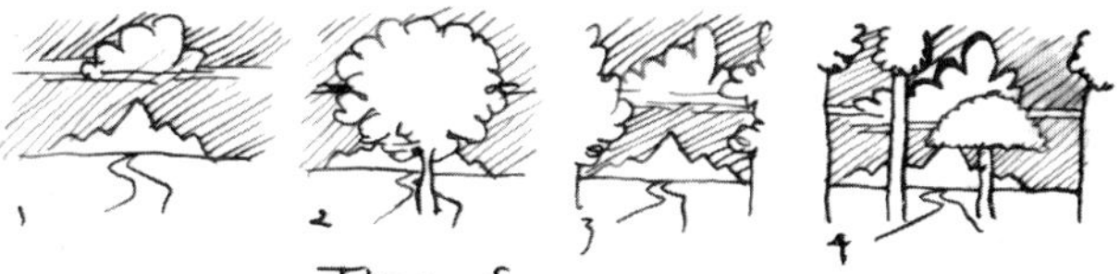

These four scenes are each seen from a spot in the middle of the road. They could all exist simultaneously but be experienced at different times at different spots along the road.

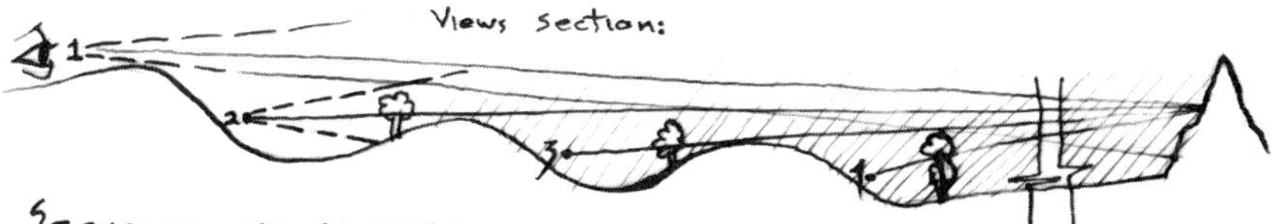

Sequence can surprise.

Since each of the views are positioned to exclude the others, they are interchangeable.

large distance compressed (removed) for sake of illustration.

These three scenes can be seen from a great variety of spots. Without the road they could be to the left or right of each other as well as to the fore or behind.

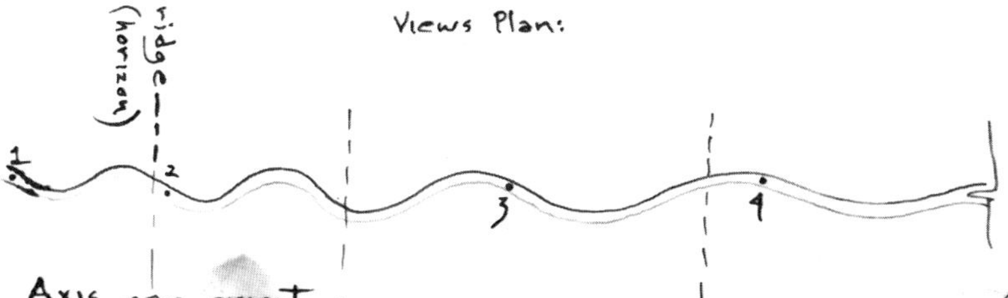

Axis can orient.

In this case the road is an axis; we can rely on it as surely as we rely on a compass needle to determine our relation to something definite.

TO SEE IS TO MOVE. TO MOVE IS TO SEQUENCE AN AXIS.

Wherever your site or situation may be, it has surrounding views which you are powerless to touch in order to make them please you. Views of this panorama are limited widths extending various distances. Choose the best vantage points.

To be enjoyed, views must be suitable to you.

After you have determined what is appropriate where and when, and you want to plan in more detail than sequence & axis...

ARE THE AXES CLEARLY FOR MOVING? HOW FOR THE EYE? HOW FOR THE BODY?

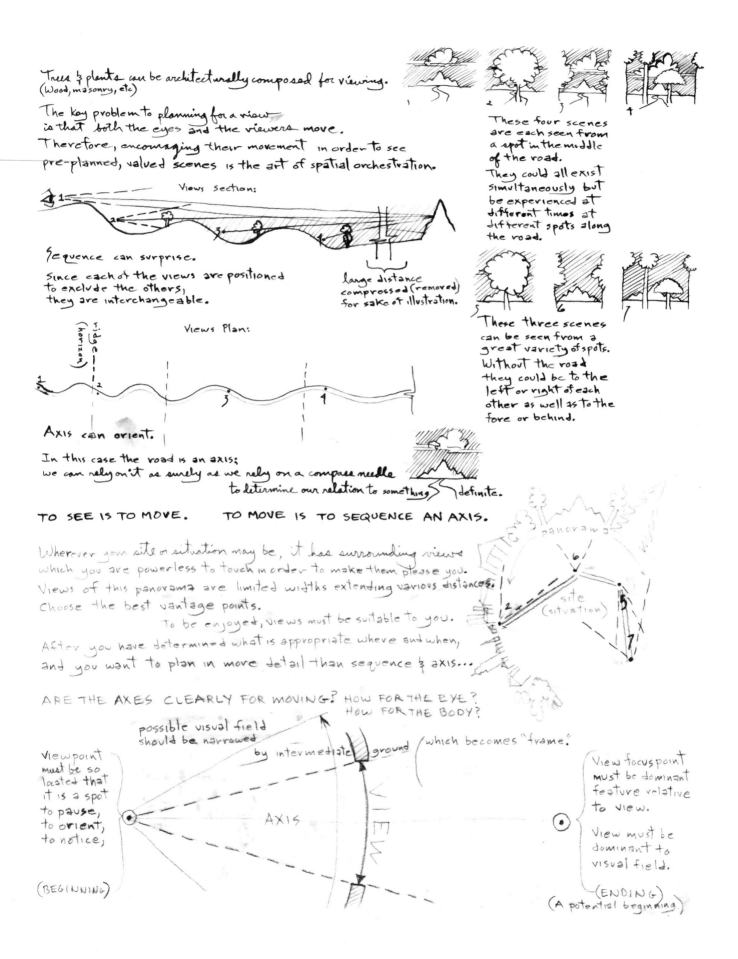

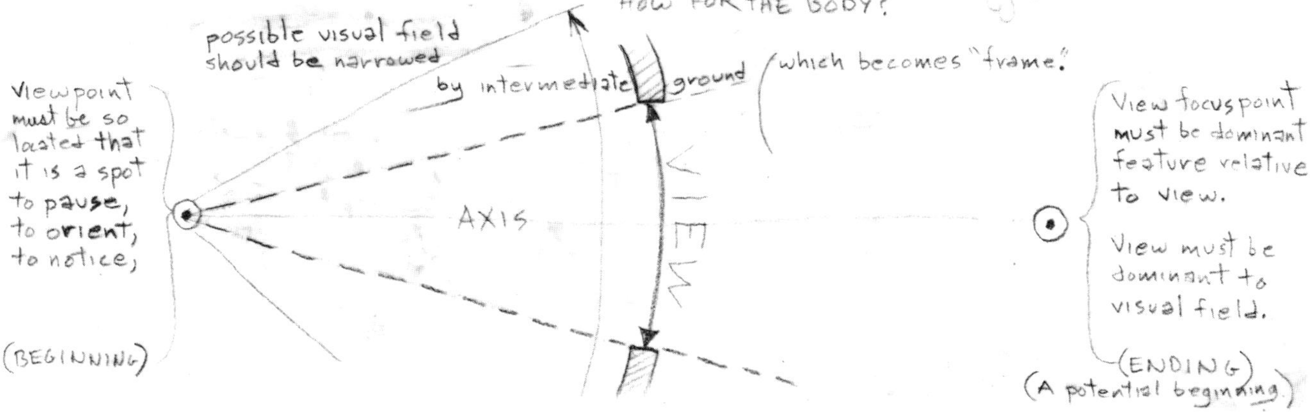

86. Finding one's way around a site is the beginning of a map.

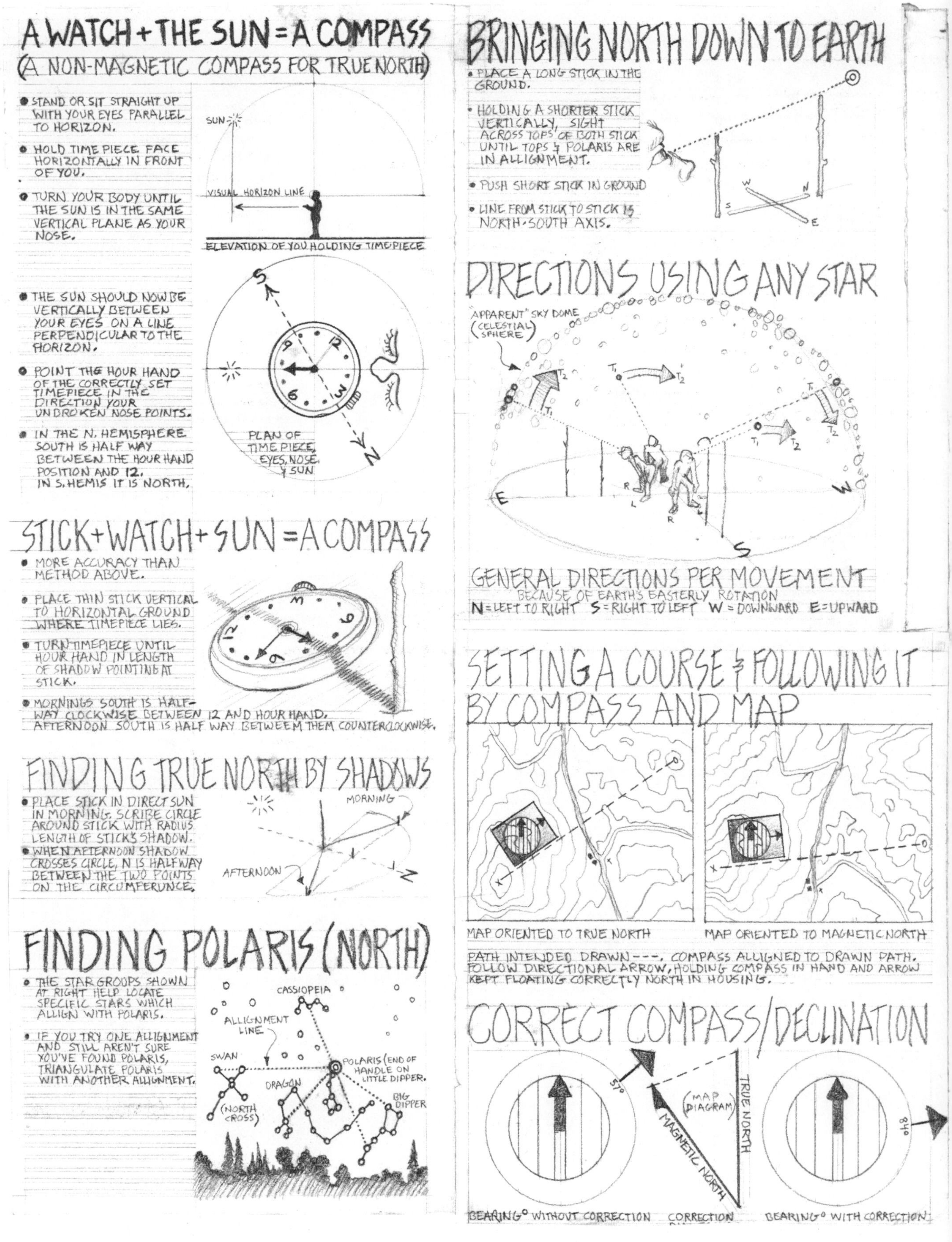
A WATCH + THE SUN = A COMPASS
(A NON-MAGNETIC COMPASS FOR TRUE NORTH)
• STAND OR SIT STRAIGHT UP WITH YOUR EYES PARALLEL TO HORIZON.
• HOLD TIME PIECE FACE HORIZONTALLY IN FRONT OF YOU.
• TURN YOUR BODY UNTIL THE SUN IS IN THE SAME VERTICAL PLANE AS YOUR NOSE.
SUN
VISUAL HORIZON LINE
ELEVATION OF YOU HOLDING TIMEPIECE
• THE SUN SHOULD NOW BE VERTICALLY BETWEEN YOUR EYES ON A LINE PERPENDICULAR TO THE HORIZON.
• POINT THE HOUR HAND OF THE CORRECTLY SET TIMEPIECE IN THE DIRECTION YOUR UNBROKEN NOSE POINTS.
• IN THE N. HEMISPHERE SOUTH IS HALF WAY BETWEEN THE HOUR HAND POSITION AND 12. IN S. HEMIS IT IS NORTH.
S
N
PLAN OF TIME PIECE, EYES, NOSE, & SUN
STICK + WATCH + SUN = A COMPASS
• MORE ACCURACY THAN METHOD ABOVE.
• PLACE THIN STICK VERTICAL TO HORIZONTAL GROUND WHERE TIMEPIECE LIES.
• TURN TIMEPIECE UNTIL HOUR HAND IN LENGTH OF SHADOW POINTING AT STICK.
• MORNINGS SOUTH IS HALF-WAY CLOCKWISE BETWEEN 12 AND HOUR HAND. AFTERNOON SOUTH IS HALF WAY BETWEEM THEM COUNTERCLOCKWISE.
FINDING TRUE NORTH BY SHADOWS
• PLACE STICK IN DIRECT SUN IN MORNING. SCRIBE CIRCLE AROUND STICK WITH RADIUS LENGTH OF STICK'S SHADOW.
• WHEN AFTERNOON SHADOW CROSSES CIRCLE, N IS HALFWAY BETWEEN THE TWO POINTS ON THE CIRCUMFERUNCE.
MORNING
AFTERNOON
N
FINDING POLARIS (NORTH)
• THE STAR GROUPS SHOWN AT RIGHT HELP LOCATE SPECIFIC STARS WHICH ALLIGN WITH POLARIS.
• IF YOU TRY ONE ALLIGNMENT AND STILL AREN'T SURE YOU'VE FOUND POLARIS, TRIANGULATE POLARIS WITH ANOTHER ALLIGNMENT.
CASSIOPEIA
ALLIGNMENT LINE
SWAN
(NORTH CROSS)
DRAGON
POLARIS (END OF HANDLE ON LITTLE DIPPER.
BIG DIPPER
BRINGING NORTH DOWN TO EARTH
• PLACE A LONG STICK IN THE GROUND.
• HOLDING A SHORTER STICK VERTICALLY, SIGHT ACROSS TOPS OF BOTH STICK UNTIL TOPS & POLARIS ARE IN ALLIGNMENT.
• PUSH SHORT STICK IN GROUND
• LINE FROM STICK TO STICK IS NORTH-SOUTH AXIS.
W
N
S
E
DIRECTIONS USING ANY STAR
"APPARENT" SKY DOME (CELESTIAL SPHERE)
E
W
S
GENERAL DIRECTIONS PER MOVEMENT
BECAUSE OF EARTH'S EASTERLY ROTATION
N = LEFT TO RIGHT S = RIGHT TO LEFT W = DOWNWARD E = UPWARD
SETTING A COURSE & FOLLOWING IT BY COMPASS AND MAP
MAP ORIENTED TO TRUE NORTH
MAP ORIENTED TO MAGNETIC NORTH
PATH INTENDED DRAWN ---. COMPASS ALLIGNED TO DRAWN PATH. FOLLOW DIRECTIONAL ARROW, HOLDING COMPASS IN HAND AND ARROW KEPT FLOATING CORRECTLY NORTH IN HOUSING.
CORRECT COMPASS/DECLINATION
57°
(MAP DIAGRAM)
TRUE NORTH
MAGNETIC NORTH
84°
BEARING° WITHOUT CORRECTION
CORRECTION
BEARING° WITH CORRECTION

87. A mental map can be verified with a compass and a length.

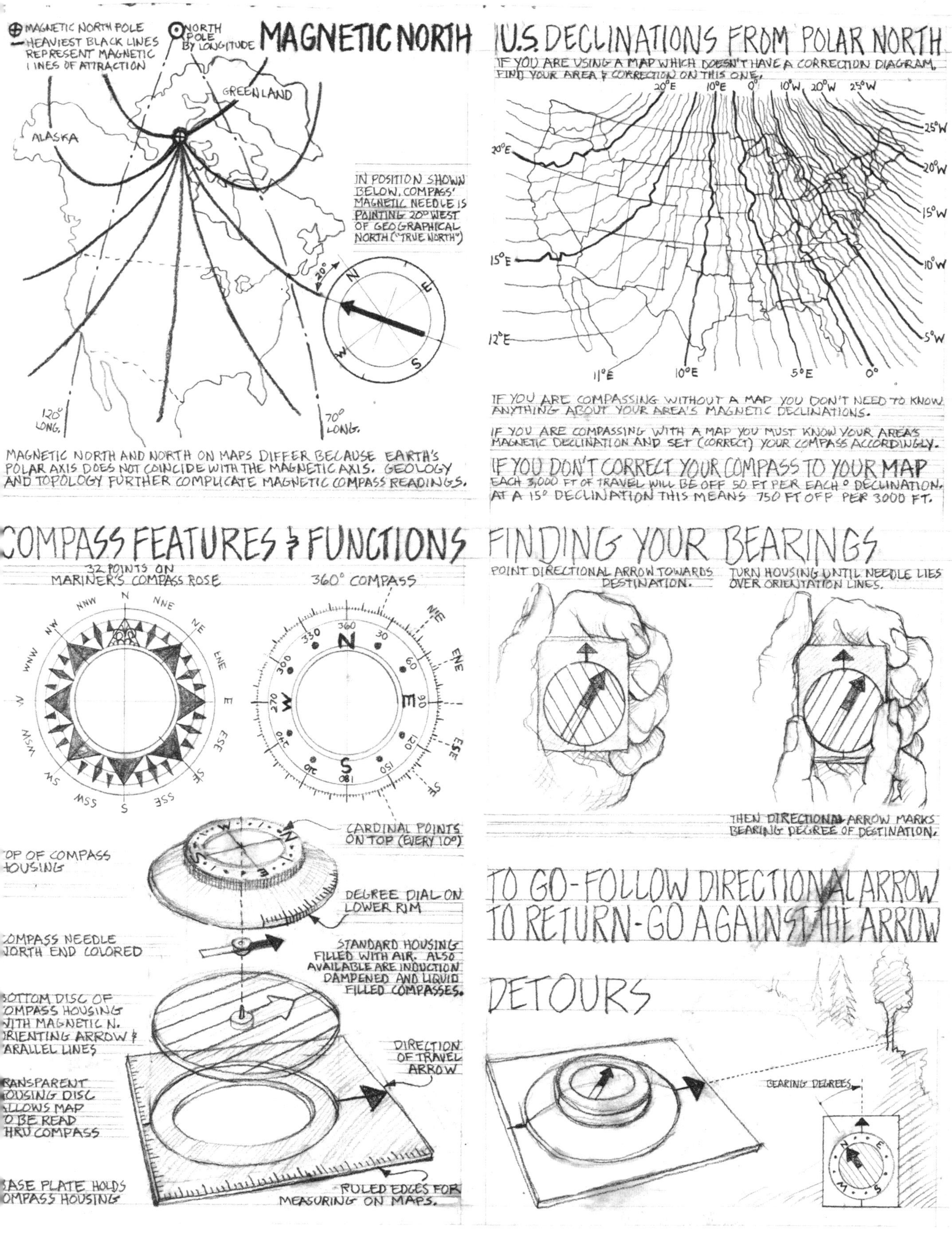
MAGNETIC NORTH
MAGNETIC NORTH POLE
HEAVIEST BLACK LINES REPRESENT MAGNETIC LINES OF ATTRACTION
NORTH POLE BY LONGITUDE
GREENLAND
ALASKA
IN POSITION SHOWN BELOW, COMPASS' MAGNETIC NEEDLE IS POINTING 20° WEST OF GEOGRAPHICAL NORTH ("TRUE NORTH")
20°
N
E
W
S
120° LONG.
70° LONG.
MAGNETIC NORTH AND NORTH ON MAPS DIFFER BECAUSE EARTH'S POLAR AXIS DOES NOT COINCIDE WITH THE MAGNETIC AXIS. GEOLOGY AND TOPOLOGY FURTHER COMPLICATE MAGNETIC COMPASS READINGS.
U.S. DECLINATIONS FROM POLAR NORTH
IF YOU ARE USING A MAP WHICH DOESN'T HAVE A CORRECTION DIAGRAM, FIND YOUR AREA & CORRECTION ON THIS ONE.
20°E
10°E
0°
10°W
20°W
25°W
15°W
5°W
15°E
12°E
11°E
5°E
IF YOU ARE COMPASSING WITHOUT A MAP YOU DON'T NEED TO KNOW ANYTHING ABOUT YOUR AREA'S MAGNETIC DECLINATIONS.
IF YOU ARE COMPASSING WITH A MAP YOU MUST KNOW YOUR AREA'S MAGNETIC DECLINATION AND SET (CORRECT) YOUR COMPASS ACCORDINGLY.
IF YOU DON'T CORRECT YOUR COMPASS TO YOUR MAP
EACH 3,000 FT OF TRAVEL WILL BE OFF 50 FT PER EACH ° DECLINATION. AT A 15° DECLINATION THIS MEANS 750 FT OFF PER 3000 FT.
COMPASS FEATURES & FUNCTIONS
32 POINTS ON MARINER'S COMPASS ROSE
N
NNE
NE
ENE
E
ESE
SE
SSE
S
SSW
SW
WSW
W
WNW
NW
NNW
360° COMPASS
360
30
60
90
120
150
180
210
240
270
300
330
CARDINAL POINTS ON TOP (EVERY 10°)
TOP OF COMPASS HOUSING
DEGREE DIAL ON LOWER RIM
COMPASS NEEDLE NORTH END COLORED
STANDARD HOUSING FILLED WITH AIR. ALSO AVAILABLE ARE INDUCTION DAMPENED AND LIQUID FILLED COMPASSES.
BOTTOM DISC OF COMPASS HOUSING WITH MAGNETIC N. ORIENTING ARROW & PARALLEL LINES
DIRECTION OF TRAVEL ARROW
TRANSPARENT HOUSING DISC ALLOWS MAP TO BE READ THRU COMPASS
BASE PLATE HOLDS COMPASS HOUSING
RULED EDGES FOR MEASURING ON MAPS.
FINDING YOUR BEARINGS
POINT DIRECTIONAL ARROW TOWARDS DESTINATION.
TURN HOUSING UNTIL NEEDLE LIES OVER ORIENTATION LINES.
THEN DIRECTIONAL ARROW MARKS BEARING DEGREE OF DESTINATION.
TO GO-FOLLOW DIRECTIONAL ARROW
TO RETURN-GO AGAINST THE ARROW
DETOURS
BEARING DEGREES

88. Circular timepieces and the circles of the sun verify the compass.

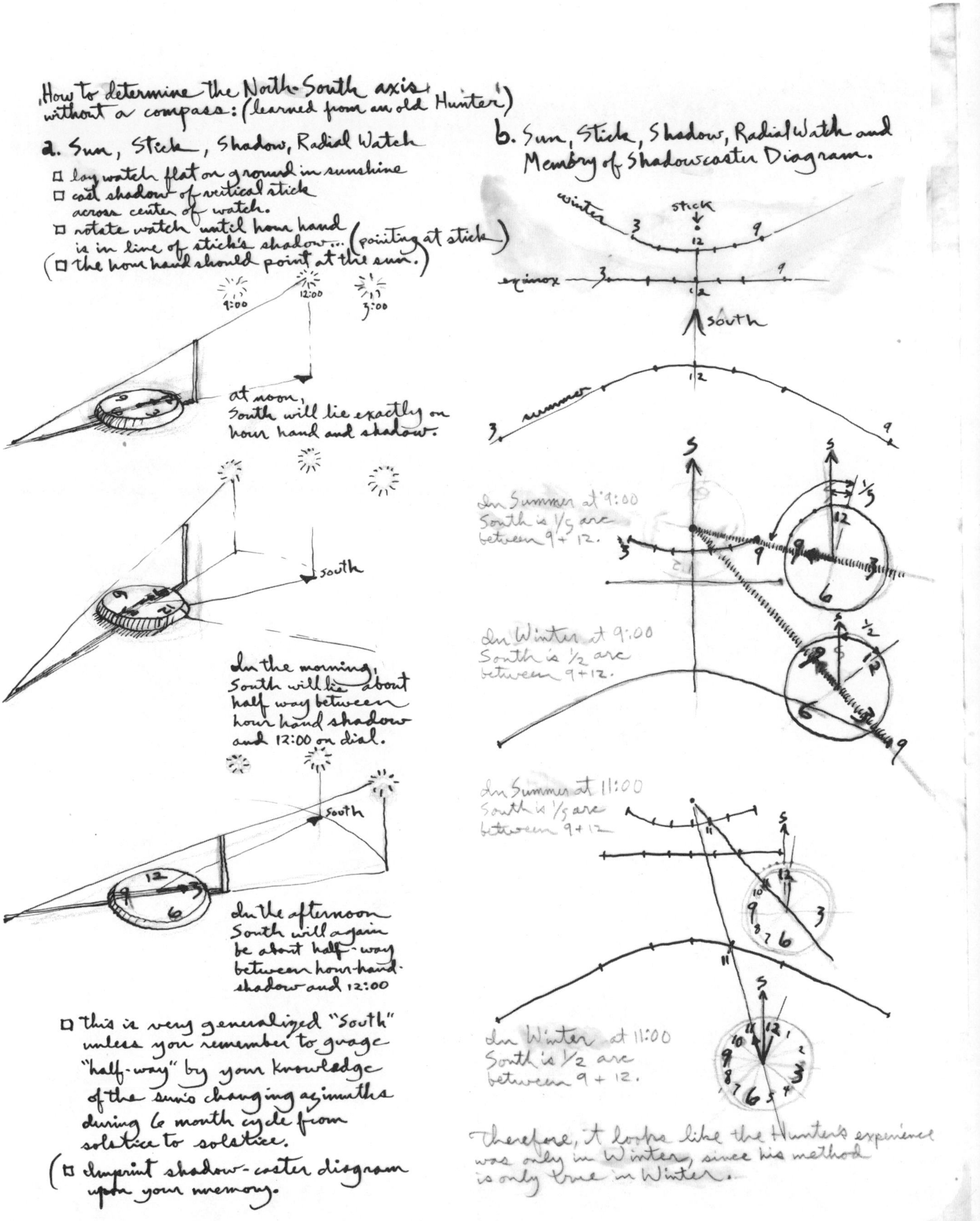

How to determine the North-South axis, without a compass: (learned from an old Hunter)
a. Sun, Stick, Shadow, Radial Watch
□ lay watch flat on ground in sunshine
□ cast shadow of vertical stick across center of watch.
□ rotate watch until hour hand is in line of stick's shadow... (pointing at stick)
(□ the hour hand should point at the sun.)
9:00
12:00
3:00
at noon, South will lie exactly on hour hand and shadow.
south
In the morning, South will lie about half way between hour hand shadow and 12:00 on dial.
South
In the afternoon South will again be about half-way between hour-hand-shadow and 12:00
□ This is very generalized "South" unless you remember to guage "half-way" by your knowledge of the sun's changing azimuths during 6 month cycle from solstice to solstice.
(□ Imprint shadow-caster diagram upon your memory.
b. Sun, Stick, Shadow, Radial Watch and Memory of Shadowcaster Diagram.
winter
stick
equinox
south
summer
In Summer at 9:00 South is 1/5 arc between 9 + 12.
In Winter at 9:00 South is 1/2 arc between 9 + 12.
In Summer at 11:00 South is 1/5 arc between 9 + 12
In Winter at 11:00 South is 1/2 arc between 9 + 12.
Therefore, it looks like the Hunter's experience was only in Winter, since his method is only true in Winter.

89. The lines and times of the shadowlabe verify the pendulum of day per season.

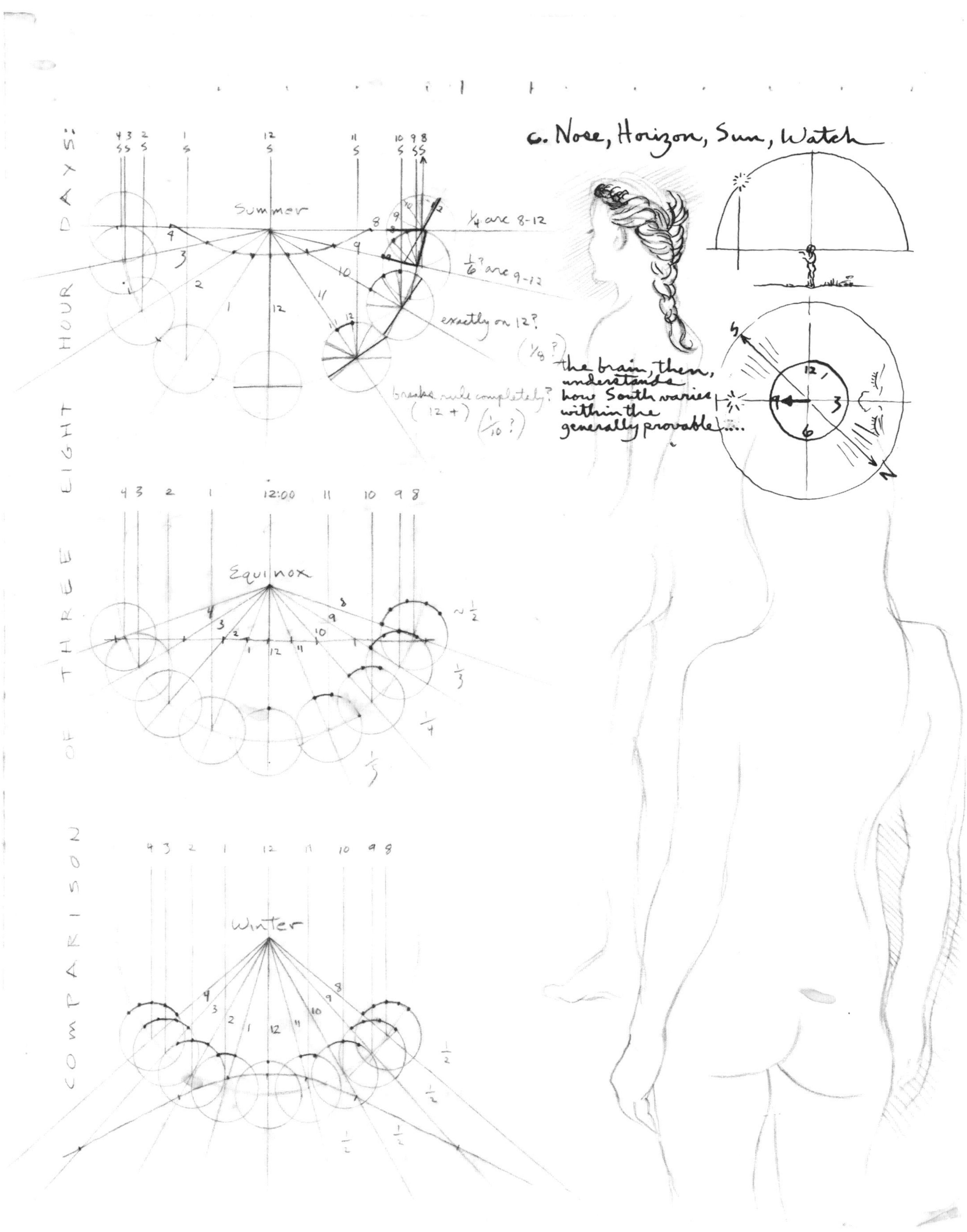

90. Surveying is the practical art of measuring changes which you consider significant in the horizon.

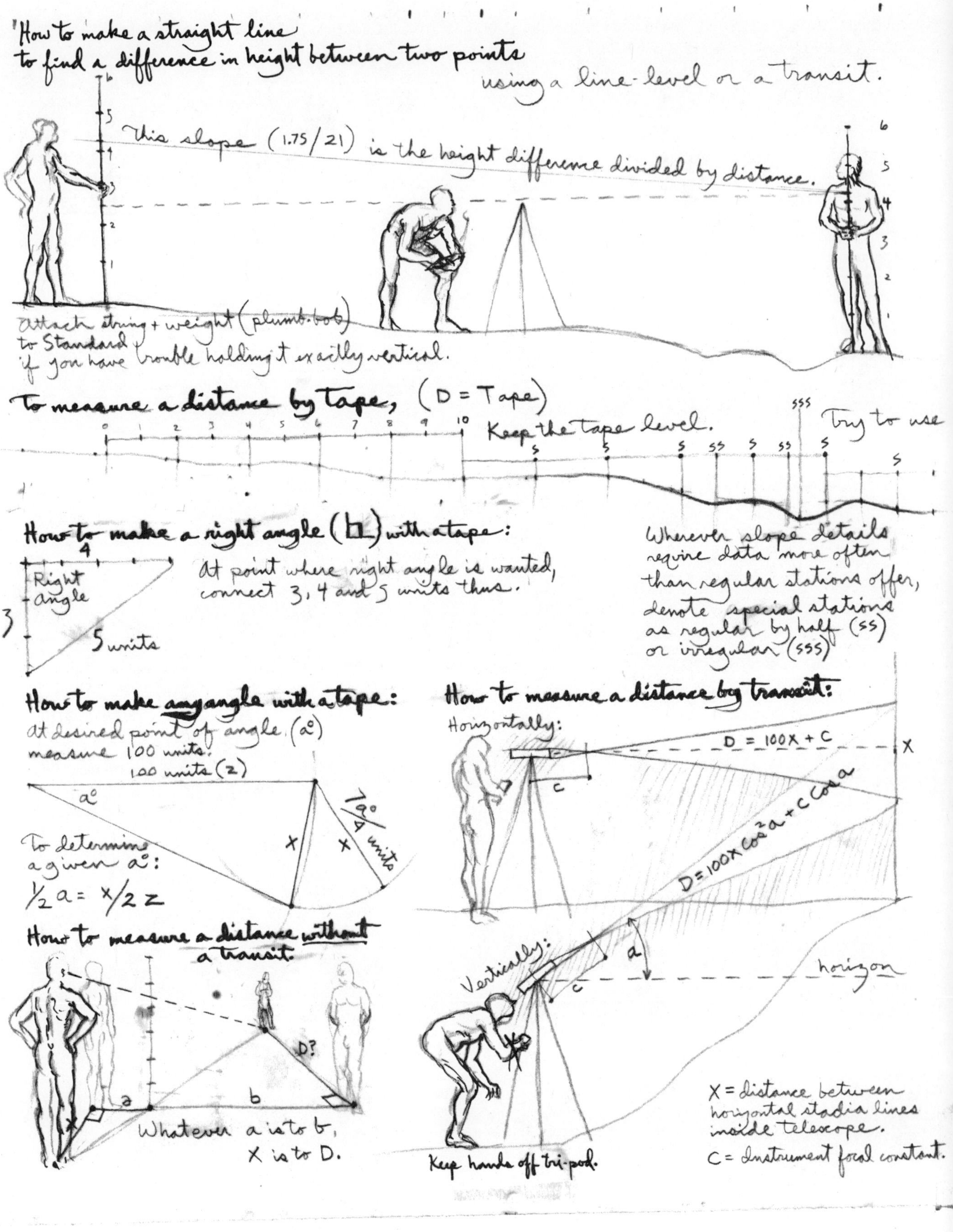
How to make a straight line
to find a difference in height between two points
using a line-level or a transit.
This slope (1.75/21) is the height difference divided by distance.
Attach string + weight (plumb-bob) to Standard if you have trouble holding it exactly vertical.
To measure a distance by tape, (D = Tape)
Keep the tape level.
Try to use
How to make a right angle (∟) with a tape:
At point where right angle is wanted, connect 3, 4 and 5 units thus.
Right angle
5 units
Wherever slope details require data more often than regular stations offer, denote special stations as regular by half (SS) or irregular (SSS)
How to make any angle with a tape:
At desired point of angle (a°) measure 100 units. 100 units (z)
To determine a given a°:
½ a = X/2 Z
How to measure a distance by transit:
Horizontally:
D = 100X + C
Vertically:
D = 100X cos² a + C cos a
horizon
Keep hands off tri-pod.
X = distance between horizontal stadia lines inside telescope.
C = Instrument focal constant.
How to measure a distance without a transit.
Whatever a is to b, X is to D.

91. What you choose to survey becomes your map of the world; that is, what isn't considered can't be part of the intended design.

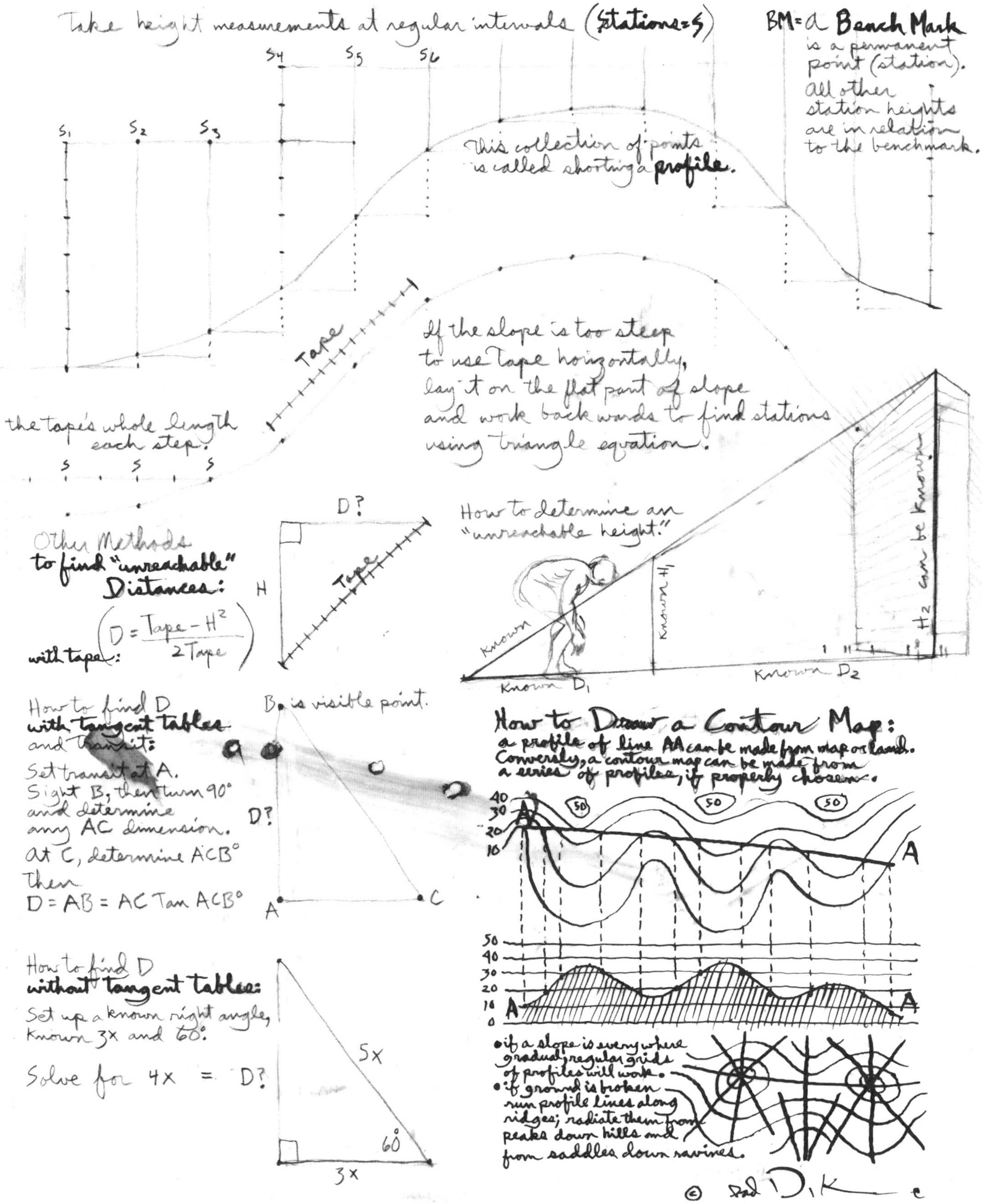

Take height measurements at regular intervals (Stations=S)
BM=a Bench Mark
is a permanent point (station).
All other station heights are in relation to the benchmark.
S1 S2 S3 S4 S5 S6
This collection of points is called shooting a profile.
Tape
If the slope is too steep to use Tape horizontally, lay it on the flat part of slope and work backwards to find stations using triangle equation.
the tape's whole length each step.
S S S
Other Methods to find "unreachable" Distances:
D?
H
Tape
with tape: $D = \frac{Tape - H^2}{2\,Tape}$
How to determine an "unreachable height."
Known H1
H2 can be known
Known
Known D1
Known D2
How to find D with tangent tables and transit:
Set transit at A.
Sight B, then turn 90° and determine any AC dimension.
At C, determine ACB°
Then
D = AB = AC Tan ACB°
B is visible point.
D?
A
C
How to Draw a Contour Map:
a profile of line AA can be made from map or land. Conversely, a contour map can be made from a series of profiles, if properly chosen.
40 30 20 10
50 50 50
A A
50 40 30 20 10 0
A A
• if a slope is everywhere gradual, regular grids of profiles will work.
• if ground is broken run profile lines along ridges; radiate them from peaks down hills and from saddles down ravines.
How to find D without tangent tables:
Set up a known right angle, known 3x and 60°.
Solve for 4x = D?
5x
60°
3x

92. Surveying instruments are better used to allow for the unplannable, rather than to master plan.

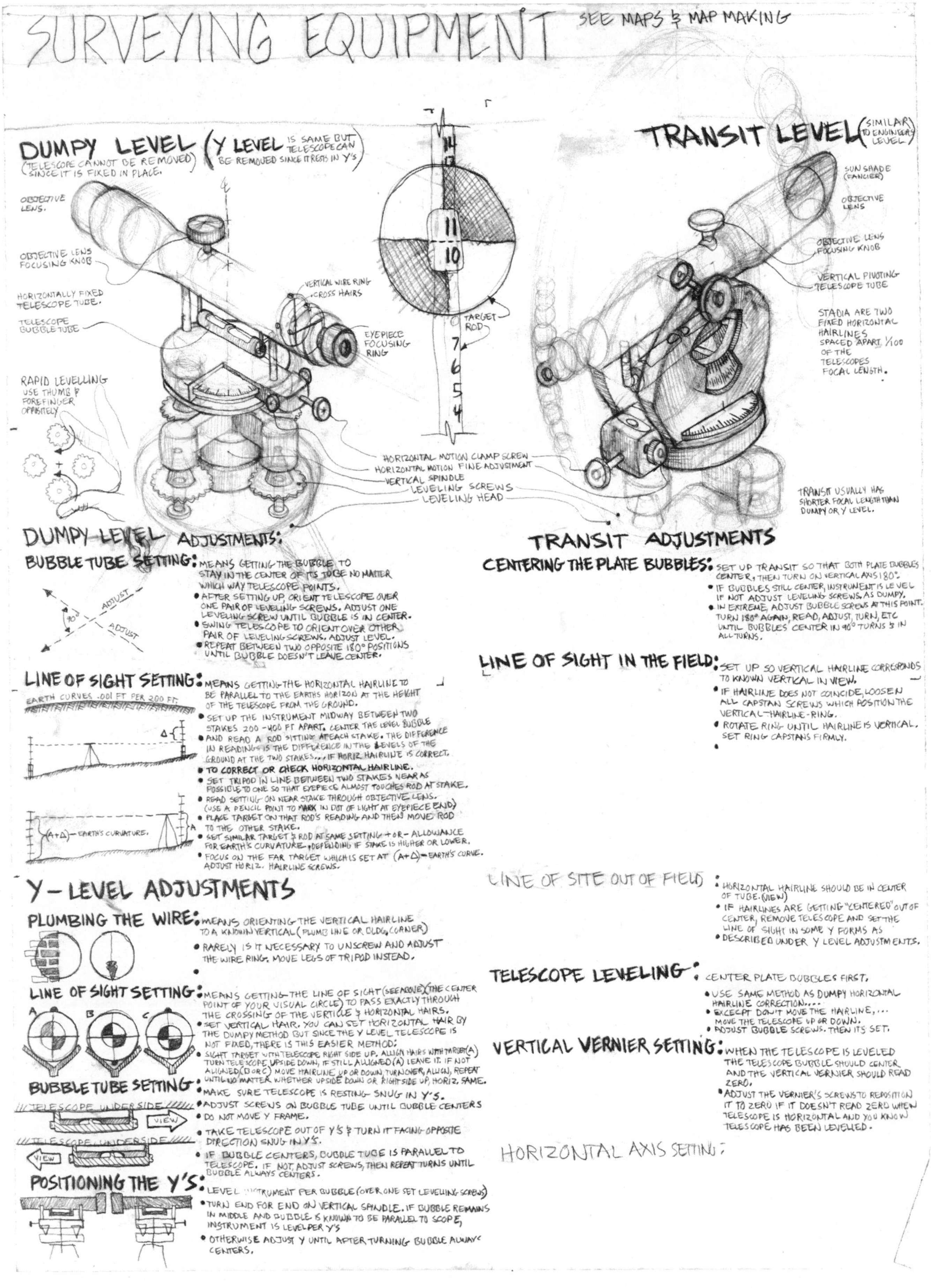
SURVEYING EQUIPMENT
SEE MAPS & MAP MAKING
DUMPY LEVEL (TELESCOPE CANNOT BE REMOVED SINCE IT IS FIXED IN PLACE.)
(Y LEVEL IS SAME BUT TELESCOPE CAN BE REMOVED SINCE IT RESTS IN Y'S)
OBJECTIVE LENS.
OBJECTIVE LENS FOCUSING KNOB
HORIZONTALLY FIXED TELESCOPE TUBE.
TELESCOPE BUBBLE TUBE
VERTICAL WIRE RING CROSS HAIRS
EYEPIECE FOCUSING RING
RAPID LEVELLING USE THUMB & FOREFINGER OPPOSITELY
TARGET ROD
HORIZONTAL MOTION CLAMP SCREW
HORIZONTAL MOTION FINE ADJUSTMENT
VERTICAL SPINDLE
LEVELING SCREWS
LEVELING HEAD
TRANSIT LEVEL (SIMILAR TO ENGINEER'S LEVEL)
SUN SHADE (FANCIER)
OBJECTIVE LENS
OBJECTIVE LENS FOCUSING KNOB
VERTICAL PIVOTING TELESCOPE TUBE
STADIA ARE TWO FIXED HORIZONTAL HAIRLINES SPACED APART 1/100 OF THE TELESCOPES FOCAL LENGTH.
TRANSIT USUALLY HAS SHORTER FOCAL LENGTH THAN DUMPY OR Y LEVEL.
DUMPY LEVEL ADJUSTMENTS:
BUBBLE TUBE SETTING: MEANS GETTING THE BUBBLE TO STAY IN THE CENTER OF ITS TUBE NO MATTER WHICH WAY TELESCOPE POINTS.
• AFTER SETTING UP, ORIENT TELESCOPE OVER ONE PAIR OF LEVELING SCREWS. ADJUST ONE LEVELING SCREW UNTIL BUBBLE IS IN CENTER.
• SWING TELESCOPE TO ORIENT OVER OTHER PAIR OF LEVELING SCREWS. ADJUST LEVEL.
• REPEAT BETWEEN TWO OPPOSITE 180° POSITIONS UNTIL BUBBLE DOESN'T LEAVE CENTER.
ADJUST
ADJUST
90°
LINE OF SIGHT SETTING: MEANS GETTING THE HORIZONTAL HAIRLINE TO BE PARALLEL TO THE EARTHS HORIZON AT THE HEIGHT OF THE TELESCOPE FROM THE GROUND.
• SET UP THE INSTRUMENT MIDWAY BETWEEN TWO STAKES 200-400 FT APART. CENTER THE LEVEL BUBBLE
• AND READ A ROD SITTING AT EACH STAKE. THE DIFFERENCE IN READINGS IS THE DIFFERENCE IN THE LEVELS OF THE GROUND AT THE TWO STAKES... IF HORIZ. HAIRLINE IS CORRECT.
• TO CORRECT OR CHECK HORIZONTAL HAIRLINE.
• SET TRIPOD IN LINE BETWEEN TWO STAKES NEAR AS POSSIBLE TO ONE SO THAT EYEPIECE ALMOST TOUCHES ROD AT STAKE.
• READ SETTING ON NEAR STAKE THROUGH OBJECTIVE LENS. (USE A PENCIL POINT TO MARK IN DOT OF LIGHT AT EYEPIECE END)
• PLACE TARGET ON THAT ROD'S READING AND THEN MOVE ROD TO THE OTHER STAKE.
• SET SIMILAR TARGET & ROD AT SAME SETTING + OR - ALLOWANCE FOR EARTH'S CURVATURE, DEPENDING IF STAKE IS HIGHER OR LOWER.
• FOCUS ON THE FAR TARGET WHICH IS SET AT (A+Δ) - EARTH'S CURVE. ADJUST HORIZ. HAIRLINE SCREWS.
EARTH CURVES .001 FT PER 200 FT.
(A+Δ) - EARTH'S CURVATURE.
Y-LEVEL ADJUSTMENTS
PLUMBING THE WIRE: MEANS ORIENTING THE VERTICAL HAIRLINE TO A KNOWN VERTICAL (PLUMB LINE OR BLDG. CORNER)
• RARELY IS IT NECESSARY TO UNSCREW AND ADJUST THE WIRE RING. MOVE LEGS OF TRIPOD INSTEAD.
LINE OF SIGHT SETTING: MEANS GETTING THE LINE OF SIGHT (SEE ABOVE) (THE CENTER POINT OF YOUR VISUAL CIRCLE) TO PASS EXACTLY THROUGH THE CROSSING OF THE VERTICLE & HORIZONTAL HAIRS.
• SET VERTICAL HAIR. YOU CAN SET HORIZONTAL HAIR BY THE DUMPY METHOD BUT SINCE THE Y LEVEL TELESCOPE IS NOT FIXED, THERE IS THIS EASIER METHOD:
• SIGHT TARGET WITH TELESCOPE RIGHT SIDE UP. ALLIGN HAIRS WITH TARGET (A) TURN TELESCOPE UPSIDE DOWN, IF STILL ALLIGNED (A) LEAVE IT. IF NOT ALIGNED (B OR C) MOVE HAIRLINE UP OR DOWN, TURN OVER, ALLIGN, REPEAT UNTIL NO MATTER WHETHER UPSIDE DOWN OR RIGHT SIDE UP, HORIZ. SAME.
A
B
C
BUBBLE TUBE SETTING: MAKE SURE TELESCOPE IS RESTING SNUG IN Y'S.
• ADJUST SCREWS ON BUBBLE TUBE UNTIL BUBBLE CENTERS
• DO NOT MOVE Y FRAME.
• TAKE TELESCOPE OUT OF Y'S & TURN IT FACING OPPOSITE DIRECTION SNUG IN Y'S.
• IF BUBBLE CENTERS, BUBBLE TUBE IS PARALLEL TO TELESCOPE. IF NOT, ADJUST SCREWS, THEN REPEAT TURNS UNTIL BUBBLE ALWAYS CENTERS.
TELESCOPE UNDERSIDE
VIEW
POSITIONING THE Y'S: LEVEL INSTRUMENT PER BUBBLE (OVER ONE SET LEVELING SCREWS)
• TURN END FOR END ON VERTICAL SPINDLE. IF BUBBLE REMAINS IN MIDDLE AND BUBBLE IS KNOWN TO BE PARALLEL TO SCOPE, INSTRUMENT IS LEVEL PER Y'S
• OTHERWISE ADJUST Y UNTIL AFTER TURNING BUBBLE ALWAYS CENTERS.
TRANSIT ADJUSTMENTS
CENTERING THE PLATE BUBBLES: SET UP TRANSIT SO THAT BOTH PLATE BUBBLES CENTER, THEN TURN ON VERTICAL AXIS 180°.
• IF BUBBLES STILL CENTER, INSTRUMENT IS LEVEL IF NOT ADJUST LEVELING SCREWS AS DUMPY.
• IN EXTREME, ADJUST BUBBLE SCREWS AT THIS POINT. TURN 180° AGAIN, READ, ADJUST, TURN, ETC UNTIL BUBBLES' CENTER IN 90° TURNS & IN ALL TURNS.
LINE OF SIGHT IN THE FIELD: SET UP SO VERTICAL HAIRLINE CORRESPONDS TO KNOWN VERTICAL IN VIEW.
• IF HAIRLINE DOES NOT COINCIDE, LOOSEN ALL CAPSTAN SCREWS WHICH POSITION THE VERTICAL-HAIRLINE-RING.
• ROTATE RING UNTIL HAIRLINE IS VERTICAL. SET RING CAPSTANS FIRMLY.
LINE OF SITE OUT OF FIELD: HORIZONTAL HAIRLINE SHOULD BE IN CENTER OF TUBE. (VIEW)
• IF HAIRLINES ARE GETTING "CENTERED" OUT OF CENTER, REMOVE TELESCOPE AND SET THE LINE OF SIGHT IN SOME Y FORMS AS
• DESCRIBED UNDER Y LEVEL ADJUSTMENTS.
TELESCOPE LEVELING: CENTER PLATE BUBBLES FIRST.
• USE SAME METHOD AS DUMPY HORIZONTAL HAIRLINE CORRECTION....
• EXCEPT DON'T MOVE THE HAIRLINE,... MOVE THE TELESCOPE UP OR DOWN.
• ADJUST BUBBLE SCREWS. THEN ITS SET.
VERTICAL VERNIER SETTING: WHEN THE TELESCOPE IS LEVELED THE TELESCOPE BUBBLE SHOULD CENTER AND THE VERTICAL VERNIER SHOULD READ ZERO.
• ADJUST THE VERNIER'S SCREWS TO REPOSITION IT TO ZERO IF IT DOESN'T READ ZERO WHEN TELESCOPE IS HORIZONTAL AND YOU KNOW TELESCOPE HAS BEEN LEVELLED.
HORIZONTAL AXIS SETTING:

93. Cutting and filling is not bulldozing, but accommodating incongruous ideas with the delights of the site.

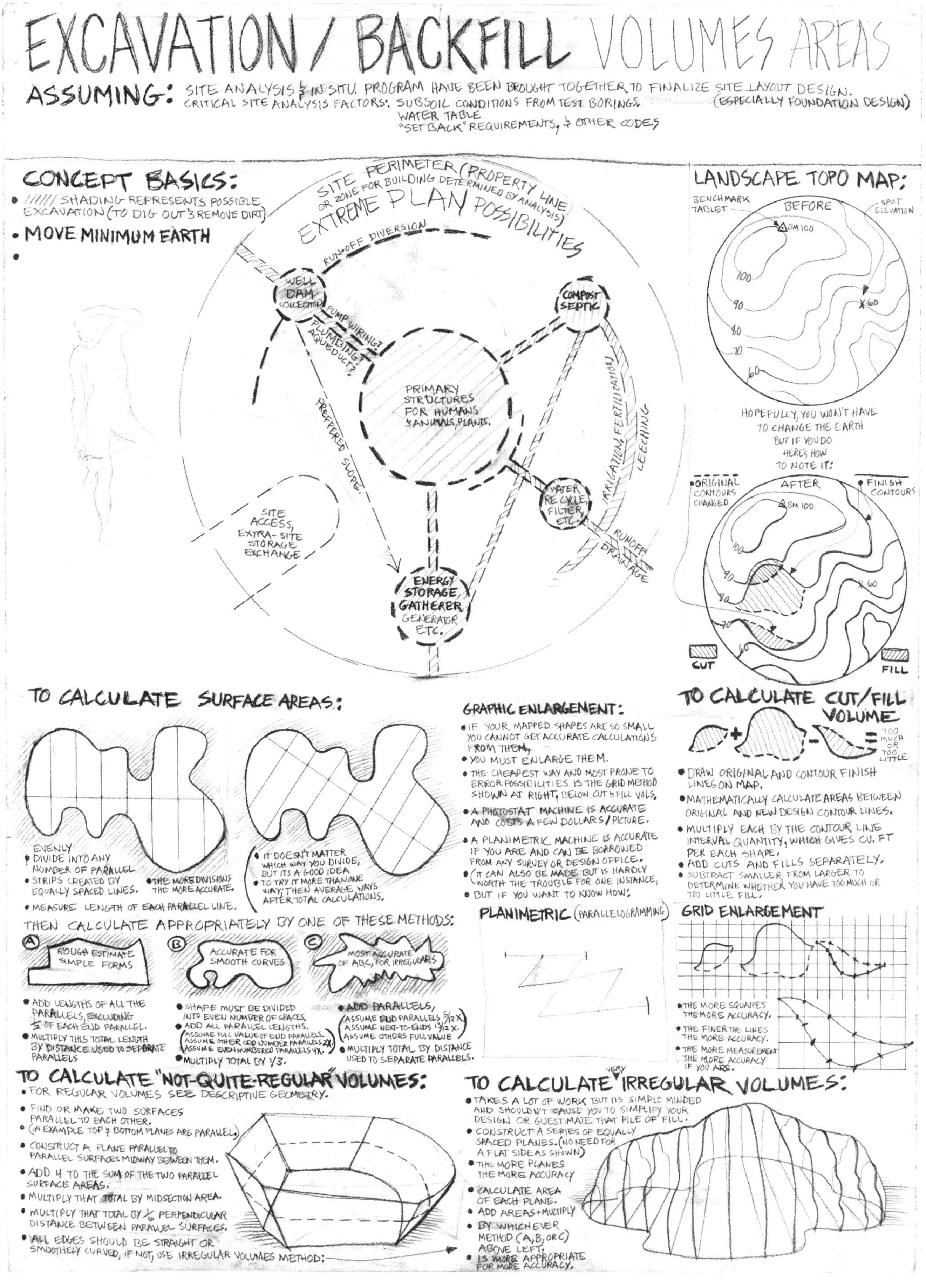

EXCAVATION / BACKFILL VOLUMES AREAS
ASSUMING: SITE ANALYSIS & IN SITU. PROGRAM HAVE BEEN BROUGHT TOGETHER TO FINALIZE SITE LAYOUT DESIGN.
CRITICAL SITE ANALYSIS FACTORS: SUBSOIL CONDITIONS FROM TEST BORINGS. (ESPECIALLY FOUNDATION DESIGN)
WATER TABLE
"SET BACK" REQUIREMENTS, & OTHER CODES
CONCEPT BASICS:
• ////// SHADING REPRESENTS POSSIBLE EXCAVATION (TO DIG OUT & REMOVE DIRT)
• MOVE MINIMUM EARTH
SITE PERIMETER (PROPERTY LINE OR ZONE FOR BUILDING DETERMINED BY ANALYSIS)
EXTREME PLAN POSSIBILITIES
RUN-OFF DIVERSION
WELL DAM COLLECTOR
PUMP WIRING
PLUMBING AQUEDUCT
COMPOST SEPTIC
PRIMARY STRUCTURES FOR HUMANS & ANIMALS, PLANTS.
PREFERRED SLOPE
IRRIGATION, FERTILIZATION
LEECHING
WATER RECYCLE, FILTER, ETC.
RUNOFF DRAINAGE
SITE ACCESS, EXTRA-SITE STORAGE EXCHANGE
ENERGY STORAGE, GATHERER GENERATOR ETC.
LANDSCAPE TOPO MAP:
BENCHMARK TABLET
BEFORE
SPOT ELEVATION
BM 100
100
90
80
70
60
X 60
HOPEFULLY, YOU WON'T HAVE TO CHANGE THE EARTH BUT IF YOU DO HERE'S HOW TO NOTE IT:
• ORIGINAL CONTOURS CHANGED
AFTER
• FINISH CONTOURS
BM 100
100
90
80
70
60
X 60
CUT
FILL
TO CALCULATE SURFACE AREAS:
• EVENLY DIVIDE INTO ANY NUMBER OF PARALLEL
• STRIPS CREATED BY EQUALLY SPACED LINES.
• THE MORE DIVISIONS THE MORE ACCURATE.
• MEASURE LENGTH OF EACH PARALLEL LINE.
• IT DOESN'T MATTER WHICH WAY YOU DIVIDE, BUT ITS A GOOD IDEA
• TO TRY IT MORE THAN ONE WAY; THEN AVERAGE WAYS AFTER TOTAL CALCULATIONS.
THEN CALCULATE APPROPRIATELY BY ONE OF THESE METHODS:
A ROUGH ESTIMATE SIMPLE FORMS
B ACCURATE FOR SMOOTH CURVES
C MOST ACCURATE OF ABC, FOR IRREGULARS
• ADD LENGTHS OF ALL THE PARALLELS, EXCLUDING ½ OF EACH END PARALLEL.
• MULTIPLY THIS TOTAL LENGTH BY DISTANCE USED TO SEPERATE PARALLELS
• SHAPE MUST BE DIVIDED INTO EVEN NUMBER OF SPACES.
• ADD ALL PARALLEL LENGTHS.
(ASSUME FULL VALUE OF END PARALLELS. ASSUME OTHER ODD NUMBER PARALLELS 2X. ASSUME EVEN NUMBERED PARALLELS 4X.)
• MULTIPLY TOTAL BY 1/3.
• ADD PARALLELS,
(ASSUME END PARALLELS 5/12X. ASSUME NEXT-TO-ENDS 13/12X. ASSUME OTHERS FULL VALUE)
• MULTIPLY TOTAL BY DISTANCE USED TO SEPARATE PARALLELS.
GRAPHIC ENLARGEMENT:
• IF YOUR MAPPED SHAPES ARE SO SMALL YOU CANNOT GET ACCURATE CALCULATIONS FROM THEM,
• YOU MUST ENLARGE THEM.
• THE CHEAPEST WAY AND MOST PRONE TO ERROR POSSIBILITIES IS THE GRID METHOD SHOWN AT RIGHT, BELOW CUT & FILL VOLS.
• A PHOTOSTAT MACHINE IS ACCURATE AND COSTS A FEW DOLLARS / PICTURE.
• A PLANIMETRIC MACHINE IS ACCURATE IF YOU ARE AND CAN BE BORROWED FROM ANY SURVEY OR DESIGN OFFICE.
• (IT CAN ALSO BE MADE BUT IS HARDLY WORTH THE TROUBLE FOR ONE INSTANCE,
• BUT IF YOU WANT TO KNOW HOW:
PLANIMETRIC (PARALLELOGRAMMING)
TO CALCULATE CUT/FILL VOLUME
TOO MUCH OR TOO LITTLE
• DRAW ORIGINAL AND CONTOUR FINISH LINES ON MAP.
• MATHEMATICALLY CALCULATE AREAS BETWEEN ORIGINAL AND NEW DESIGN CONTOUR LINES.
• MULTIPLY EACH BY THE CONTOUR LINE INTERVAL QUANTITY, WHICH GIVES CU. FT PER EACH SHAPE.
• ADD CUTS AND FILLS SEPARATELY.
• SUBTRACT SMALLER FROM LARGER TO DETERMINE WHETHER YOU HAVE TOO MUCH OR TOO LITTLE FILL.
GRID ENLARGEMENT
• THE MORE SQUARES THE MORE ACCURACY.
• THE FINER THE LINES THE MORE ACCURACY.
• THE MORE MEASUREMENT THE MORE ACCURACY IF YOU ARE.
TO CALCULATE "NOT-QUITE-REGULAR" VOLUMES:
• FOR REGULAR VOLUMES SEE DESCRIPTIVE GEOMETRY.
• FIND OR MAKE TWO SURFACES PARALLEL TO EACH OTHER.
• (IN EXAMPLE TOP & BOTTOM PLANES ARE PARALLEL.)
• CONSTRUCT A PLANE PARALLEL TO PARALLEL SURFACES MIDWAY BETWEEN THEM.
• ADD 4 TO THE SUM OF THE TWO PARALLEL SURFACE AREAS.
• MULTIPLY THAT TOTAL BY MIDSECTION AREA.
• MULTIPLY THAT TOTAL BY 1/6 PERPENDICULAR DISTANCE BETWEEN PARALLEL SURFACES.
• ALL EDGES SHOULD BE STRAIGHT OR SMOOTHLY CURVED, IF NOT, USE IRREGULAR VOLUMES METHOD:
TO CALCULATE VERY IRREGULAR VOLUMES:
• TAKES A LOT OF WORK BUT ITS SIMPLE MINDED AND SHOULDN'T CAUSE YOU TO SIMPLIFY YOUR DESIGN OR GUESTIMATE THAT PILE OF FILL.
• CONSTRUCT A SERIES OF EQUALLY SPACED PLANES. (NO NEED FOR A FLAT SIDE AS SHOWN)
• THE MORE PLANES THE MORE ACCURACY
• CALCULATE AREA OF EACH PLANE.
• ADD AREAS + MULTIPLY
• BY WHICHEVER METHOD (A, B, OR C) ABOVE LEFT,
• IS MORE APPROPRIATE FOR MORE ACCURACY.

 Structure originates in sun/slope.

STRUCTURE

STANDARD VARIABLES ANY CLIMATE

POLAR & TROPIC ZONES HAVE LESS RANGE BETWEEN EXTREMES MOST OF THE TIME. BELOW SHOWN FOR N. HEMIS TEMPERATE; REVERSE FOR S. HEMIS.

COMMON SENSE
RAD DIKE 8·74

SUN & SLOPE (CLIMATE) ARE INHERENT IN ALL THESE VARIABLES AS DEMONSTRATED HERE.

PLAN & LAT. & LONG. SECTIONS SHOWN AT RIGHT ARE FOR BASIC EVALUATION WITHOUT SUBTLETIES.

ANY SITE HAS THESE VARIABLES WHICH CAN FORM STRUCTURE, DEPENDENT ON THEIR INTEGRATION.

MONEY THE BIGGER THE SPACE, THE SLOWER IT HEATS UP, THE SLOWER IT COOLS OFF AT NIGHT. ETC. ETC. THE MORE THE FOLLOWING VARIABLES CAN BE BALANCED TO ADVANTAGE NATURALLY, THE CHEAPER THE SPACE.

REAL ESTATE ALL LAND HAS ITS PECULIARITIES. EVEN FLAT LAND VARIES BY ITS PROXIMITY TO OTHER FORMS, GROUND COVER, GEOLOGY, LATITUDE, ETC. DEPRESSIONS ARE APT TO ACT LIKE SINKS BY HOLDING COLD OR HOT AIR, STAYING DAMP AND THEREFORE FOGGY IN THE MORNINGS. SEE SLOPE ANALYSIS TO SEE WHY A LITTLE DISTANCE UP A SLOPE CAN BE MORE ADVANTAGEOUS.

AIR OF COURSE WARMS BY DAYLIGHT & COOLS BY NIGHT; RISES WHEN WARM, LOWERS WHEN COOL. BREEZES CAN BE CREATED BY PLACEMENT OF OPENINGS. EVERY LIVING SPACE CAN BE A REGULATABLE CHIMNEY FOR DRAFTS. SEE AIR, WIND, VENTILATION, DETAILS. IN COMBINATION WITH PROPER INSULATION, COOLER AIR TEMPERATURES ARE MORE PLEASANT, LESS STUFFY, LESS DRY, LESS DRAFTY AND GENERALLY MUCH HEALTHIER.

PLANTS & TREES MAKE EXCELLENT RADIATION & WIND DIFFUSERS.

EARTH A FEW FEET BENEATH THE GROUND SURFACE, THE SOIL IS COOL IN SUMMER AND WARM IN WINTER. SOIL IS NEARLY THE SAME TEMPERATURE YEAR ROUND BUT COMPARATIVELY WITH AIR TEMPERATURE AND BODY NEEDS (SEE HUMAN TEMP) SOIL IS COOLER + WARMER.

WATER

FIRE THE BIGGER THE SOUTH WINDOWS, THE QUICKER ACTING HEAT GENERATOR MUST BE, WHEN CLOUDS BLOCK THE SUN. USE SMALL OPENINGS FOR WINDOWS IN WINTER IF FIRE IS USED, AND WEATHER GENERALLY CLOUDY. OR STORE HEAT WITH WATER. ANIMAL HEAT IS HUMAN'S REAL FURNACE. SEE HUMAN.

COLOR, LIGHT SILVER & WHITE SURFACES REFLECT LIGHT BEST. THIS KEEPS HEAT FROM ENTERING WALL OR ROOF MATERIAL. IN WINTER WHEN WARM WALLS ARE DESIRABLE, IN WINTER, SUN IS LOWER AND LESSER ANGLE IS NOT GOOD FOR SURFACE ABSORBTION; WALLS IN WINTER ARE MORE DIRECTLY HIT BY SUN THAN ROOF. DARK WALLS ARE APPROPRIATE BUT DIRECT SUNSHINE IS BETTER, BUT NO STORAGE. BEST TRANSPARENT HEAT STORAGE MATERIAL IS WATER. SEE SOLAR ENERGY.

TEXTURE, DENSITY, INSULATION WITHOUT INSULATING FACTORS OUTSIDE COLD COMES THROUGH SURFACES SO THAT INSIDE SURFACES ARE COOL. THIS CAUSE THE HEAT RADIATED FROM THE BODY TO BE USED UP ON THESE COOL SURFACES THEREBY KEEPING THE BODY LESS WARM. EITHER THE BODY MUST EXPEND MORE ENERGY TO KEEP WARM OR THE AIR TEMP MUST BE RAISED. WITH INSULATION, THE BODY IS MORE COMFORTABLE AT COOLER AIR TEMPS.

FLOORS, ROOFS MUST RESIST PRECIPITATION ETC. SEE DETAILS. WIDE EAVES ON SOUTH, SHADE SOUTH WALL IN SUMMER, ALLOW WARM SOUTH WALL IN WINTER,

WALLS, ARE SURFACES WHICH COLLECT HEAT. THE WEST WALL COLLECTS THE MOST HEAT BECAUSE OF SETTING SUN THROUGH ALREADY WARMED AIR. THIS DOESN'T HAPPEN TO EAST WALL BECAUSE AIR IS COOL FROM NIGHT.

WINDOWS, DOORS SHOULD BE LOCATED WITH PARTICULAR DISCRETION TO SUNSHINE & WEATHER (BESIDES VIEWS, ACCESS ETC.) NORTH WINDOWS GET MORE SHINE MORN + NITE THAN OTHERS. THE COLDER THE WINTER, THE MORE SOUTH WINDOWS IS MOST ADVANTAGEOS. THE HOTTER THE SUMMER, THE SMALLER WEST WINDOWS ARE MOST ADVANTAGEOUS.

ENERGY

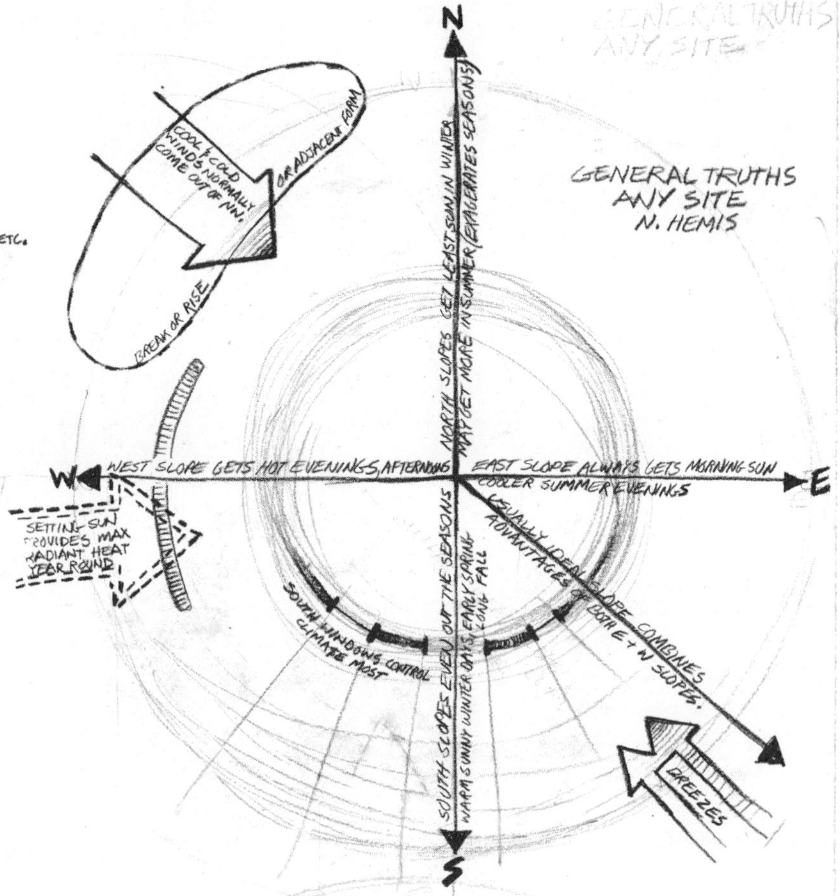
N
S
W
E
GENERAL TRUTHS
ANY SITE
N. HEMIS
COOL & COLD WINDS NORMALLY COME OUT OF NW.
OR ADJACENT FORM
BREAK OR RISE
NORTH SLOPES GET LEAST SUN IN WINTER MAY GET MORE IN SUMMER (EXAGGERATES SEASONS)
WEST SLOPE GETS HOT EVENINGS, AFTERNOONS
EAST SLOPE ALWAYS GETS MORNING SUN COOLER SUMMER EVENINGS
SETTING SUN PROVIDES MAX RADIANT HEAT YEAR ROUND
SOUTH WINDOWS CONTROL CLIMATE MOST
SOUTH SLOPES EVEN OUT THE SEASONS WARM SUNNY WINTER DAYS, EARLY SPRING, LONG FALL
USUALLY IDEAL SLOPE COMBINES ADVANTAGES OF BOTH E + N SLOPES.
BREEZES

LONG SETTING DIRECT SUMMER
WALL OR SPACE
LIVING SPACE
W
WEST-EAST SECTION
E

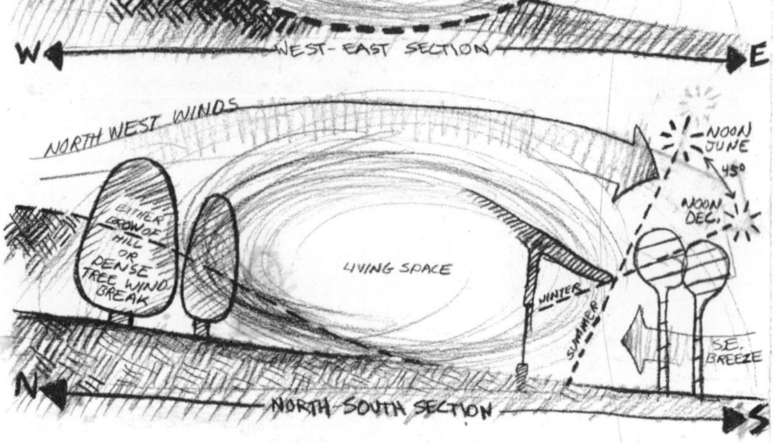
NORTH WEST WINDS
NOON JUNE
45°
NOON DEC.
EITHER BROW OF HILL OR DENSE TREE WIND BREAK
LIVING SPACE
WINTER
SUMMER
SE. BREEZE
N
NORTH-SOUTH SECTION
S

95. Each sun/slope suggests its own true comfortable architecture.

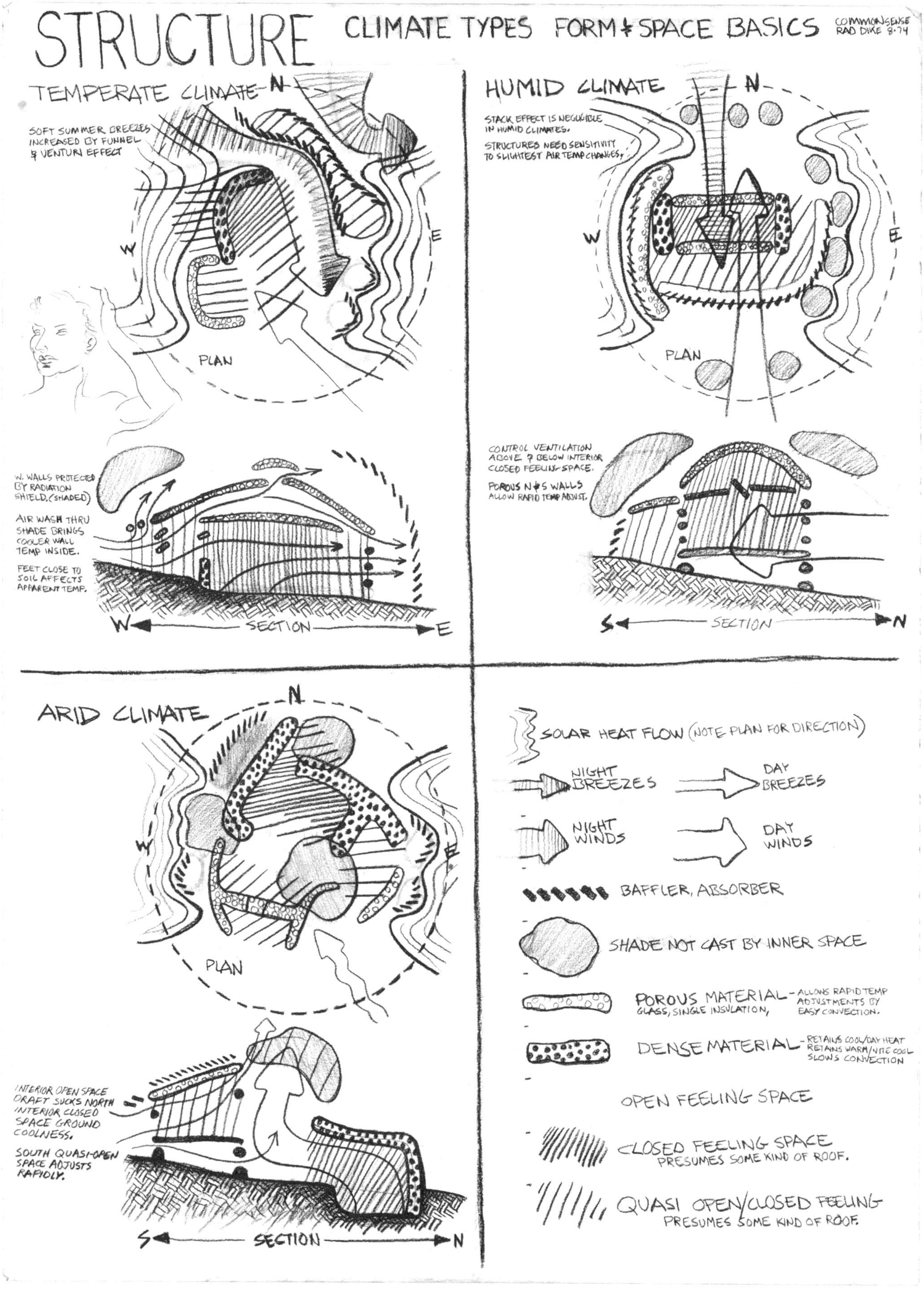
STRUCTURE
CLIMATE TYPES FORM & SPACE BASICS
COMMONSENSE
RAD DIKE 8·74
TEMPERATE CLIMATE
N
SOFT SUMMER BREEZES INCREASED BY FUNNEL & VENTURI EFFECT
W
E
PLAN
W. WALLS PROTECTED BY RADIATION SHIELD. (SHADED)
AIR WASH THRU SHADE BRINGS COOLER WALL TEMP INSIDE.
FEET CLOSE TO SOIL AFFECTS APPARENT TEMP.
W
SECTION
E
HUMID CLIMATE
N
STACK EFFECT IS NEGLIGIBLE IN HUMID CLIMATES.
STRUCTURES NEED SENSITIVITY TO SLIGHTEST AIR TEMP CHANGES.
W
E
PLAN
CONTROL VENTILATION ABOVE & BELOW INTERIOR CLOSED FEELING SPACE.
POROUS N & S WALLS ALLOW RAPID TEMP ADJUST.
S
SECTION
N
ARID CLIMATE
N
W
E
PLAN
INTERIOR OPEN SPACE DRAFT SUCKS NORTH INTERIOR CLOSED SPACE GROUND COOLNESS.
SOUTH QUASI-OPEN SPACE ADJUSTS RAPIDLY.
S
SECTION
N
SOLAR HEAT FLOW (NOTE PLAN FOR DIRECTION)
NIGHT BREEZES
DAY BREEZES
NIGHT WINDS
DAY WINDS
BAFFLER, ABSORBER
SHADE NOT CAST BY INNER SPACE
POROUS MATERIAL - ALLOWS RAPID TEMP ADJUSTMENTS BY EASY CONVECTION.
GLASS, SINGLE INSULATION,
DENSE MATERIAL - RETAINS COOL/DAY HEAT RETAINS WARM/NITE COOL SLOWS CONVECTION
OPEN FEELING SPACE
CLOSED FEELING SPACE PRESUMES SOME KIND OF ROOF.
QUASI OPEN/CLOSED FEELING PRESUMES SOME KIND OF ROOF.

96. Truth-finding originated in practicality, and had faith in the unanswerable, which diverged religion and science.

Philosophy comes from the Greek words philos(loving) & sophos(wise).

Nearly all of western Civilization's known philosophical development has occurred since the Greeks. Eastern Civilization's thought is assumed to have been developing for a longer known period, not agreed upon. Though eastern thought certainly has been a "love of the wise", it has not paralleled the western distinctions and definitions:

Western philosophy has differed from both Science and Religion in self concious, distinct ways. Philosophy differs with religion because it searches for underlying causes and principles of being and thinking and does not rely on dogma or faith. Philosophy differs with science since it does not depend only on fact but relies heavily on speculation.

Historically, philosophy developed from religion when thinkers sought truth independent of theological considerations; science then developed from philosophy and recently branches of science broke from the embrace of philosophy , notably psychology, sociology, ethnology et cetera. Scientology, theosophy et cetera are breaks from not only from science or religion or philosophy, but from combinations and permutations of those areas.

The combination or permutation area of philosophy and religion, or philosophy and science or science and religion, or any or all with art or all together are new hard tp define areas none of which are contained within the distinctions of philosophy defined above. Certainly these areas are "philosophical" and represent parts of humanity's mind concerning being and thinking. My intentions are to cover those areas if I am capable at another place and time.

Philosophy alone,without the many other areas of thought, has developed a multitude of viewpoints which to the layman can be an overwhelming whole. The problem solver or designer is possibly one philosophical layman whose process and solution may or may not reflect his philosophical tendencies or convictions.

If one desires to carry one's integrity through all scales of his existence, then one may want a concious articulation of who and why philosophically. For those who for some reason have difficulty in locating philosophical information appropriate to their needs, I offer this basic array of statements which attempt to articulate facets of philosophy.

Corresponding to the number of the statement are one or more words and/or names which can be effective as appropriate search leads into the mass of information available. By separating the statements from the categorical leads, I have intended to provide these usable opportunities: Skimming or reading over the statements, one is free to act & react to the ideas rather than Names or Words representative of the ideas. This may lead to restating the ideas in ones own words or discussion et cetera, responses developing the idea. Therefore one may establish a group of statements which together or each in some way provokes a desire to search further without knowing if their representative names or words contradict, complement or otherwise affect one.another. Rather, when correlating the statements to the leads, one is encouraged to use them as a string of adjectives describing one' own curiosity.

THESE STATEMENTS ARE MEANT ONLY TO REPRESENT VARIOUS POINTS OF VIEW. NO SUMMARY OF PHILOSOPHICAL SYSTEMS IS INTENDED. PERHAPS SOME STATEMENT(S) WILL GIVE YOU REFERENCE(S).

The attempt herein is to respect the distinctions of philosophy from religion et al, these generalizations are still subject to great interpretations as wellas the words employed. Basically they will refer to either ethics(the study of human conduct) and/or epistemology(the study of theories of knowledge) and/or metaphysics(the study of the universe).

This compilation is and will be forever incomplete and inadequate; it is being reworked and developed. Contributions are welcome.

97. I believe that what a being creates is shaped in part by his reason for living and creating, difficult as it may be to verbalize.

1. The effect which a course of action produces in this world makes it good or bad.

2. The goodness or badness of a course of action must be judged by standards dictated from an other world external to man(from God or from some force for good)

3. In each human there is a still small voice which guides him as to right or wrong.

4. Before acting in a certain way, each individual must ask himself, "Act only according to a maxim by which you can at the same time will that it shall become general law."

5. Pleasure is the beginning and end of the good life, but pleasure is good only when moderate or passive; dynamic pleasure which causes painful after-effects is not good.

6. Pain and pleasure being sovereign masters governing human conduct, the greatest happiness of the greatest number is the measure of right and wrong.

7. There are different qualities of pain and pleasure and some kinds of pleasure are more valuable than others.

8. "Whatever works is right" is not as shallow as it might seem since it depends on what standards are used to decide what works.

9. There is an ultimate reality in which all differences are reconciled.

10. The ultimate answers to all inquiries is that we do not know.

11. Goodness is to live and act in the interests of others rather than oneself.

12. Withdrawel from the physical world into the inner world of the spirit is the highest good attainable.

13. The only workable hypothesis in this world is to reject the concept of God.

14. The entire universe is composed of distinct and indivisible units.

15. Universal ideas are neither created by finite(human) minds, nor entirely apart from from absolute mind(God).

16. Humans cannot determine whether there is anything beyond each's own experience.

17. Reality is tri-partite; in addittion to the mental and physical aspects of reality, there is a third aspect called essences.

18. The path to knowledge lies midway between dogmatism and scepticism.

19. The universe follows a fixed or pre-determined pattern.

20. Reality is strictly material and is based on a struggle between opposing forces, with occasional interludes of harmony.

21. One may believe anything and espouse it without any authoritive support.

22. The world consists of two radically independent and absolute elements(paradoxes, e.g. good & evil, spirit & matter, etc)

23. Serving one's own interests is the highest end.

24. The most workable hypothesis in this world is to reject a priori knowledge in favor of experience and induction.

25. The uhiverse is a continuous progression of inter-related phenomena.

98. If there is a meaning to life, it seems most likely to be found in the ordinary basics of life.

26. Humans must create values for themselves through action; self is the ultimate reality.

27. Human interests and the human mind are are paramount in the universe.

28. Thought or idea is the basis either of knowledge or existence(the search for best,highest)

29. Ideas are instruments and tools rather than goals for living.

30. Perception of truth is by intuition, not analysis.

31. The independent spirit does not exist,the only substance is matter, physical well-being is paramount.

32. The world is capable of improvement, and humas have the power of helping in its betterment.

33. There is only one ultimate reality, whatever its nature.

34. The ultimate real lies in direct contact with the divine.

35. All phenomena can be explained by strictly natural categories(as opposed to supernatural)

36. Reality is neither physical nor spiritual but capable of expressing itself as either.

37. General terms have no corresponding reality either in or out of the mind, and are in effect no nothing more than words.

38. The universe is the best of all possible ones; all will work out for the best.

39. God and the universe are one and the same.

40. Ultimate reality consists of a plurality of spiritual beings or independent persons.

41. The universe is the worst possible and all is doomed to evil.

42. Reality is only appearance.

43. There are more than two irreducable components of reality.

44. Humans can have no knowledge except of phenomena, and the knowledge of phenomena is relative, not absolute.

45. Practical consequences are the test of truth.

46. Reason alone, without the aid of experience , can arrive at the basic reality of the universe.

47. There are no absolutes; everything is relative.

48. No facts can be certainly known.

49. The only workable hypothesis in this world is the concept of God.

50. There is an ultimate reality which transcends human experience.

51. Will is the determining factor in the universe.

52. The one permanent thing is change; there is no one ultimate reality.

53. Plurality and change are appearances, not realities.

54. It is reasonable to doubt the human's ability to attain absolute truth.

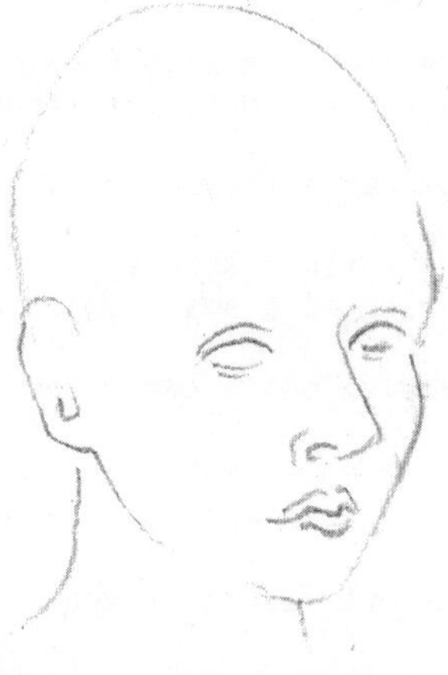

55/ The four factors in causation are: the interrelated factors of form & matter; motive cause, which produces change; and the end, for which a process of change occurs.

56. Mankind's role is to accept nature and all it offers, good or bad.

99. This provocative list of statements may help to relate design intention to beliefs, or at least to stretch the imaginative use of hardware.

57. God transcends human comprehension.

58. Faith and reason are in agreement.

59. The State of Government is paramount in human affairs.

60. Sense and reason are mutually deceptive; truth lies between dogmatism & skepticism.

61. Material things exist only in being perceived.

62. Existence precedes essence; only existence has reality and each individual is unique.

63. The "will to power" is basic in life; the spontaneous is to be preferred to the orderly; Christianity is a system which fosters the weak; the function of evolution is to evolve supermen.

64. Overall reality is not orderly and to live appropriately one must be subjective and irrational.

TO FIND WHICH OF THE REFERENCES BELOW ARE APPROPRIATE TO YOUR INTERESTS, MOST PEOPLE BROWSE THROUGH THE POINTS OF VIEW ABOVE BEFORE EVER LOOKING AT THE CATEGORIES & NAMES BELOW.

Since the statements-above are not mutually exclusive, more than one will most likely help to describe your point of view since it is doubtful that your view is accurately represented here.

1. Utilitarian, Mill, Bentham,
2. Idealism, Plato, Socrates, Aristotle
3. Practical Idealism, Kant
4. Practical Idealism, Kant
5. Epicurus
6. Bentham, Utilitarian
7. Mill, Utilitarianism
8. Pragmatism, James, Sociology
9. Absolutism
10. Agnosticism
11. Altruism
12. Asceticism
13. Atheism
14. Atomism
15. Conceptualism
16. Critical Idealism
17. Critical realism
18. Criticism
19. Determinism
20. Dialectical Materialism
21. Dogmatism
22. Dualism
23. Egoism
24. Empiricism
25. Evolutionism
26. Existentialism
27. Humanism
28. Idealism
29. Instrumentalism
30. Intuitionism
31 Materialism
32. Meliorism
33. Monism,
34. Mysticism
35. Naturalism
36. Neutral Monism
37. Nominalism
38 Optimism
39. Pantheism
40. Personalism
41. Pessimism
42. Phenomenalism
43. Pluralism
44. Positivism
45. Pragmatism
46. Rationalism
47. Relativism
48. Skepticism
49. Theism
50. Transeendentalism
51. Voluntarism
52. Heraclitus
53. Zeno of Elea
54. Protagoras
55. Aristotle
56. Zeno of Citrium
57. Augustine
58. Aquinas
59. Machiavelli
60. Pascal
61. Berkeley
62. Kierkegaard
63. Nietzsche
64. Sartre

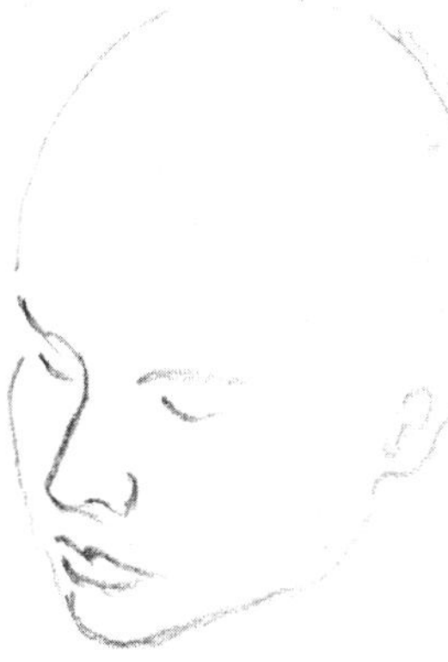

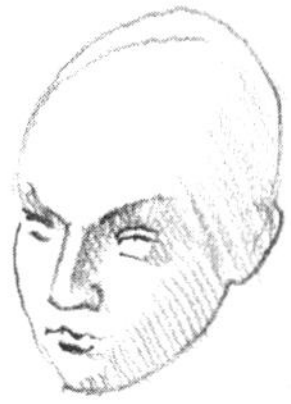

100. It does not satisfy me to add the parts of technology together to make architecture or metaphors for organisms.

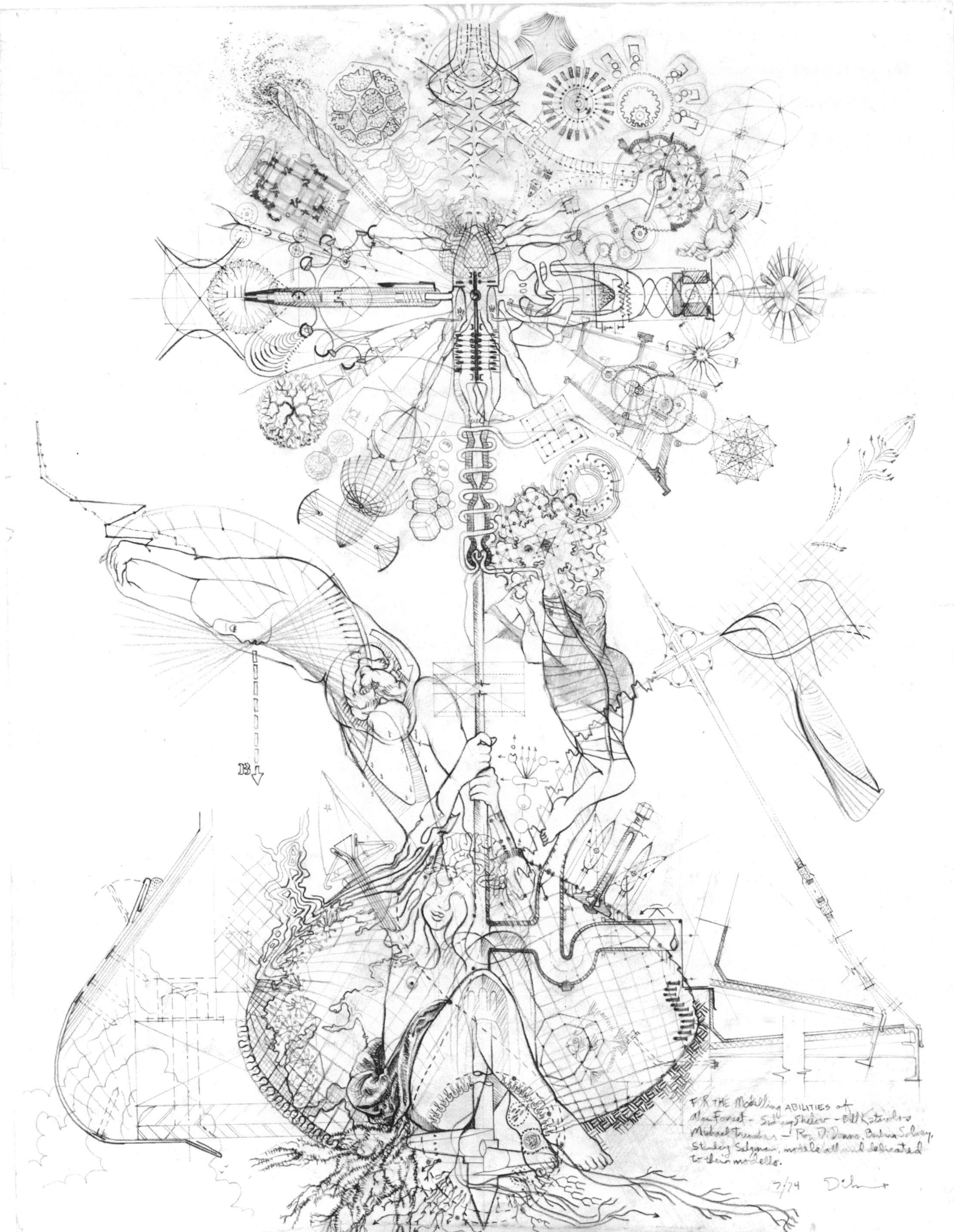

101. The psyche and its processes may prove yet that the parts we don't see are the parts we use most.

The physical birth of biology lasts and is seen for only a short while compared to the physical birth of psychology.

This second birth, whatever we may call it (religous etc), we all continue to feel long after we are first breathing. I am refering to our developable abilities to work with our range of feelings from "separateness" to "one-ness" with the world.

I am not familiar with the large body of knowledge developed by professional psychoanalysts, but many of their reference words to describe the process also describe to me self to site situations:

THE SEPARATION-INDIVIDUATION PROCESS
is making sense of forms as separate things...

IN ARCHITECTURE
we often have difficulty making sense of the separate factors forming the multitude of features comprising the whole space and one-ness of form.

THE SELF'S OWN BODY & THE PRIMARY LOVE OBJECT
are the whole world to the newborn world of the infant.

IN ARCHITECTURE
our concept of our ideal building is initially identical to all that we love, which at this stage is not noticed to be an undifferentiated mass of individual requirements.

THE SEPARATION-INDIVIDUATION PHASE
is when the main achievements of this process occur.

IN ARCHITECTURE
the main events develop the factors of the self, site and situation to achieve livable forms. Life after "moving-in" continues the process.

We want to live functionally and emotionally separate from our home but rely on it to support us when we need it.

102. However, truth is of the whole human, not only of the intellect.

APPRAISING $\frac{\text{WHAT YOU WANT}}{\text{WHAT YOU DESIGN}}$

SUN SITE & SELF
decision-making factors

	TOLERABLE	DESIRABLE	TOLERABLE
SPACE TEMPERATURE : comfortable in day in JANUARY (example)	55°	70°	75°
" " night " "	40	50°	60°
" " day " JULY			
" " night " "			
in any other monthes			

In order to visualize simply and comparatively YOUR IDEALS PER YOUR DESIGN'S PREDICTED PERFORMANCE,

Graph the ideals as a constant, against which predicted & actual performance can be seen to deviate or coincide.

FACTORS are as numerous as your concerns.
The factors included in this volume are restricted to those circumscribing the nature of both site & man.
Time & space further limit the selections to those I understand most and with which I am most concerned.
They are
those factors of site and man, which, when interacted, result in a visual proportional layer between man & site called architecture

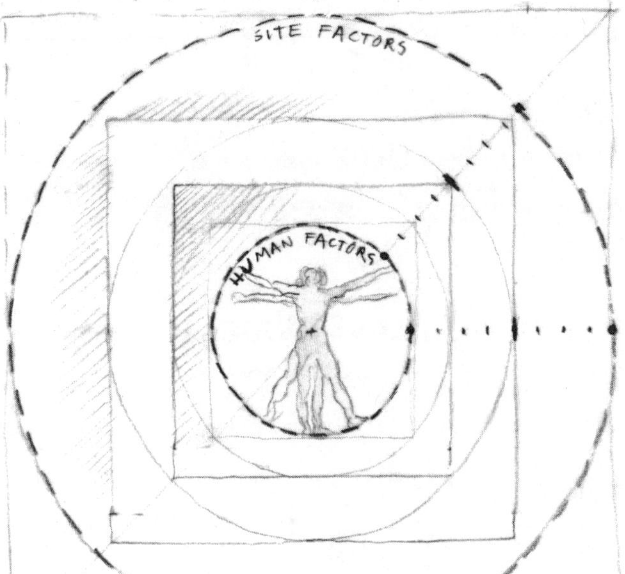

You know what is desirable for you and what is tolerable.
Between these constant concepts, actual quantifiable data appropriate to your needs will make such a chart meaningful.
However,
each self or group of selves has its own preferences. I know of no universally applicable quantities, only factors.
Therefore, in this book only the factors are explained and brief suggestions are made about quantifying the range.

An architectural factor resulting at man-site interface is

OPERATION/ATTENTION: daily controlls
monthly "
annual "

Such factors,
mostly dependent on design choices heavily weighted by preference derived from other than man site interactions, are not included in this volume.

103. Thermodynamics, the mechanical relations of heat, is certainly the basis of our comfortable existence.

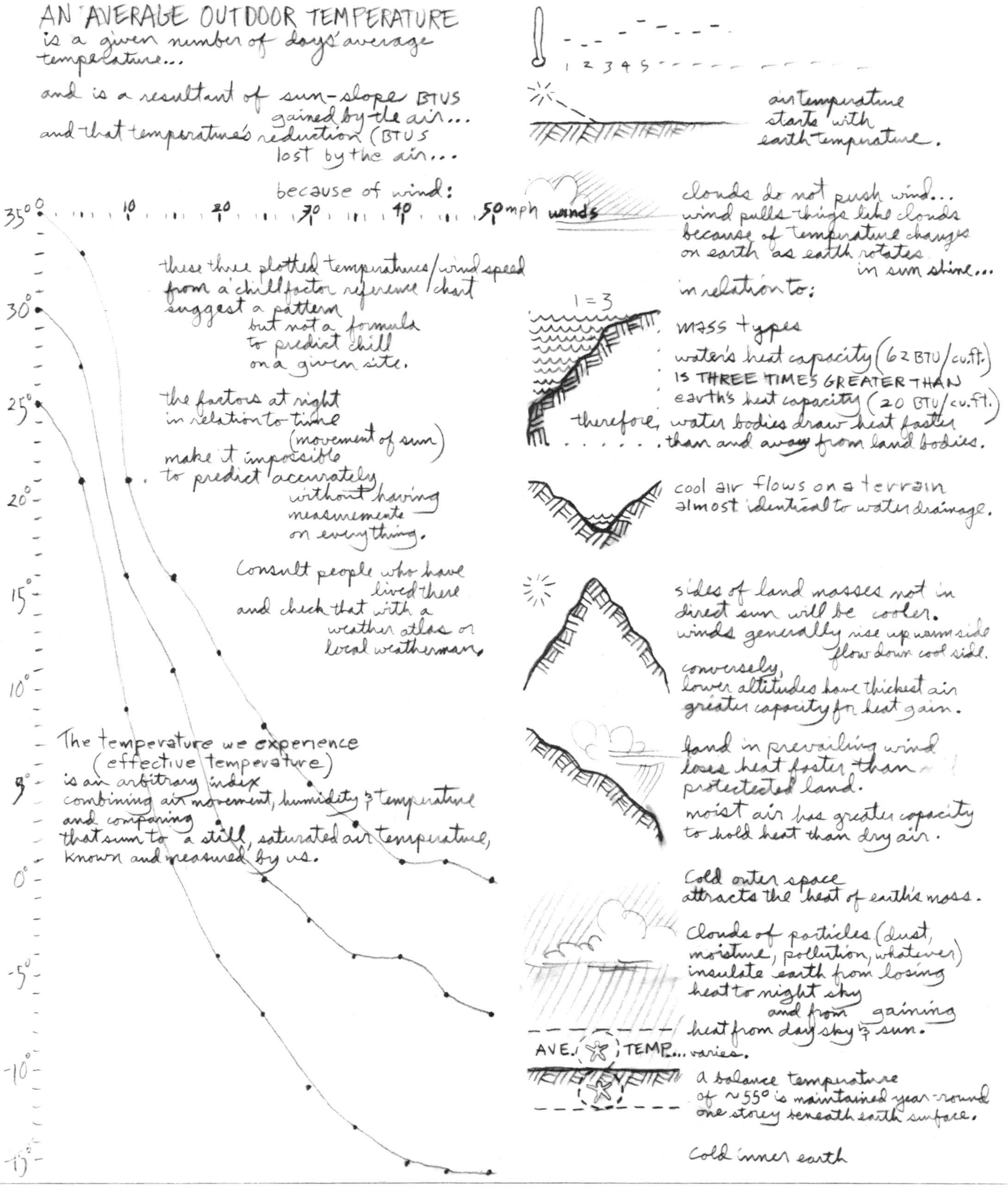

AN AVERAGE OUTDOOR TEMPERATURE
is a given number of days' average temperature...
and is a resultant of sun-slope BTUS gained by the air...
and that temperature's reduction (BTUS lost by the air...
because of wind:
0 10 20 30 40 50 mph winds
35° 30° 25° 20° 15° 10° 5° 0° -5° -10° -15°
these three plotted temperatures/wind speed from a chillfactor reference chart suggest a pattern but not a formula to predict chill on a given site.
the factors at right in relation to time (movement of sun) make it impossible to predict accurately without having measurements on everything.
Consult people who have lived there and check that with a weather atlas or local weatherman.
The temperature we experience (effective temperature) is an arbitrary index combining air movement, humidity & temperature and comparing that sum to a still, saturated air temperature, known and measured by us.
1 2 3 4 5
air temperature starts with earth temperature.
clouds do not push wind... wind pulls things like clouds because of temperature changes on earth as earth rotates in sun shine...
in relation to:
1 = 3
mass types
water's heat capacity (62 BTU/cu.ft.) IS THREE TIMES GREATER THAN earth's heat capacity (20 BTU/cu.ft.)
therefore, water bodies draw heat faster than and away from land bodies.
cool air flows on a terrain almost identical to water drainage.
sides of land masses not in direct sun will be cooler. winds generally rise up warm side flow down cool side.
conversely, lower altitudes have thickest air greater capacity for heat gain.
land in prevailing wind loses heat faster than protected land.
moist air has greater capacity to hold heat than dry air.
Cold outer space attracts the heat of earth's mass.
Clouds of particles (dust, moisture, pollution, whatever) insulate earth from losing heat to night sky and from gaining heat from day sky & sun.
AVE. TEMP... varies.
a balance temperature of ~55° is maintained year-round one storey beneath earth surface.
Cold inner earth

104. Each living body is its own source of heat energy, emmeshed in the heat flow of earth and sun.

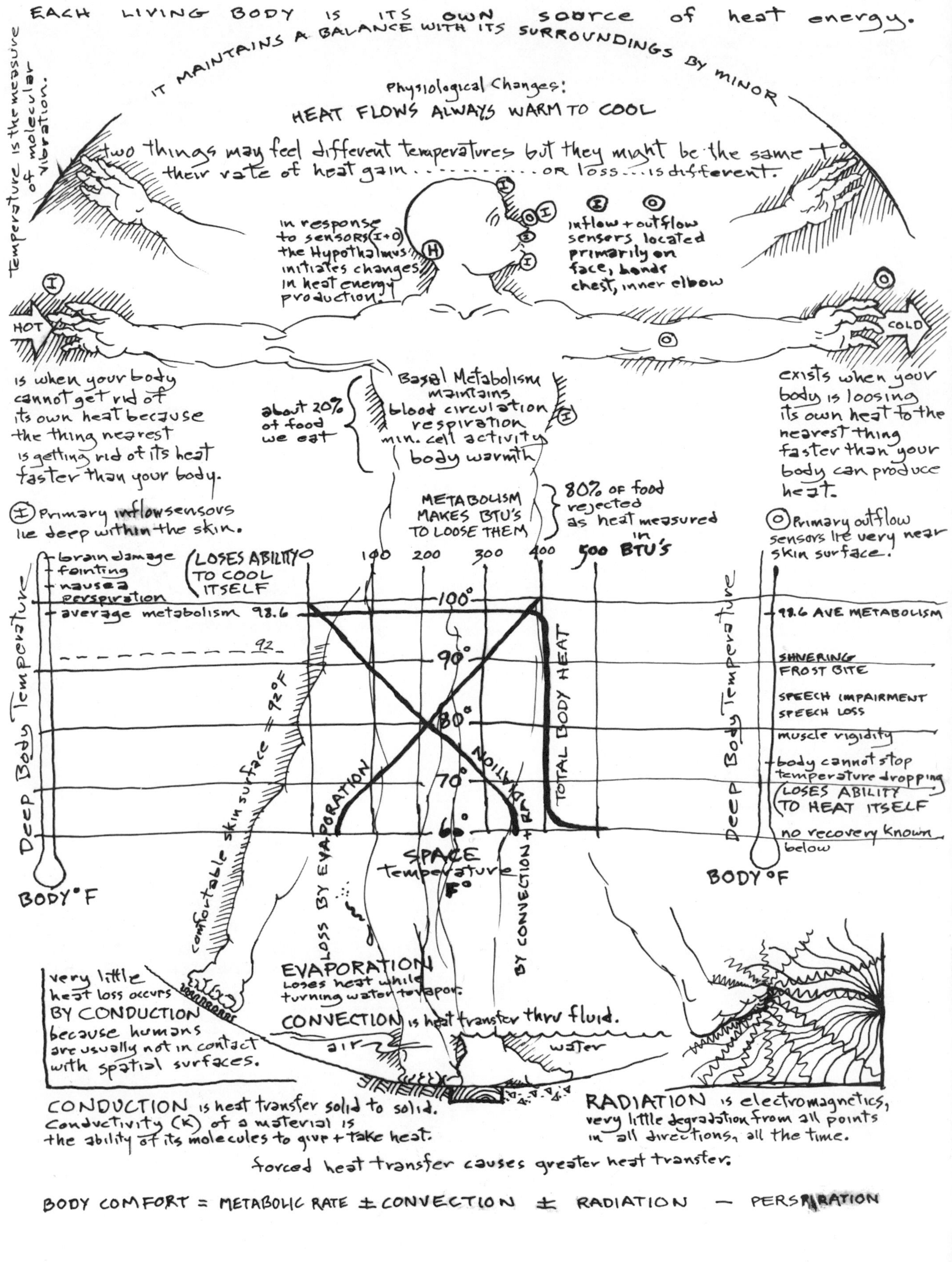
EACH LIVING BODY IS ITS OWN source of heat energy.
IT MAINTAINS A BALANCE WITH ITS SURROUNDINGS BY MINOR
Physiological Changes:
HEAT FLOWS ALWAYS WARM TO COOL
two things may feel different temperatures but they might be the same their rate of heat gain .... OR loss....is different.
Temperature is the measure of molecular vibration.
in response to sensors (I+O) the Hypothalmus initiates changes in heat energy production.
inflow + outflow sensers located primarily on face, hands chest, inner elbow
HOT
COLD
is when your body cannot get rid of its own heat because the thing nearest is getting rid of its heat faster than your body.
exists when your body is loosing its own heat to the nearest thing faster than your body can produce heat.
Basal Metabolism maintains blood circulation respiration min. cell activity body warmth
about 20% of food we eat
METABOLISM MAKES BTU'S TO LOOSE THEM
80% OF food rejected as heat measured in
100 200 300 400 500 BTU'S
Primary inflow sensors lie deep within the skin.
Primary outflow sensors lie very near skin surface.
brain damage
fainting
nausea
perspiration
LOSES ABILITY TO COOL ITSELF
average metabolism 98.6
92
100°
90°
80°
70°
60°
98.6 AVE METABOLISM
SHIVERING
FROST BITE
SPEECH IMPAIRMENT
SPEECH LOSS
muscle rigidity
body cannot stop temperature dropping
LOSES ABILITY TO HEAT ITSELF
no recovery known below
Deep Body Temperature
BODY °F
EVAPORATION
RADIATION
TOTAL BODY HEAT
comfortable skin surface = 92°F
LOSS BY EVAPORATION
BY CONVECTION + RADIATION
SPACE temperature F°
very little heat loss occurs BY CONDUCTION because humans are usually not in contact with spatial surfaces.
EVAPORATION Loses heat while turning water to vapor.
CONVECTION is heat transfer thru fluid.
air
water
CONDUCTION is heat transfer solid to solid. Conductivity (K) of a material is the ability of its molecules to give + take heat.
RADIATION is electromagnetics, very little degradation from all points in all directions, all the time.
forced heat transfer causes greater heat transfer.
BODY COMFORT = METABOLIC RATE ± CONVECTION ± RADIATION − PERSPIRATION

105. The sun is the source of greenhouse-effect heat energy, which architects the heat flow of earth and sun.

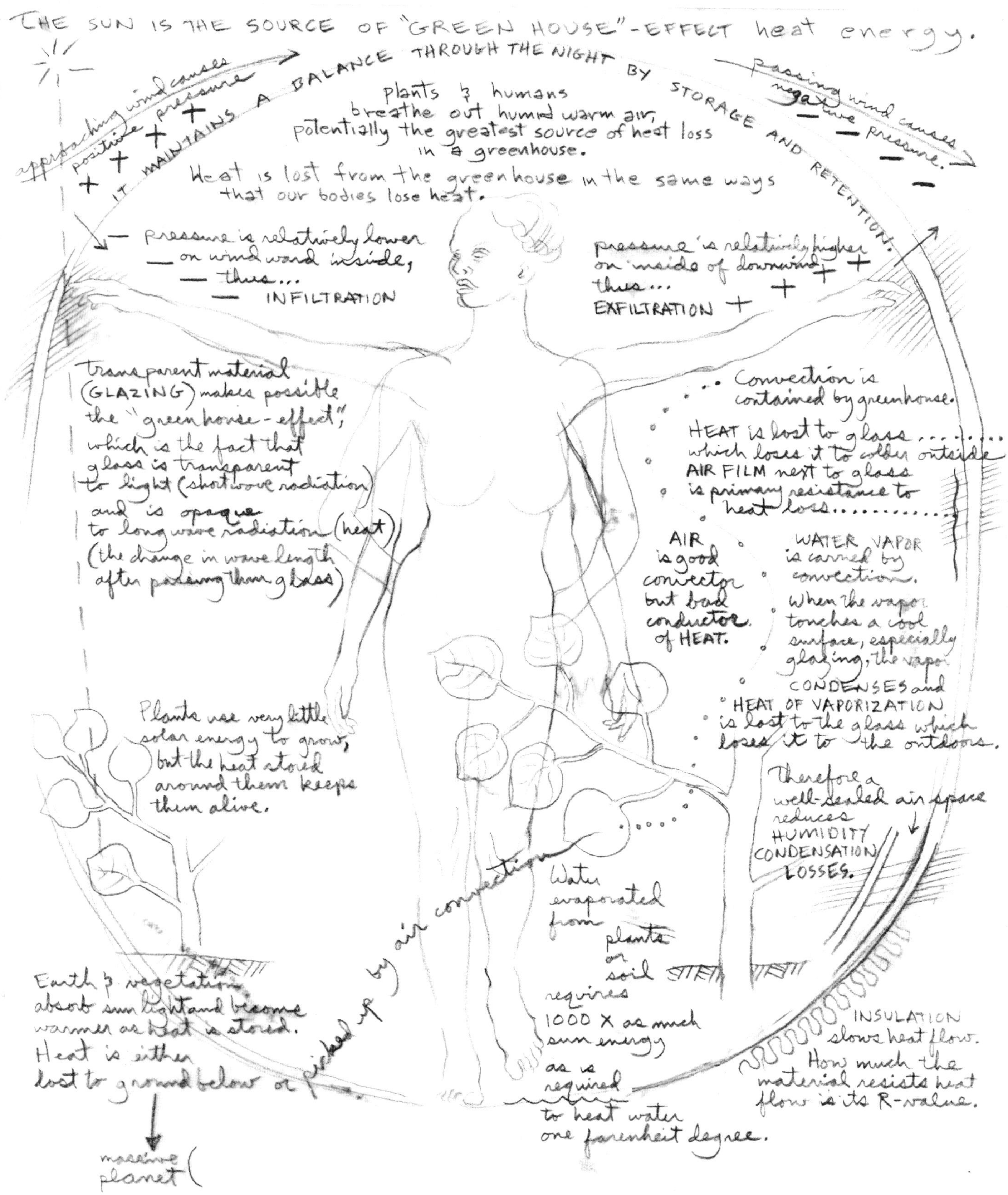

THE SUN IS THE SOURCE OF "GREEN HOUSE"-EFFECT heat energy.
IT MAINTAINS A BALANCE THROUGH THE NIGHT BY STORAGE AND RETENTION.
approaching wind causes positive pressure
passing wind causes negative pressure
plants & humans breathe out humid warm air, potentially the greatest source of heat loss in a greenhouse.
Heat is lost from the greenhouse in the same ways that our bodies lose heat.
pressure is relatively lower on wind ward inside, thus... INFILTRATION
pressure is relatively higher on inside of downwind thus... EXFILTRATION
transparent material (GLAZING) makes possible the "greenhouse-effect", which is the fact that glass is transparent to light (shortwave radiation) and is opaque to long wave radiation (heat) (the change in wave length after passing thru glass)
Convection is contained by greenhouse.
HEAT is lost to glass which loses it to colder outside
AIR FILM next to glass is primary resistance to heat loss
AIR is good convector but bad conductor of HEAT.
WATER VAPOR is carried by convection. When the vapor touches a cool surface, especially glazing, the vapor CONDENSES and HEAT OF VAPORIZATION is lost to the glass which loses it to the outdoors.
Plants use very little solar energy to grow, but the heat stored around them keeps them alive.
Therefore a well-sealed air space reduces HUMIDITY CONDENSATION LOSSES.
picked up by air convection
Water evaporated from plants on soil requires 1000 X as much sun energy as is required to heat water one farenheit degree.
Earth & vegetation absorb sunlight and become warmer as heat is stored. Heat is either lost to ground below or
massive planet
INSULATION slows heat flow. How much the material resists heat flow is its R-value.

HEAT CONTAINED = SOLAR GAIN + FILTRATION, CONVECTION, CONDUCTION, RADIATION LOSSES.

106. The ever changing sky, by clouds or seasons, is the primary stimulation of living organisms' receptor organs, photosynthesis, sight, and sensation.

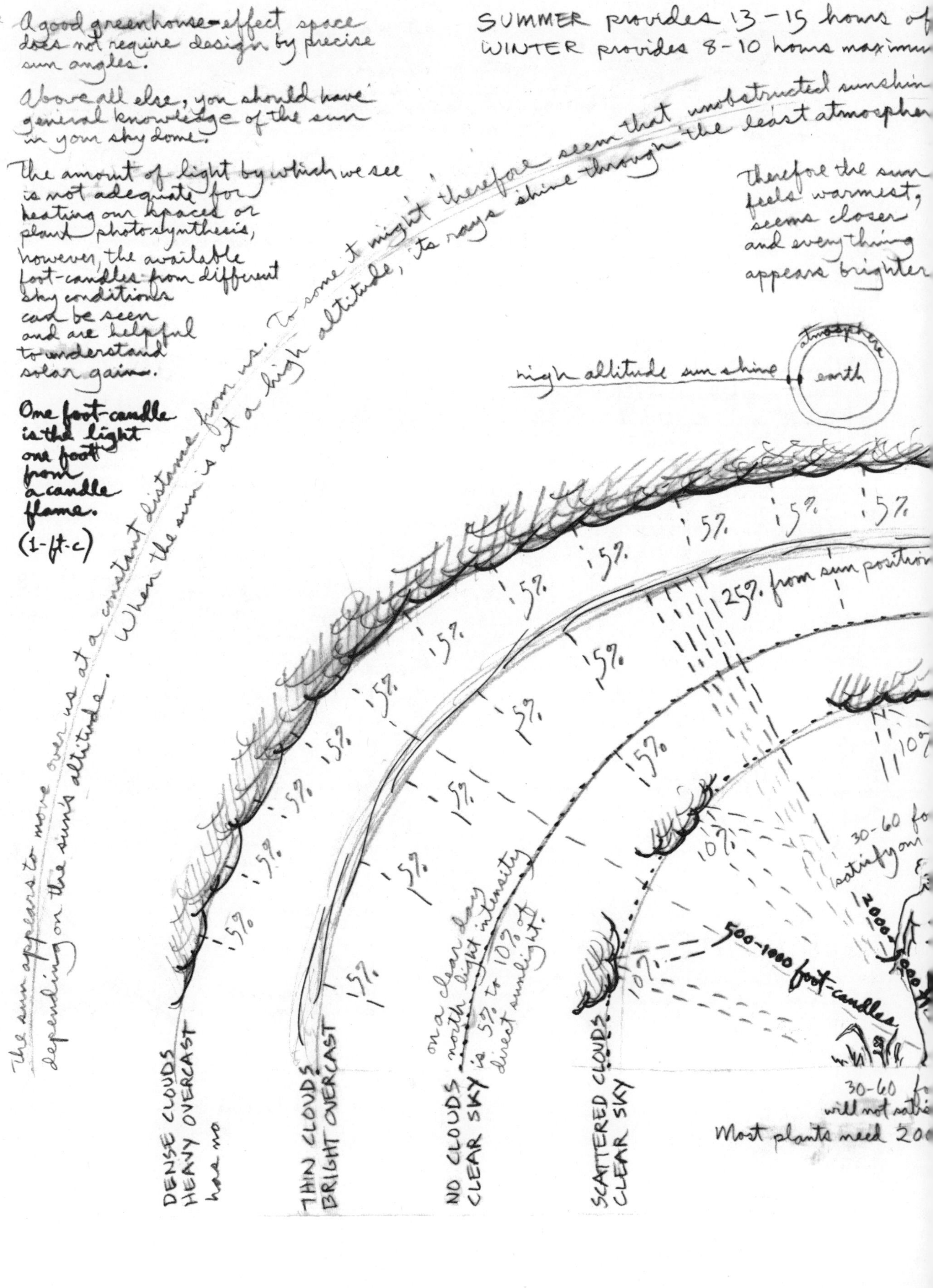

A good greenhouse-effect space does not require design by precise sun angles.
Above all else, you should have general knowledge of the sun in your sky dome.
The amount of light by which we see is not adequate for heating our spaces or plant photosynthesis; however, the available foot-candles from different sky conditions can be seen and are helpful to understand solar gain.
One foot-candle is the light one foot from a candle flame. (1-ft-c)
SUMMER provides 13-15 hours of
WINTER provides 8-10 hours maximum
The sun appears to move over us at a constant distance from us. To some it might therefore seem that unobstructed sunshine depending on the sun's altitude. When the sun is at a high altitude, its rays shine through the least atmosphere
therefore the sun feels warmest, seems closer and everything appears brighter
high altitude sun shine
atmosphere
earth
5%
25% from sun position
10%
30-60 foot-candles satisfy our
500-1000 foot-candles
2000-
On a clear day north light intensity is 5% to 10% of direct sunlight.
DENSE CLOUDS HEAVY OVERCAST has no
THIN CLOUDS BRIGHT OVERCAST
NO CLOUDS CLEAR SKY
SCATTERED CLOUDS CLEAR SKY
30-60 foot-candles will not satisfy
Most plants need 20

107. The value of refelection can be seen on a cloudy day with unobstructed sun, for that sky is brighter than a clear sky. (How to get sunburned on a cloudy day.)

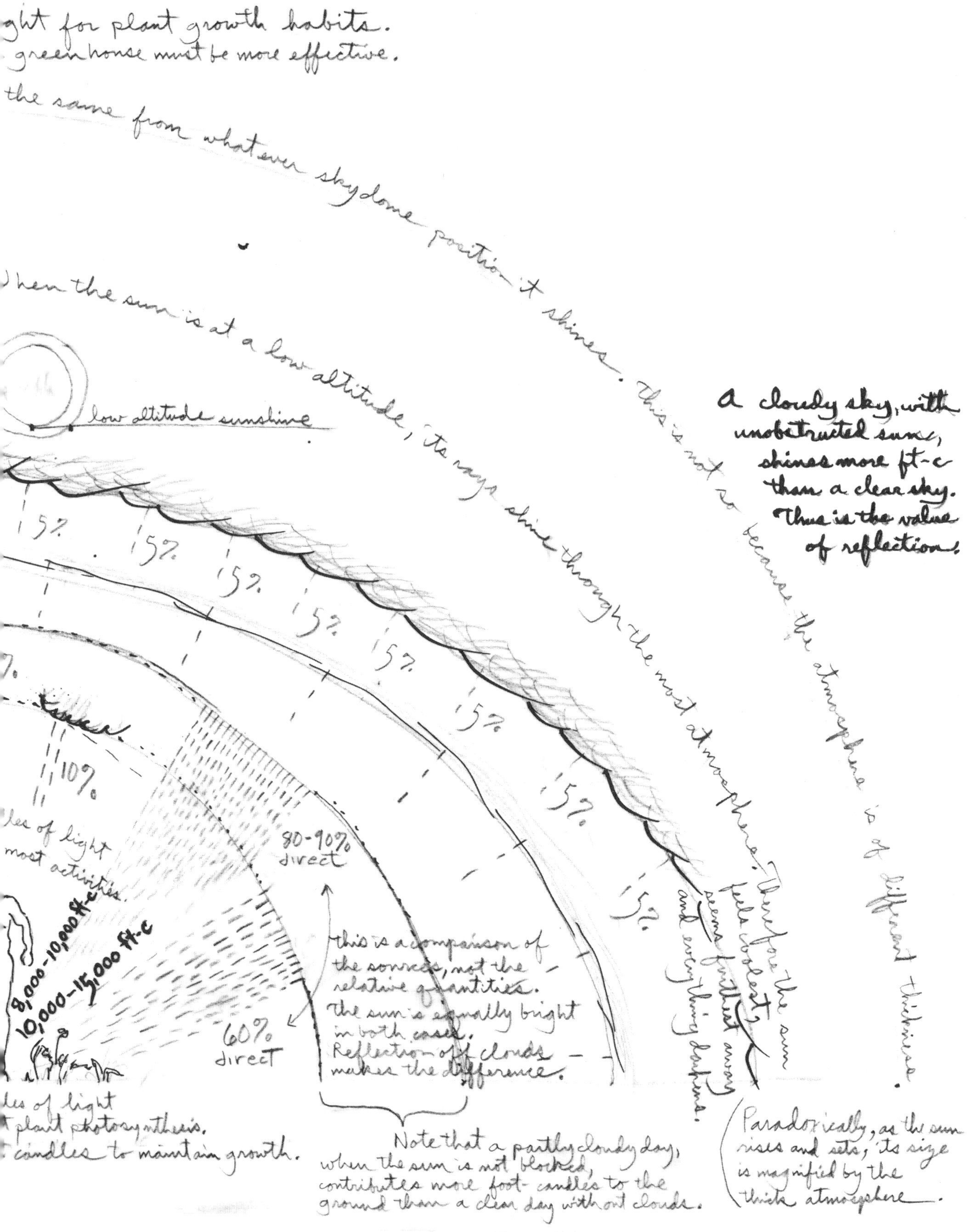
ght for plant growth habits.
greenhouse must be more effective.
the same from whatever skydome position it shines. This is not so because the atmosphere is of different thickness.
When the sun is at a low altitude, its rays shine through the most atmosphere. Therefore the sun feels closest, seems brightest, and everything darkens.
low altitude sunshine
A cloudy sky, with unobstructed sun, shines more ft-c than a clear sky. Thus is the value of reflection.
5%
10%
80-90% direct
60% direct
les of light most activities.
8,000-10,000 ft-c
10,000-15,000 ft-c
This is a comparison of the sources, not the relative quantities. The sun is equally bright in both cases. Reflection off clouds makes the difference.
les of light plant photosynthesis.
candles to maintain growth.
Note that a partly cloudy day, when the sun is not blocked, contributes more foot-candles to the ground than a clear day without clouds.
Paradoxically, as the sun rises and sets, its size is magnified by the thick atmosphere.

108. The sun's angle of incidence is the angle that the sun ray makes with the perpendicular to the slope.

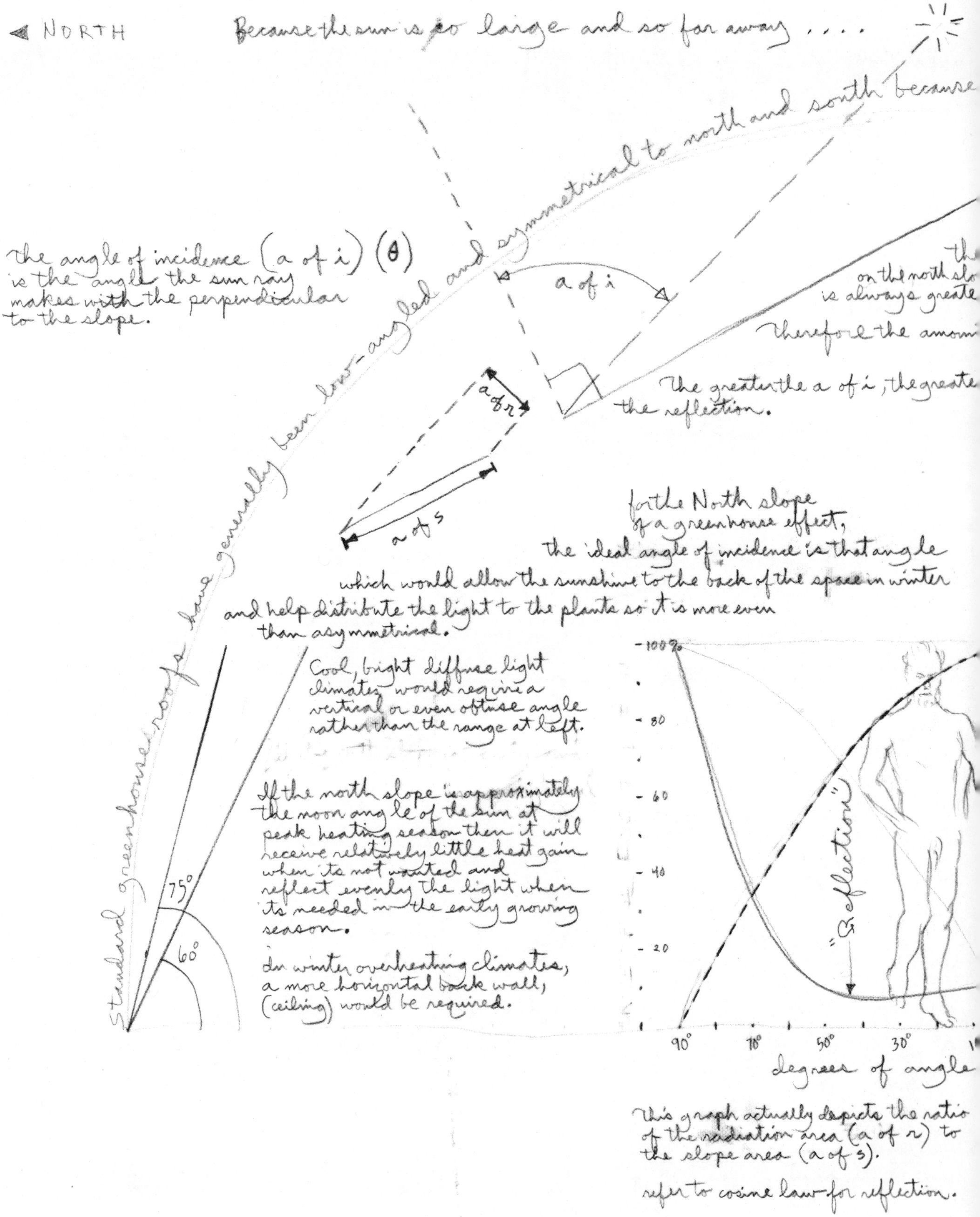
◄ NORTH
Because the sun is so large and so far away . . .
Standard greenhouse roofs have generally been low-angled and asymmetrical to north and south because
the angle of incidence (a of i) (θ) is the angle the sun ray makes with the perpendicular to the slope.
a of i
on the north slo
is always greate
therefore the amou
The greater the a of i, the greate
the reflection.
a of r
a of s
for the North slope of a greenhouse effect,
the ideal angle of incidence is that angle
which would allow the sunshine to the back of the space in winter
and help distribute the light to the plants so it is more even
than asymmetrical.
Cool, bright diffuse light climates would require a vertical or even obtuse angle rather than the range at left.
If the north slope is approximately the noon angle of the sun at peak heating season then it will receive relatively little heat gain when its not wanted and reflect evenly the light when its needed in the early growing season.
In winter overheating climates, a more horizontal back wall, (ceiling) would be required.
75°
60°
100%
80
60
40
20
"Reflection"
90°
70°
50°
30°
degrees of angle
This graph actually depicts the ratio of the radiation area (a of r) to the slope area (a of s).
refer to cosine law for reflection.

109. When a slope is perpendicular to the sun ray, the area of the slope equals the area of the solar radiation.

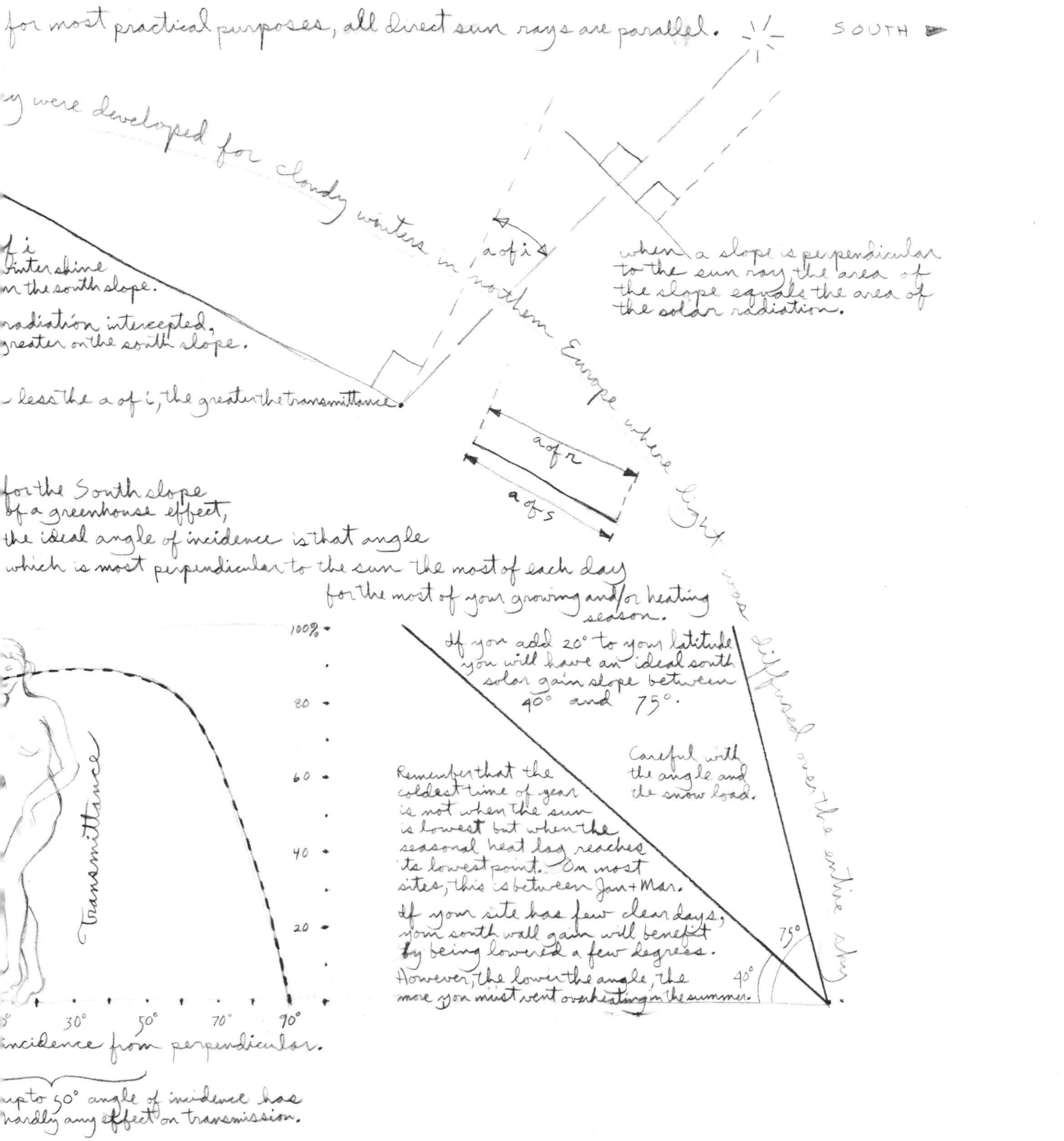

for most practical purposes, all direct sun rays are parallel.
SOUTH
y were developed for cloudy winters in northern Europe where light was diffused over the entire sky
a of i
when a slope is perpendicular to the sun ray, the area of the slope equals the area of the solar radiation.
f i
intershine
n the south slope.
radiation intercepted,
greater on the south slope.
less the a of i, the greater the transmittance.
a of r
a of s
for the South slope of a greenhouse effect, the ideal angle of incidence is that angle which is most perpendicular to the sun the most of each day for the most of your growing and/or heating season.
If you add 20° to your latitude you will have an ideal south solar gain slope between 40° and 75°.
Careful with the angle and the snow load.
Remember that the coldest time of year is not when the sun is lowest but when the seasonal heat lag reaches its lowest point. On most sites, this is between Jan + Mar.
If your site has few clear days, your south wall gain will benefit by being lowered a few degrees.
However, the lower the angle, the more you must vent overheating in the summer.
75°
40°
100%
80
60
40
20
transmittance
30°
50°
70°
90°
incidence from perpendicular.
up to 50° angle of incidence has hardly any effect on transmission.

110. The heat gain benefits of the greenhouse effect are understood by calculating the expected heat losses.

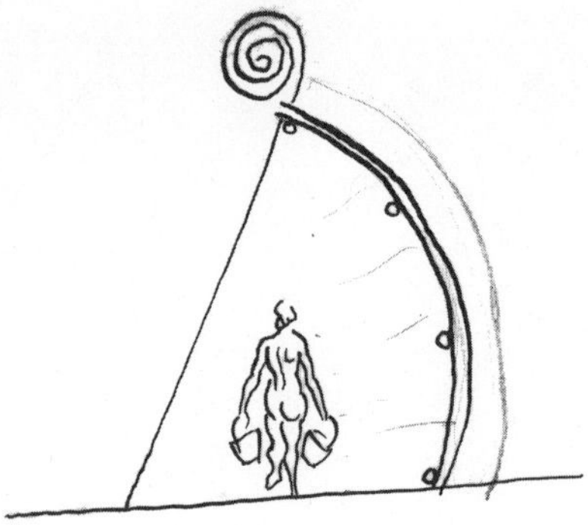

My experience with large areas of south-facing glass has been that they tend to overheat on the sunniest winter days and freeze on the coldest clearest nights.

Where average winter temperatures are temperate (35° - 45° F) occupants have found that indoor temperatures can be sun heated 65° - 70° most of the winter if their space has 1/4 square ft of southern glass for every 1 square ft of floor.

HOWEVER, this is a rough unreliable guideline and if you can tolerate doing some basic calculations, you will be able to predict the space's performance.

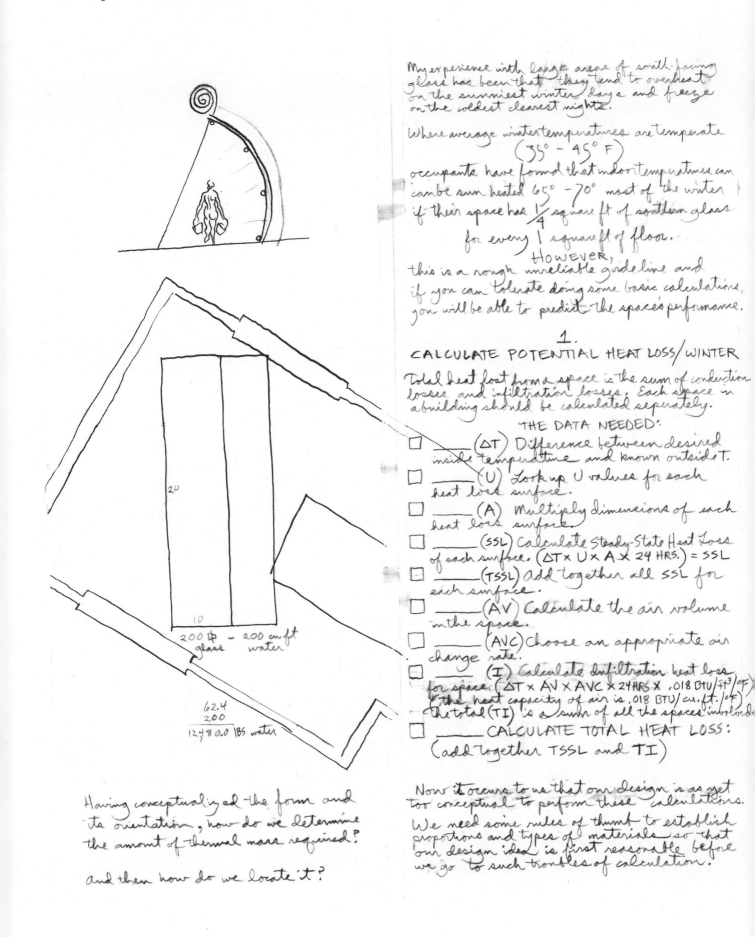

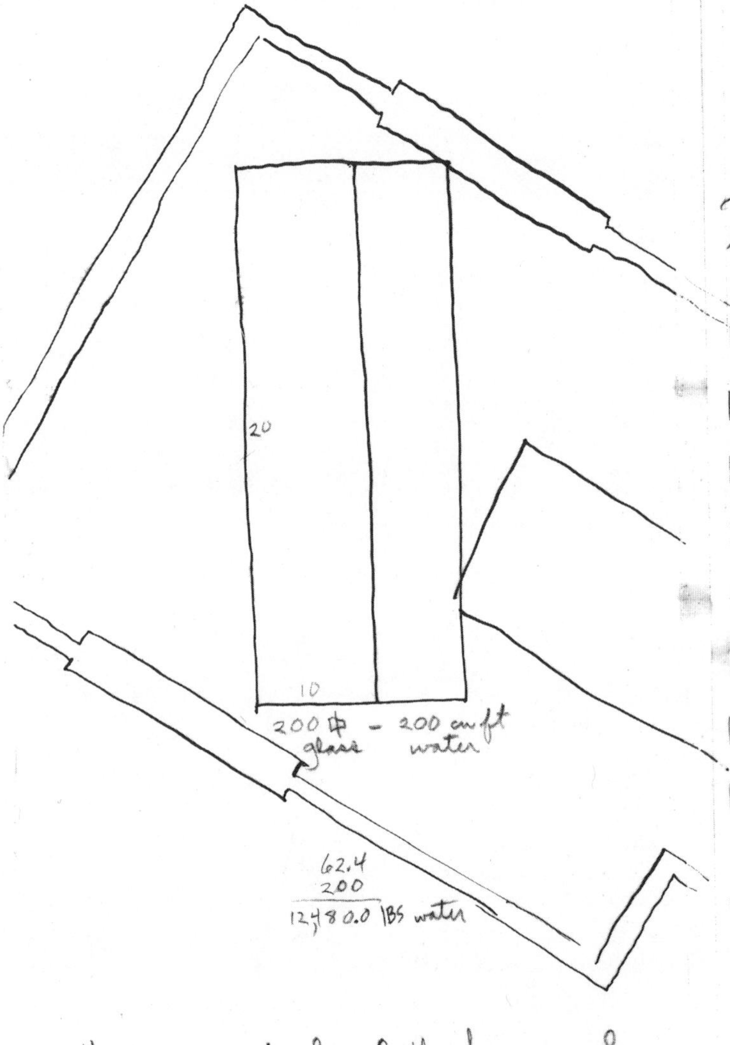

1.

CALCULATE POTENTIAL HEAT LOSS/WINTER

Total heat lost from a space is the sum of conduction losses and infiltration losses. Each space in a building should be calculated separately.

THE DATA NEEDED:

- ☐ _____ (ΔT) Difference between desired inside temperature and known outside T.
- ☐ _____ (U) Look up U values for each heat loss surface.
- ☐ _____ (A) Multiply dimensions of each heat loss surface.
- ☐ _____ (SSL) Calculate Steady-State Heat Loss of each surface. (ΔT × U × A × 24 HRS.) = SSL
- ☐ _____ (TSSL) add together all SSL for each surface.
- ☐ _____ (AV) Calculate the air volume in the space.
- ☐ _____ (AVC) Choose an appropriate air change rate.
- ☐ _____ (I) Calculate Infiltration heat loss for space. (ΔT × AV × AVC × 24 HRS × .018 BTU/ft³/°F) (the heat capacity of air is .018 BTU/cu.ft./°F) The total (TI) is a sum of all the spaces involved.
- ☐ _____ CALCULATE TOTAL HEAT LOSS: (add together TSSL and TI)

Having conceptualized the form and its orientation, how do we determine the amount of thermal mass required?

and then how do we locate it?

Now it occurs to us that our design is as yet too conceptual to perform these calculations. We need some rules of thumb to establish proportions and types of materials so that our design idea is first reasonable before we go to such troubles of calculation.

111. The atmosphere of the earth is its greenhouse glass and its clouds are its nighttime insulation.

If we patiently continue to look at the calculations which precise design prediction requires, then we will know what basic rules of thumb we should look for.

Therefore, we continue to establish this horrendous looking list knowing that we don't need to know what it's all about yet.....

2.

CALCULATE POTENTIAL HEAT GAIN/WINTER

Total heat gained by a space is a combination of the sunlight transmitted directly into a space and the heat stored in the materials of the space.

DIRECT

☐ ___(S) Look up sunshine available per square foot of gain surface. (BTU/day/sq. ft.) Multiply BTU/day/sq. ft by .94 which reduces the total by its 6% absorption loss.

☐ ___(SA) Calculate the area of sunshine on the glass. (the unshaded portion of the glass.)

☐ ___(SAS) The total heat gained through an area of glass in one day is SA multiplied by S. (that is, the available sunshine area by the sunshine)

INDIRECT

☐ ___(%SAS) the percentage of sunshine on a material which is transferred beyond or back into the space. This calculation is beyond the scope of this work and begs once again for common sense rules of thumb.

If Heat Loss (BTU/DAY/FLOOR FT²/°F)

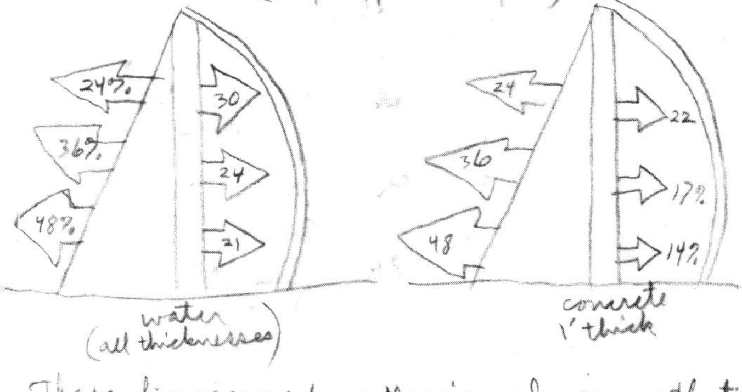

These figures are from Mazria and assume that the thermal wall is 1-1 proportion to glazing.

Obviously, such a wall defeats the use of the back wall as reflector. Also, how do we know how big a thermal mass we need?

Therefore, the following rules of thumb are useful.

The size of a water thermal wall is generally recommended to be the same wall area as the window, or one cubic ft of water for each one square foot of window.

For cloudy day storage, double or triple the cubic feet of water.

In climates where clouds prevail, cloudy day storage is not usually possible since the thermal mass requires several sunny days to store up excess heat.

One cubic foot of water is 7.5 gallons. and weighs 62.4 lbs.

A 55 gal Oil Drum contains 455 lbs water.

7.5) 550 = 7.3
525
250

62.4
× 7.3
1872
4368
455.52
× 10
4,555
× 3
13,665

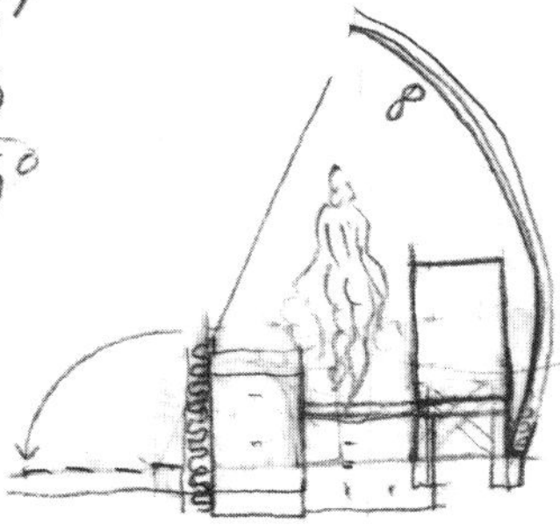

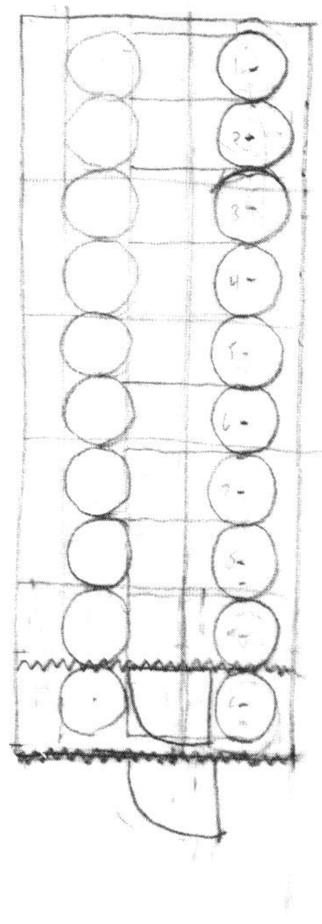

112. Earthen materials buffer our heat capacity with their own.

Specific Heat is the ratio of

the quantity of heat required to raise the temperature of a body one degree

per

the quantity of heat required to raise the temperature of an equal mass of water one degree

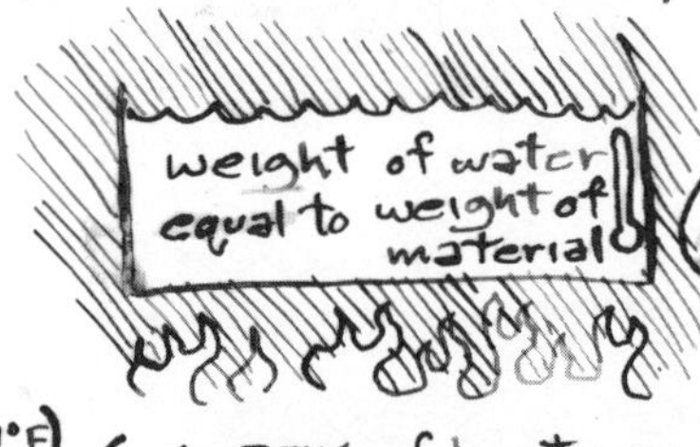

(size relationship is irrelevant)

Heat Capacity:

specific Heat of a material (BTU/1 LB/1°F)
multiplied by its density (LBs/cu. ft.)
equals
heat capacity per volume (BTU/cu.ft./1°F)

(1 BTU of heat raises 1 LB. of water 1° F.) (British Thermal Unit)

therefore heat capacity per volume is found by multiplying the density by the specific heat.

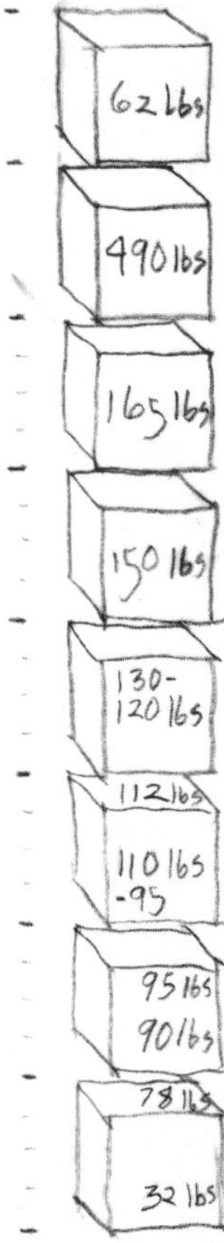

	Density		Specific Heat		Heat capacity
62 lbs	cu. ft. water	×	1.00	=	62 BTUS
490 lbs	cu. ft. steel	×	.12	=	59 BTUS
165 lbs	cu. ft. stone	×	.21	=	34 BTUS
150 lbs	cu. ft. concrete	×	.23	=	34 BTUS
130–120 lbs	cu. ft brick	×	.20	=	24 BTUS
112 lbs	cu. ft tile	×	.20	=	22 BTUS
110 lbs–95	cu. ft sand	×	.20		
95 lbs	cu. ft earth	×	.21	=	20 BTUS
90 lbs	cu. ft adobe	×	.20		
78 lbs	cu. ft gypsum board		.26		
32 lbs	cu ft wood	×	.33		10 BTUS

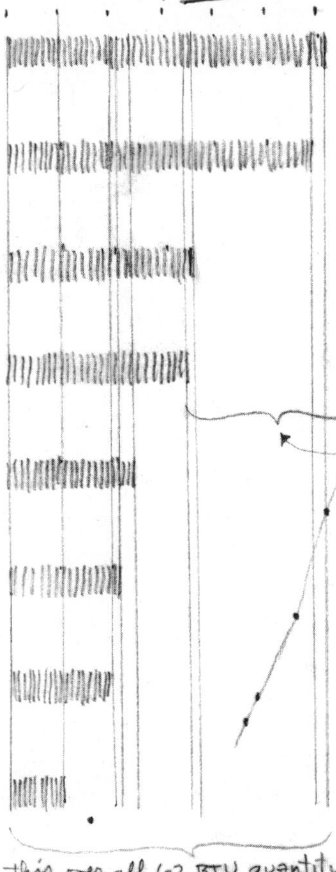

water must be contained from leaking & rusting and steel is used with least density/space.

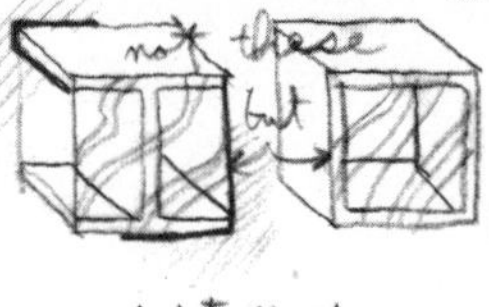

50-100 lbs/ft³ (~10 BTUS)

weights (•) of the heavy common building materials are each in relatively similar ratios to their heat capacities.

with concrete having the greatest capacity in proportion to its weight.

for same daily heat gain & loss data a space storing heat with brick, sand or adobe, needs 3x the storage volume that water needs. This is comparing capacities only.

wood's capacity makes it a better insulator than storer.

one cubic foot is always the same for a human to look at; for a human to lift...

this over all 62 BTU quantity is arbitrarily visualized but the relative proportions are accurate.

113. Water materials have the greatest heat capacity and the most uniform distribution of that capacity.

Water gains heat and loses heat throughout its volume nearly uniformly. (more uniformly than masonry.)

Compare two freshly filled containers at equal temperatures:

as soon as heat equally reaches both containers, both heated surfaces begin to warm.

at first both surfaces equally gain BTUS until...

no sooner have both surfaces absorbed their first BTUS...

the water begins to move as warmed water rises...

unheated water moves to replace risen water while

heated masonry remains immovable adding heat to itself.

Because of the increased surface temperature, the same available heat cannot be absorbed as fast as the first heat and some is lost.

Some is being lost equally from both all along by reflection et cetera.

The surface temperature of the slower conductor (masonry) is higher than the surface temperature of the faster conductor (water).

It becomes apparent why water stores 60 BTU/°F and masonry stores 30 BTU/°F

Heating surface to interior seems to be half as fast as heating by interior circulation.

However, steel, which is eight times as dense as water, has nearly equal capacity.

The rate at which the materials lose their heat is the same as the rate at which they gained it per degree change in temperature.

Water has greatest loss because of highest surface temperature.

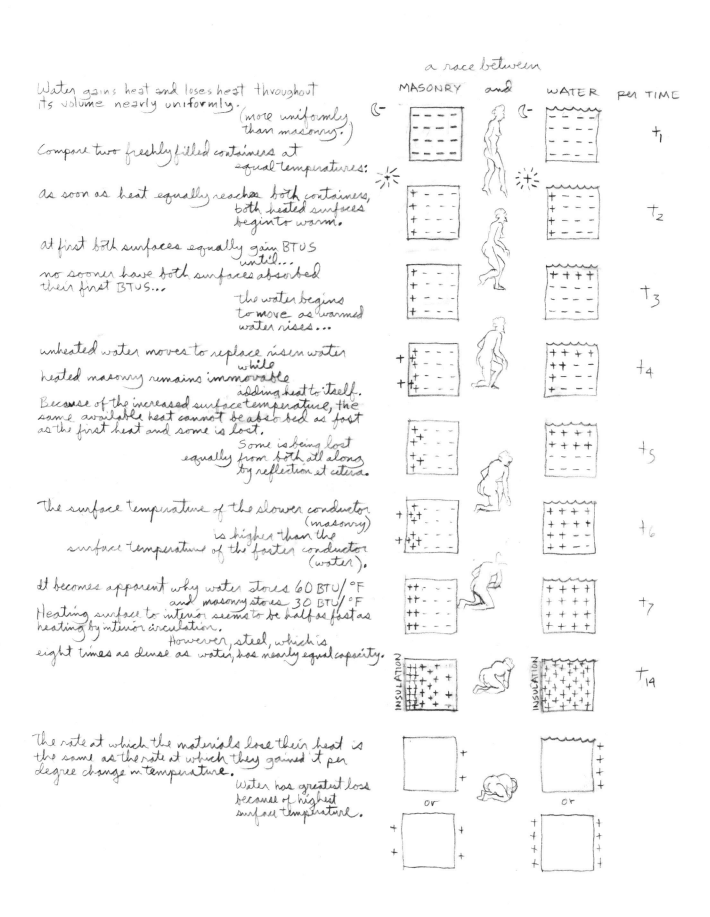

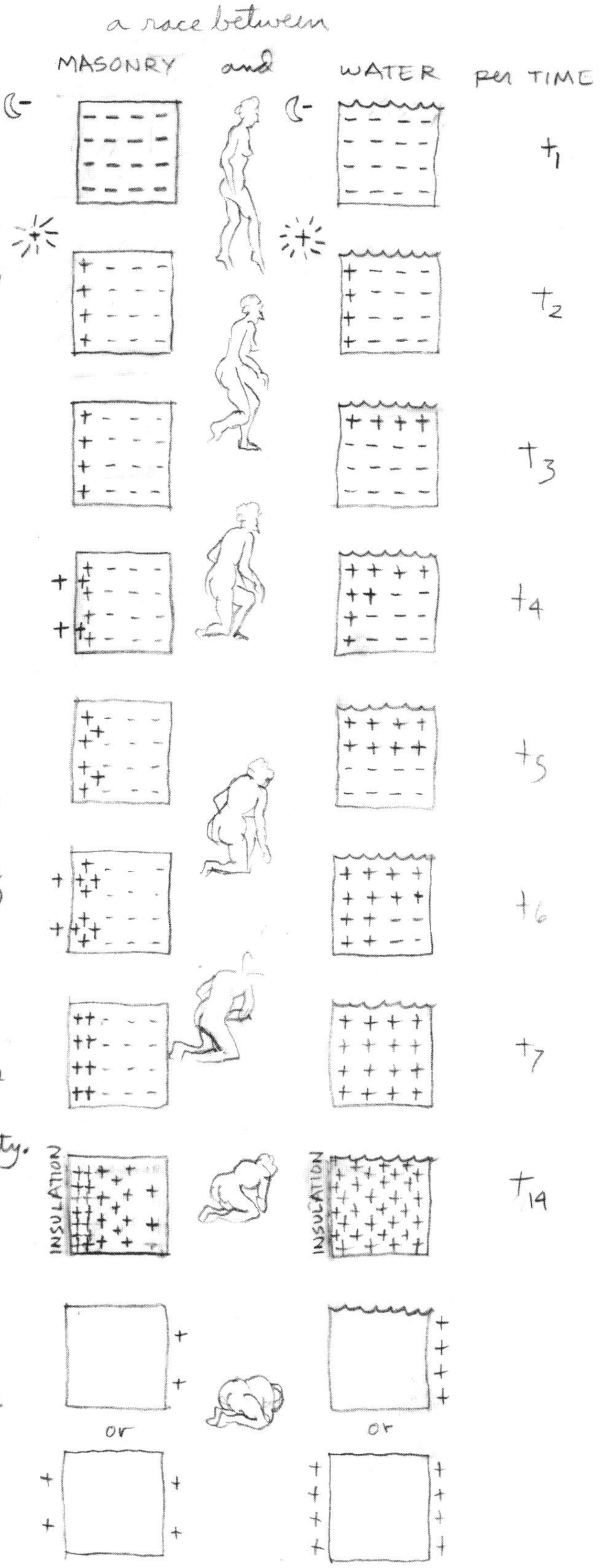

114. Out comfort range is 65 to 70°F; fluctuations above and below this can be minimized or encouraged by proportions of glass to material mass.

Since the average comfort range is 65° to 70°, fluctuation, above and below, day and night, will occur behind glass per the presence and absense of sun.

In order to keep this fluctuation to a minimum, an appropriate amount of mass per surface sun-gain-area will absorb extra heat and keep the space from overheating. When the space begins to cool below the temperature of the mass, the mass will lose heat to the space and keep the space from over cooling.

In generally due-south-oriented vertical glass walls with interior dark masses (>60% sun absorption) if mass receives direct sunlight from 10:00 - 2:00 (or equivalent BTU's) then 7.5 gallons / 1 sq.ft. glass (1 cu.ft./1 sq.ft.) will keep the fluctuations reasonable, especially if glass is insulated at night. (assuming the mass to be water.)

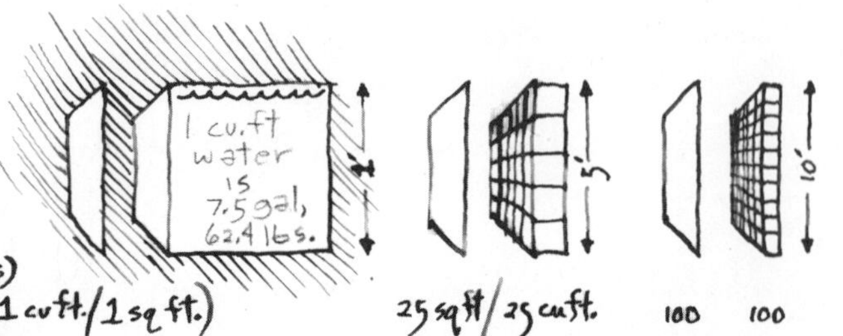

the greater the area the thinner the relative thickness.

What is the proof of this?

Effective Storage Mass ( ) = exposed surface area of storage ( ) × mass storage thickness ( ) × density of the storage material (lb.s / cu.ft.) × heat capacity of storage material (BTU / lb. / °F) = BTU/°F/cu.ft.

A space passively heat-tempered by a mass contained within the insulated space can overheat if the storage mass is too small. (Venting excess solar gain or furnace gain lowers performance rating.

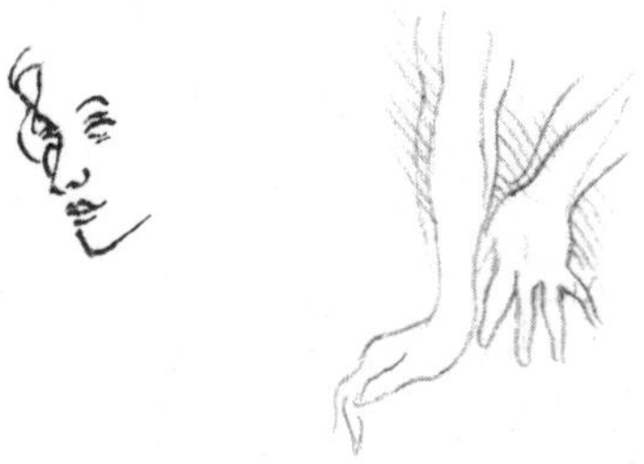

if the storage mass is too large, the space is underheated. (Any time the sun shines through glass and the air temperature is therefore raised, if a properly sized mass is not there to store the gained heat, the mass can be said to be too large.

THE OPTIONS OF HEAT GAINED:

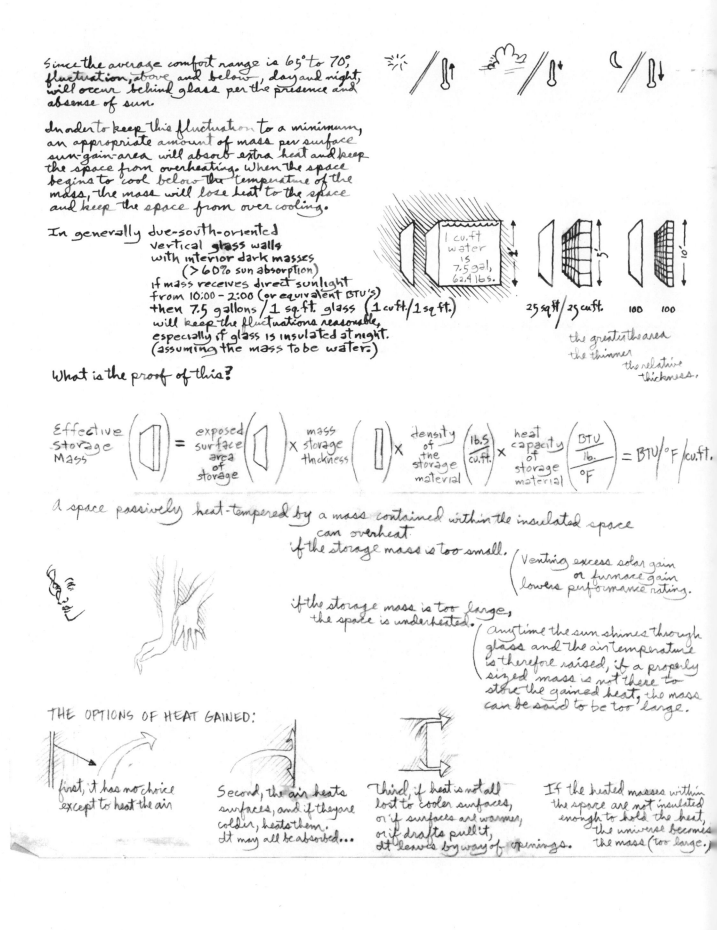

first, it has no choice except to heat the air

Second, the air heats surfaces, and if they are colder, heats them. It may all be absorbed...

Third, if heat is not all lost to cooler surfaces, or if surfaces are warmer, or if drafts pull it, it leaves by way of openings.

If the heated masses within the space are not insulated enough to hold the heat, the universe becomes the mass (too large.)

115. Overheating and underheating are prevented by design based on the division of effective mass by sunshine area.

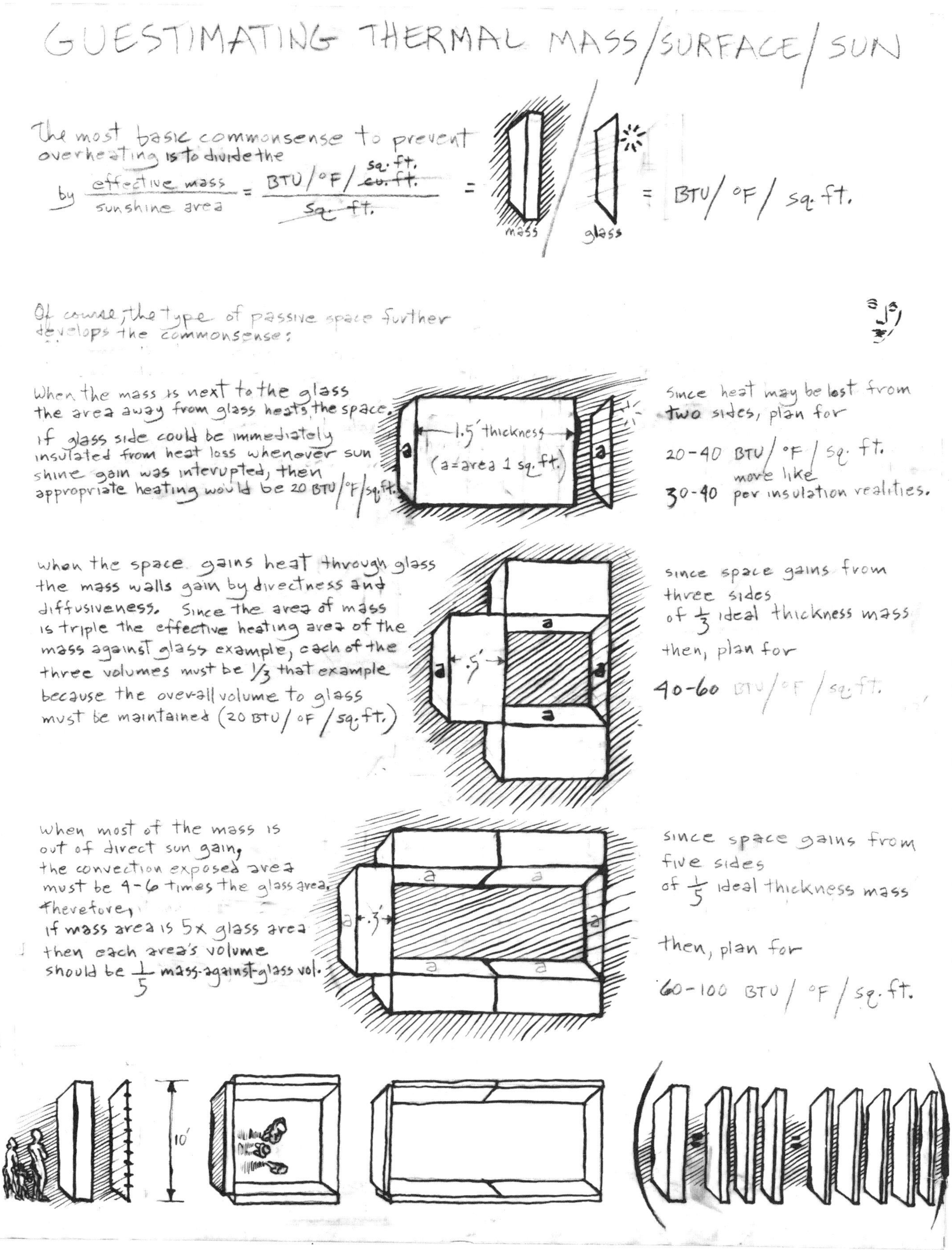
GUESTIMATING THERMAL MASS/SURFACE/SUN
The most basic commonsense to prevent overheating is to divide the
effective mass by sunshine area = BTU/°F/sq. ft. / sq. ft. = BTU/°F/sq. ft.
mass
glass
Of course, the type of passive space further develops the commonsense:
When the mass is next to the glass the area away from glass heats the space.
If glass side could be immediately insulated from heat loss whenever sun shine gain was interupted, then appropriate heating would be 20 BTU/°F/sq. ft.
1.5' thickness
(a=area 1 sq. ft.)
a
Since heat may be lost from two sides, plan for
20-40 BTU/°F/sq. ft.
more like
30-40 per insulation realities.
when the space gains heat through glass the mass walls gain by directness and diffusiveness. Since the area of mass is triple the effective heating area of the mass against glass example, each of the three volumes must be 1/3 that example because the overall volume to glass must be maintained (20 BTU/°F/sq. ft.)
.5'
since space gains from three sides of 1/3 ideal thickness mass
then, plan for
40-60 BTU/°F/sq. ft.
when most of the mass is out of direct sun gain, the convection exposed area must be 4-6 times the glass area.
therefore,
if mass area is 5x glass area then each area's volume should be 1/5 mass-against-glass vol.
.3'
since space gains from five sides of 1/5 ideal thickness mass
then, plan for
60-100 BTU/°F/sq. ft.
10'

116. Generally, the greater the wall thickness, the less the air temperature fluctuation.

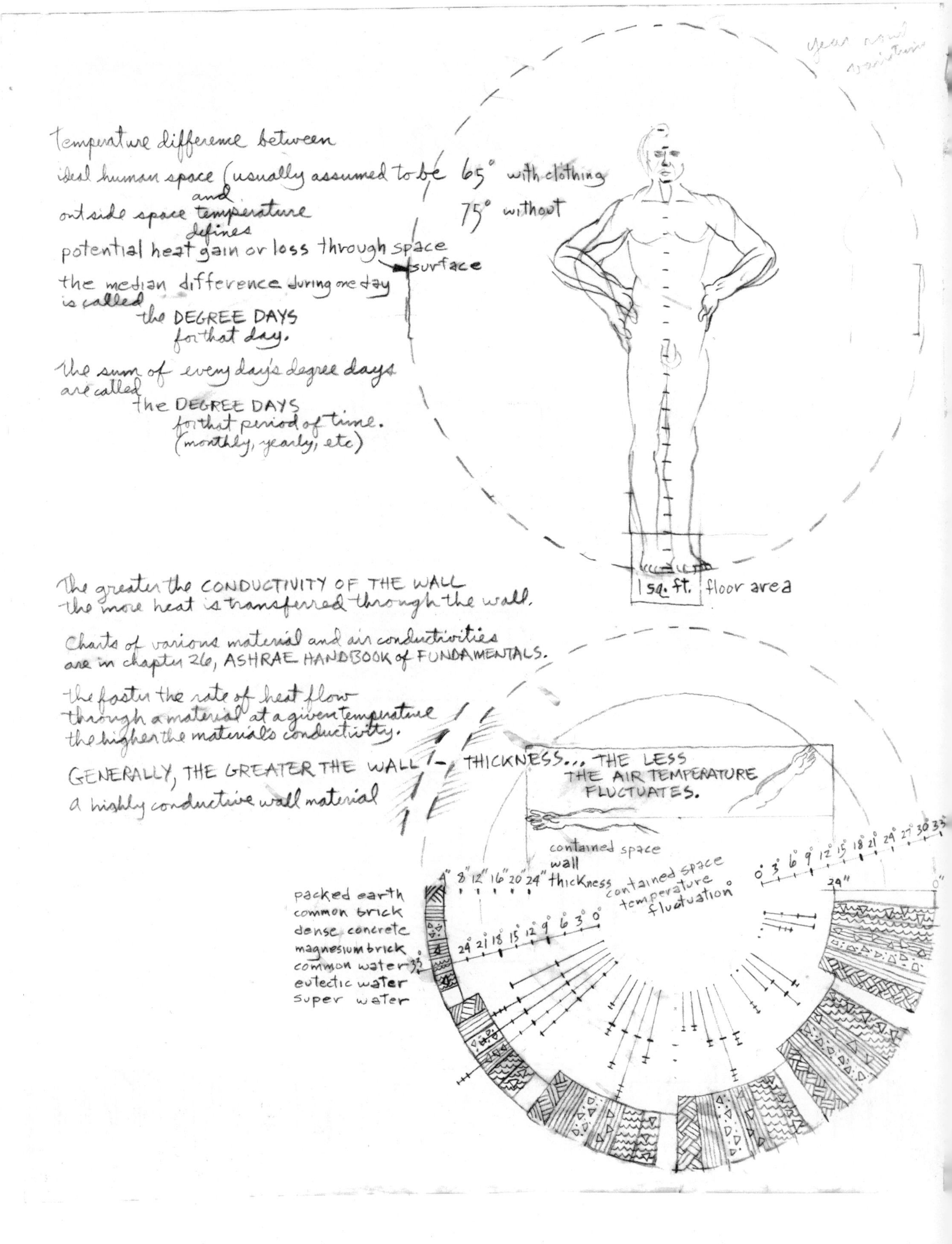
temperature difference between
ideal human space (usually assumed to be 65° with clothing
75° without
and outside space temperature
defines potential heat gain or loss through space surface
the median difference during one day is called the DEGREE DAYS for that day.
the sum of every day's degree days are called the DEGREE DAYS for that period of time. (monthly, yearly, etc)
1 sq. ft. floor area
The greater the CONDUCTIVITY OF THE WALL the more heat is transferred through the wall.
Charts of various material and air conductivities are in chapter 26, ASHRAE HANDBOOK of FUNDAMENTALS.
the faster the rate of heat flow through a material at a given temperature the higher the material's conductivity.
GENERALLY, THE GREATER THE WALL THICKNESS... THE LESS THE AIR TEMPERATURE FLUCTUATES.
a highly conductive wall material
contained space
wall thickness
contained space temperature fluctuation
4" 8" 12" 16" 20" 24"
0° 3° 6° 9° 12° 15° 18° 21° 24° 27° 30° 33°
24° 21° 18° 15° 12° 9° 6° 3° 0°
packed earth
common brick
dense concrete
magnesium brick
common water
eutectic water
super water

117. The difference in heights of the sums of % of wall area/1 sq ft floor between a water wall and a masonry wall is the area of a man's chest, the human heat regulator.

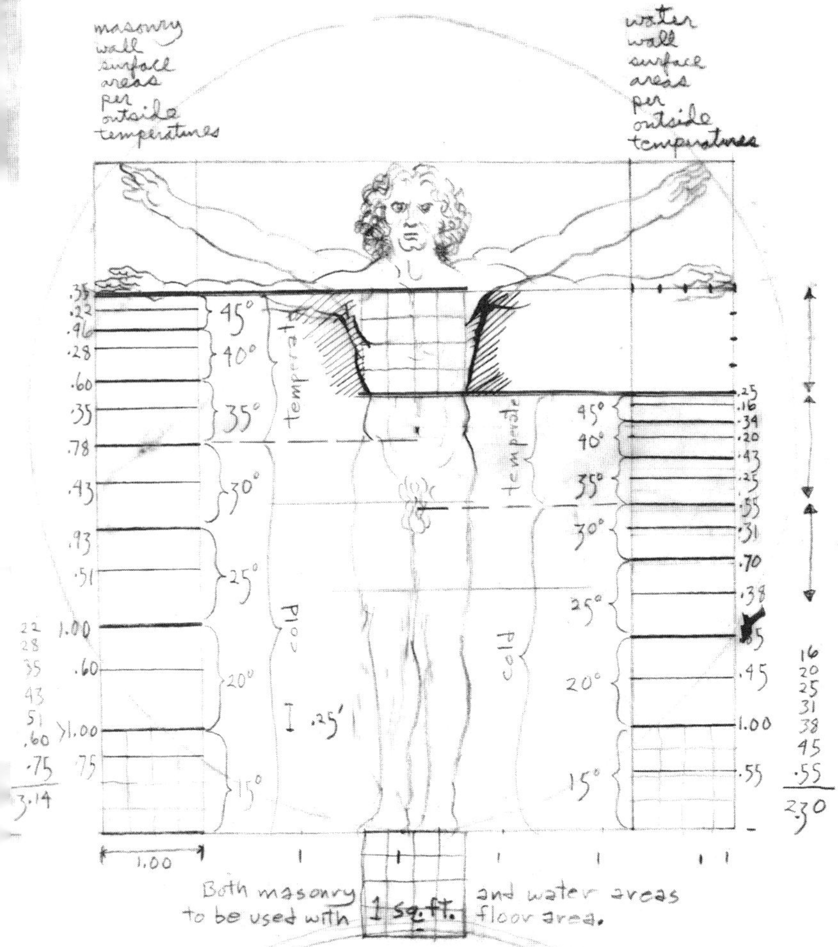

The diagram at left compares 1 ft wide proportions of 1 sq ft areas for seven most profable outdoor temperature medians added together as height and situated in scale to Vitruvius Man.

The data used can be found on page 156 of The Passive Solar Energy Book by Mazria.

One square foot, about the chest area of Vitruvius Man, containing the human BTU regulator, the heart.

One square foot, about the stomach area of Vitruvius man, is the difference between maximum water area and minimum masonry area, the center of BTU input.

< One square foot, about the thigh area of Vitruvius Man, is the difference between minimum water area and minimum masonry area, the combined center of waste & locomotion, BTU output.

These three observations are perhaps useful as a visual ordering canon to express heat storage, but the profound proportion is the range of area size shown as two percentages (minimum & max.) per each possible outside temperature.

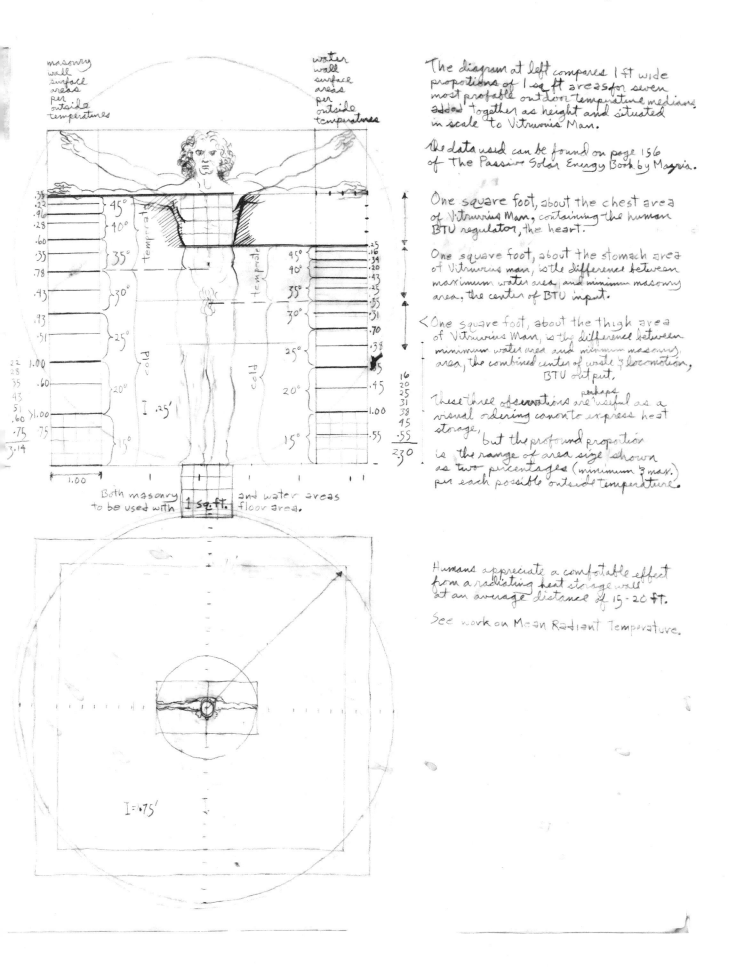

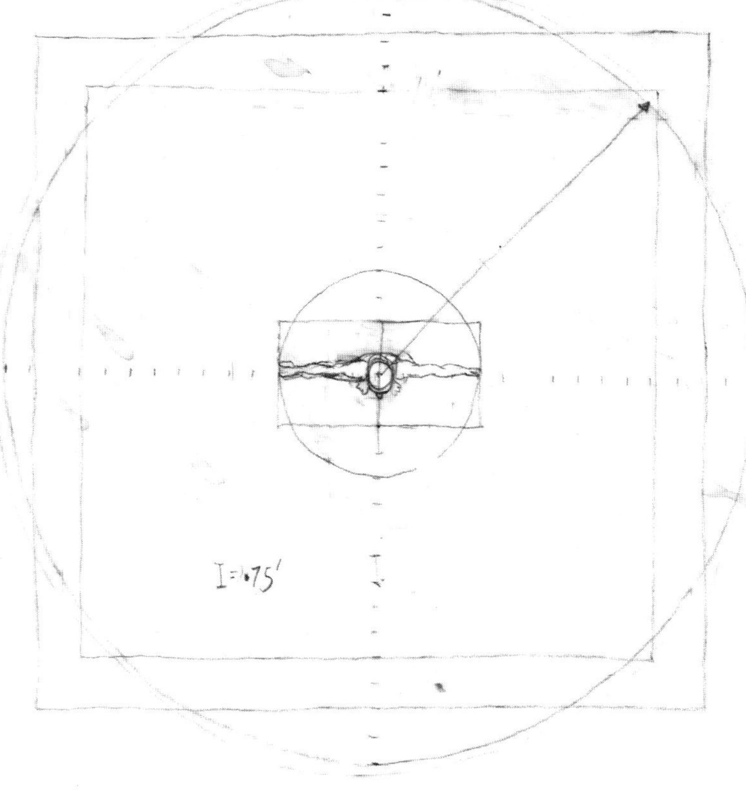

Humans appreciate a comfortable effect from a radiating heat storage wall at an average distance of 15-20 ft.

See work on Mean Radiant Temperature.

118. Radiation exchange between us and our walls affects us more than the air temperature of the room.

HOW TO TAKE THE TEMPERATURE OF A POINT IN SPACE,

Not the temperature of the air, but the temperature of electromagnetic radiation

FOR EACH QUESTIONABLE POINT IN TIME

At least in plan, if not also in section

PLAN

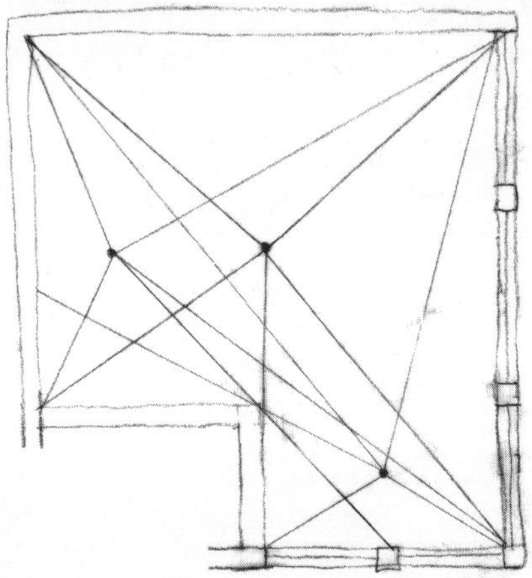

locate at least three points in the space which are centers of human use.

from each of these points, draw straight lines to all the edges of all homogenous surfaces and bends in all homogenous surfaces.

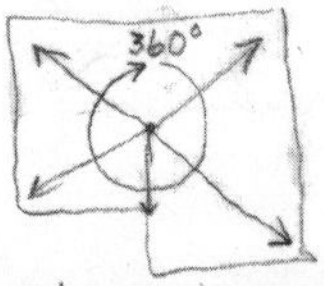

all the angles between these lines should add up to 360° (a full circle)

SECTION

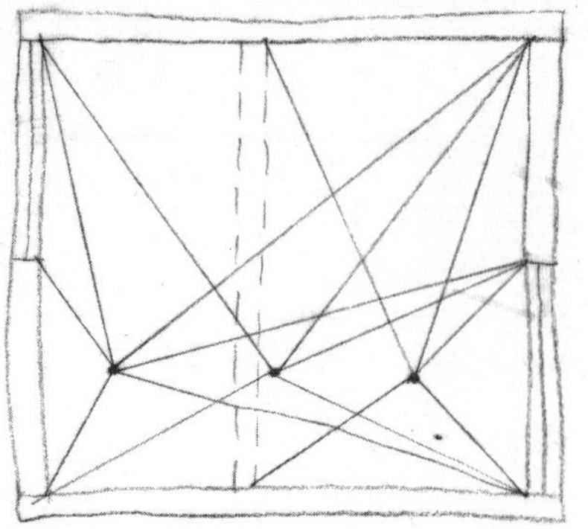

Each of these homogenous surfaces has a temperature; multiply each t° by its angle. Add together all the products of all the temperatures by their angles. Divide the sum by 360°

IF DONE ONLY IN PLAN
most professionals use this method and assume the human to be a column from floor to ceiling.

IF DONE IN PLAN & SECT
must divide total by 720° because data is summed 360° in two planes. this is the most thorough method but hardest due to air stratification.

IF DONE ONLY IN SECTION
see air stratification calcs for high ceilings especially. its best not to ignore the horizontal radiations. (plan calcs)

RADIATION FROM WALLS AFFECT US MORE THAN AIR TEMPERATURE

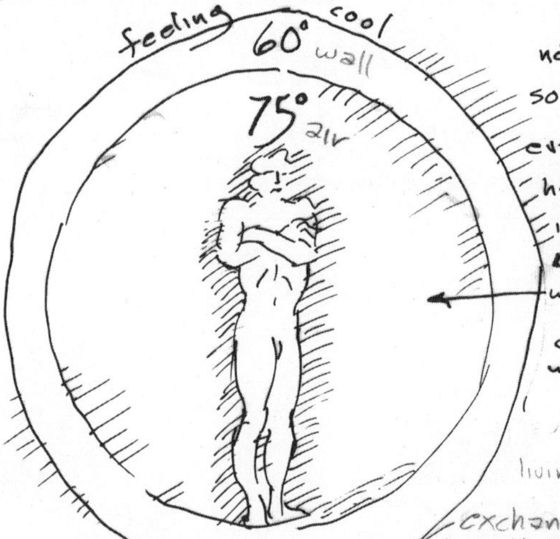

not considering solar radiation & evaporative effects, heat sensation in a closed space can be cool when air is warm OR can be warm when air is cool because body surfaces and living space surfaces primarily exchange with each other.

when the air is warm Surfaces must take our heat. when the air is cool Surfaces must give us heat.

from any given observation point ----radiations are proportionate to their distance from their surfaces.

HEAT SENSATION = AMBIENT AIR TEMPERATURE ± MEAN RADIENT TEMPERATURE = COMFORT IS AN INVERSE PROPORTION BETWEEN MRT & AAT

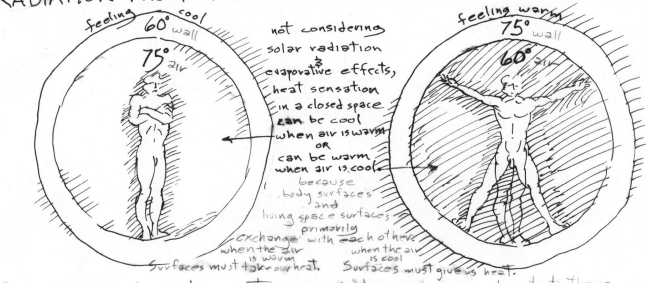

119. The mean radiant temperature that we feel from ourselves to our space surfaces is the average impact of all their radiations.

THE MEAN RADIANT TEMPERATURE (MRT) OF A POINT IN SPACE AND TIME IS ....
THE AVERAGE IMPACT OF ALL THE RADIATIONS FROM ALL THE SURROUNDING SURFACES.

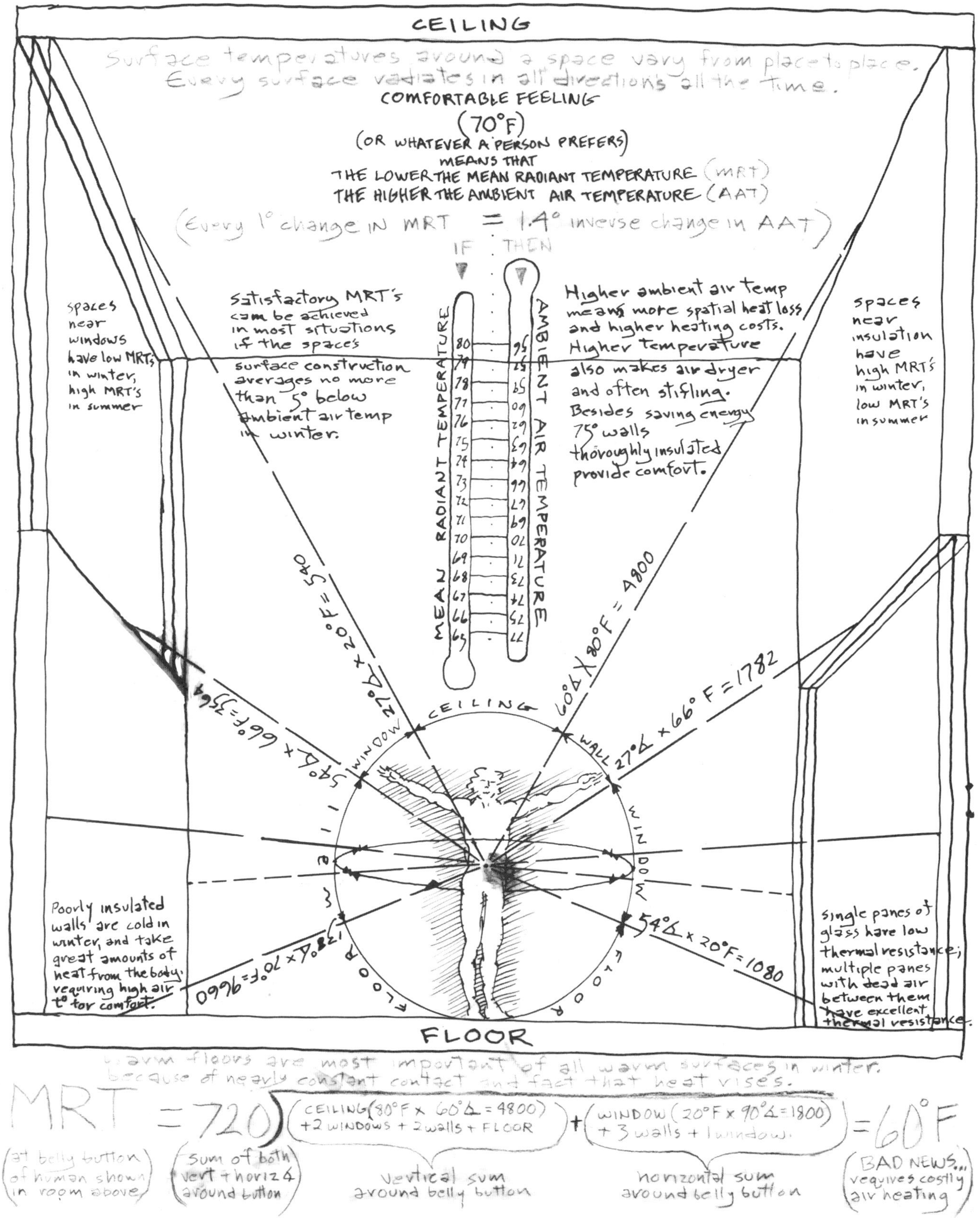
CEILING
Surface temperatures around a space vary from place to place.
Every surface radiates in all directions all the time.
COMFORTABLE FEELING
(70°F)
(OR WHATEVER A PERSON PREFERS)
MEANS THAT
THE LOWER THE MEAN RADIANT TEMPERATURE (MRT)
THE HIGHER THE AMBIENT AIR TEMPERATURE (AAT)
(Every 1° change in MRT = 1.4° inverse change in AAT)
IF
THEN
spaces near windows have low MRT's in winter, high MRT's in summer
Satisfactory MRT's can be achieved in most situations if the space's surface construction averages no more than 5° below ambient air temp in winter.
MEAN RADIANT TEMPERATURE
80 79 78 77 76 75 74 73 72 71 70 69 68 67 66 65
AMBIENT AIR TEMPERATURE
56 58 60 62 63 64 66 67 69 70 71 73 74 75 77
Higher ambient air temp means more spatial heat loss and higher heating costs. Higher temperature also makes air dryer and often stifling. Besides saving energy 75° walls thoroughly insulated provide comfort.
spaces near insulation have high MRT's in winter, low MRT's in summer
27°∠ x 20°F = 540
60°∠ X 80°F = 4800
27°∠ x 66° F = 1782
54°∠ + 66°F = 3564
CEILING
WINDOW
WALL
WINDOW
WALL
FLOOR
FLOOR
54°∠ x 20°F = 1080
138°∠ x 70°F = 9660
Poorly insulated walls are cold in winter, and take great amounts of heat from the body, requiring high air t° for comfort.
Single panes of glass have low thermal resistance; multiple panes with dead air between them have excellent thermal resistance.
FLOOR
warm floors are most important of all warm surfaces in winter because of nearly constant contact and fact that heat rises.
MRT = 720 (CEILING (80°F x 60°∠ = 4800) +2 WINDOWS + 2 walls + FLOOR) + (WINDOW (20°F x 90°∠ = 1800) + 3 walls + 1 window) = 60°F
(at belly button of human shown in room above)
(sum of both vert + horiz ∠ around button)
vertical sum around belly button
horizontal sum around belly button
(BAD NEWS... requires costly air heating)

120. The radiation we see is a different vision cone when we move or do not move.

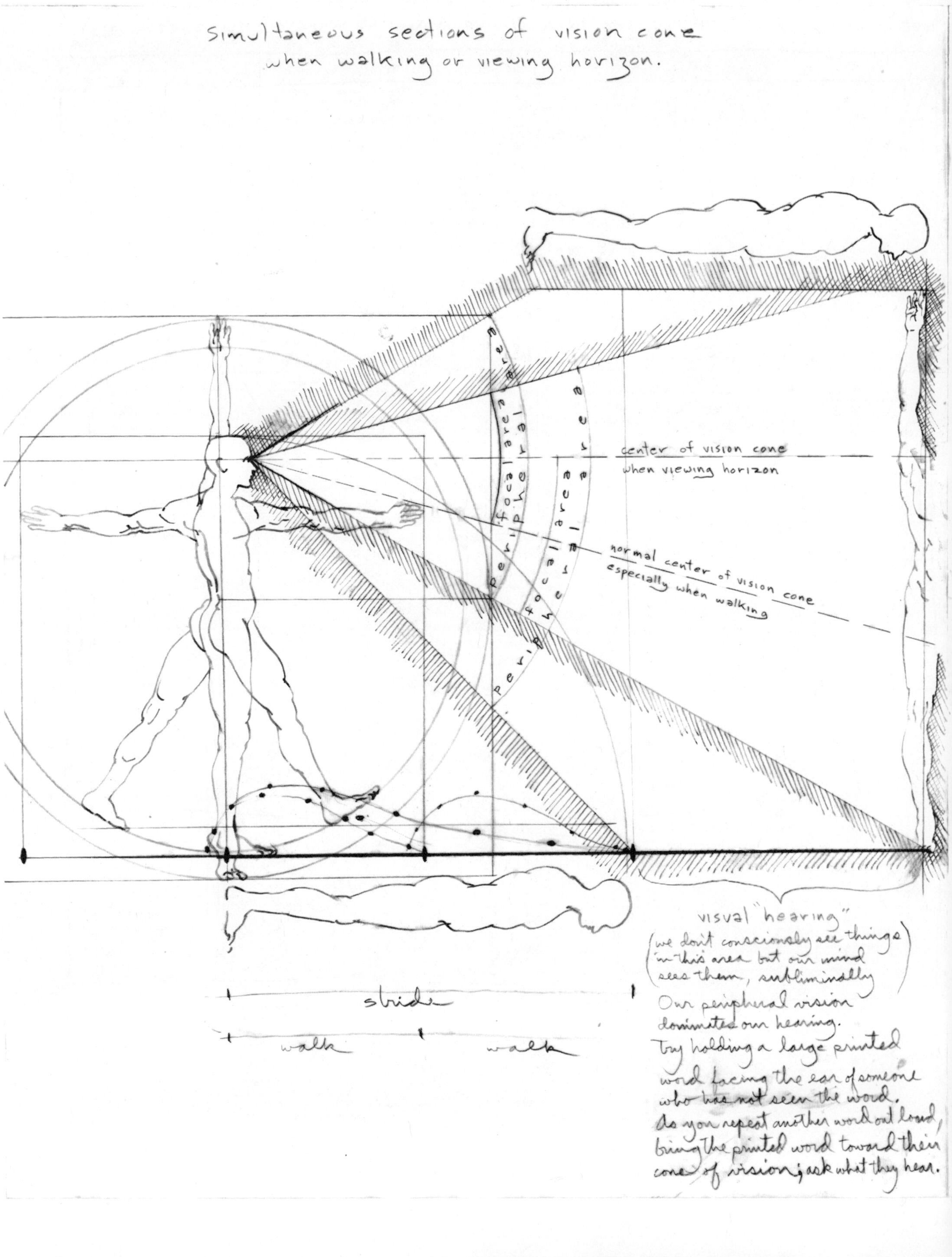

Simultaneous sections of vision cone
when walking or viewing horizon.
center of vision cone
when viewing horizon
normal center of vision cone
especially when walking
stride
walk
walk
visual "hearing"
(we don't consciously see things in this area but our mind sees them, subliminally)
Our peripheral vision dominates our hearing.
Try holding a large printed word facing the ear of someone who has not seen the word.
As you repeat another word out loud, bring the printed word toward their cone of vision; ask what they hear.

121. The plan and section of our vision cones on the floor, ceiling, and walls of a space resemble the lines on a shadowlabe.

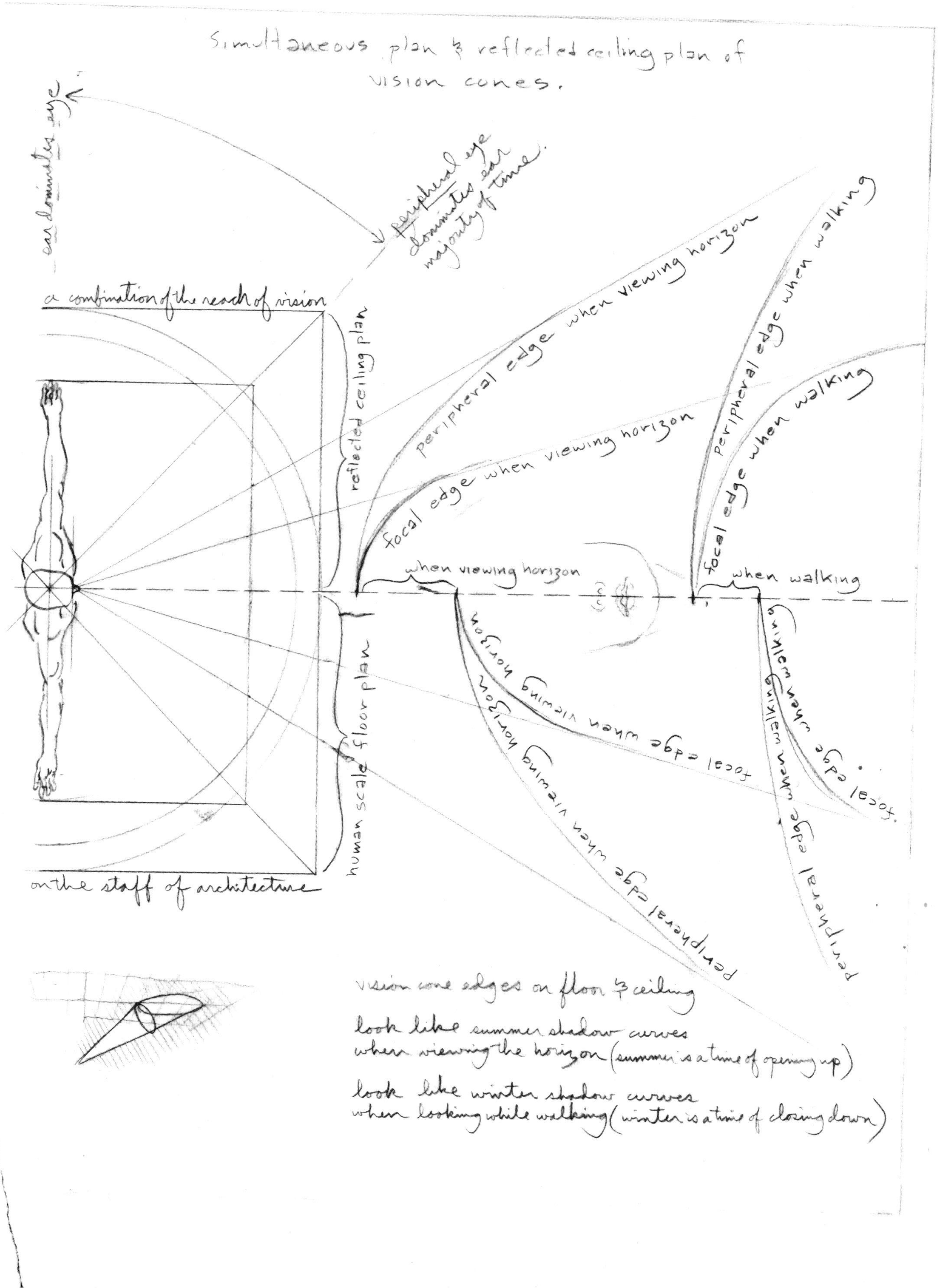

122. Space is not a single note; space is a span of time; space is the movement of sight.

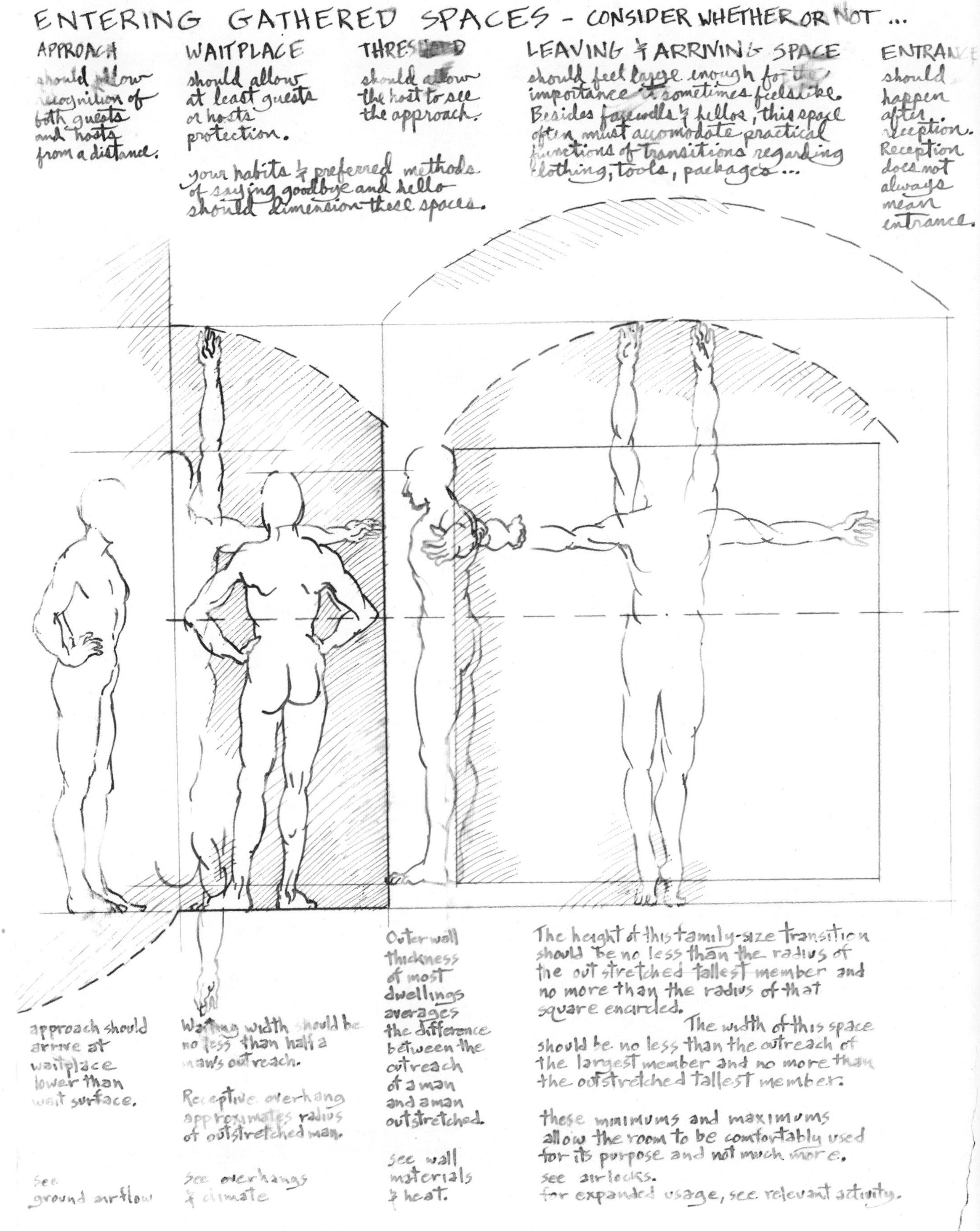
ENTERING GATHERED SPACES - CONSIDER WHETHER OR NOT ...
APPROACH
should allow recognition of both guests and hosts from a distance.
WAITPLACE
should allow at least guests or hosts protection.
THRES
should allow the host to see the approach.
LEAVING & ARRIVING SPACE
should feel large enough for the importance it sometimes feels like. Besides farewells & hellos, this space often must accomodate practical functions of transitions regarding clothing, tools, packages ...
ENTRAN
should happen after reception. Reception does not always mean entrance.
your habits & preferred methods of saying goodbye and hello should dimension these spaces.
approach should arrive at waitplace lower than wait surface.
see ground airflow
Waiting width should be no less than half a man's outreach.
Receptive overhang approximates radius of outstretched man.
see overhangs & climate
Outer wall thickness of most dwellings averages the difference between the outreach of a man and a man outstretched.
see wall materials & heat.
The height of this family-size transition should be no less than the radius of the outstretched tallest member and no more than the radius of that square encircled.
The width of this space should be no less than the outreach of the largest member and no more than the outstretched tallest member.
these minimums and maximums allow the room to be comfortably used for its purpose and not much more.
see airlocks.
for expanded usage, see relevant activity.

123. Simultaneous verticals and horizontals counter each other in a vocabular akin to music.

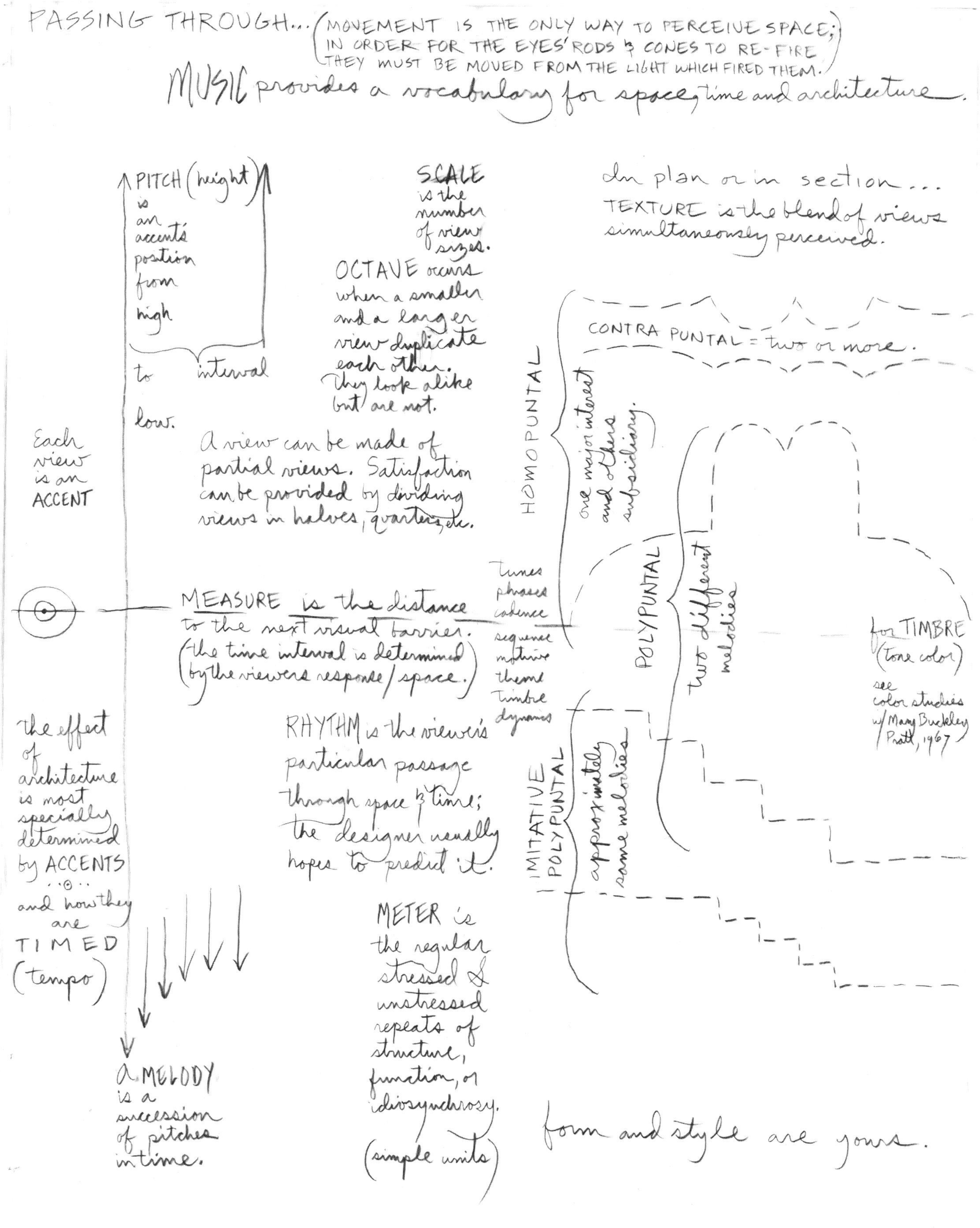
PASSING THROUGH... (MOVEMENT IS THE ONLY WAY TO PERCEIVE SPACE; IN ORDER FOR THE EYES' RODS & CONES TO RE-FIRE THEY MUST BE MOVED FROM THE LIGHT WHICH FIRED THEM.)
MUSIC provides a vocabulary for space, time and architecture.
PITCH (height) is an accent's position from high to low.
interval
SCALE is the number of view sizes.
OCTAVE occurs when a smaller and a larger view duplicate each other. They look alike but are not.
In plan or in section...
TEXTURE is the blend of views simultaneously perceived.
CONTRA PUNTAL = two or more.
HOMOPUNTAL
one major interest and others subsidiary.
Each view is an ACCENT
A view can be made of partial views. Satisfaction can be provided by dividing views in halves, quarters, etc.
tunes
phrases
cadence
sequence
motive
theme
timbre
dynamics
POLYPUNTAL
two different melodies
MEASURE is the distance to the next visual barrier. (the time interval is determined by the viewers response/space.)
for TIMBRE (tone color)
see color studies w/ Mary Buckley Pratt, 1967
the effect of architecture is most specially determined by ACCENTS and how they are TIMED (tempo)
RHYTHM is the viewer's particular passage through space & time; the designer usually hopes to predict it.
IMITATIVE POLYPUNTAL
approximately same melodies
METER is the regular stressed & unstressed repeats of structure, function, or idiosynchrosy.
(simple units)
A MELODY is a succession of pitches in time.
form and style are yours.

124. Balance is an economy of motion; a musical octave is the harmonic combination of two tones an octave apart, apparently blending as one.

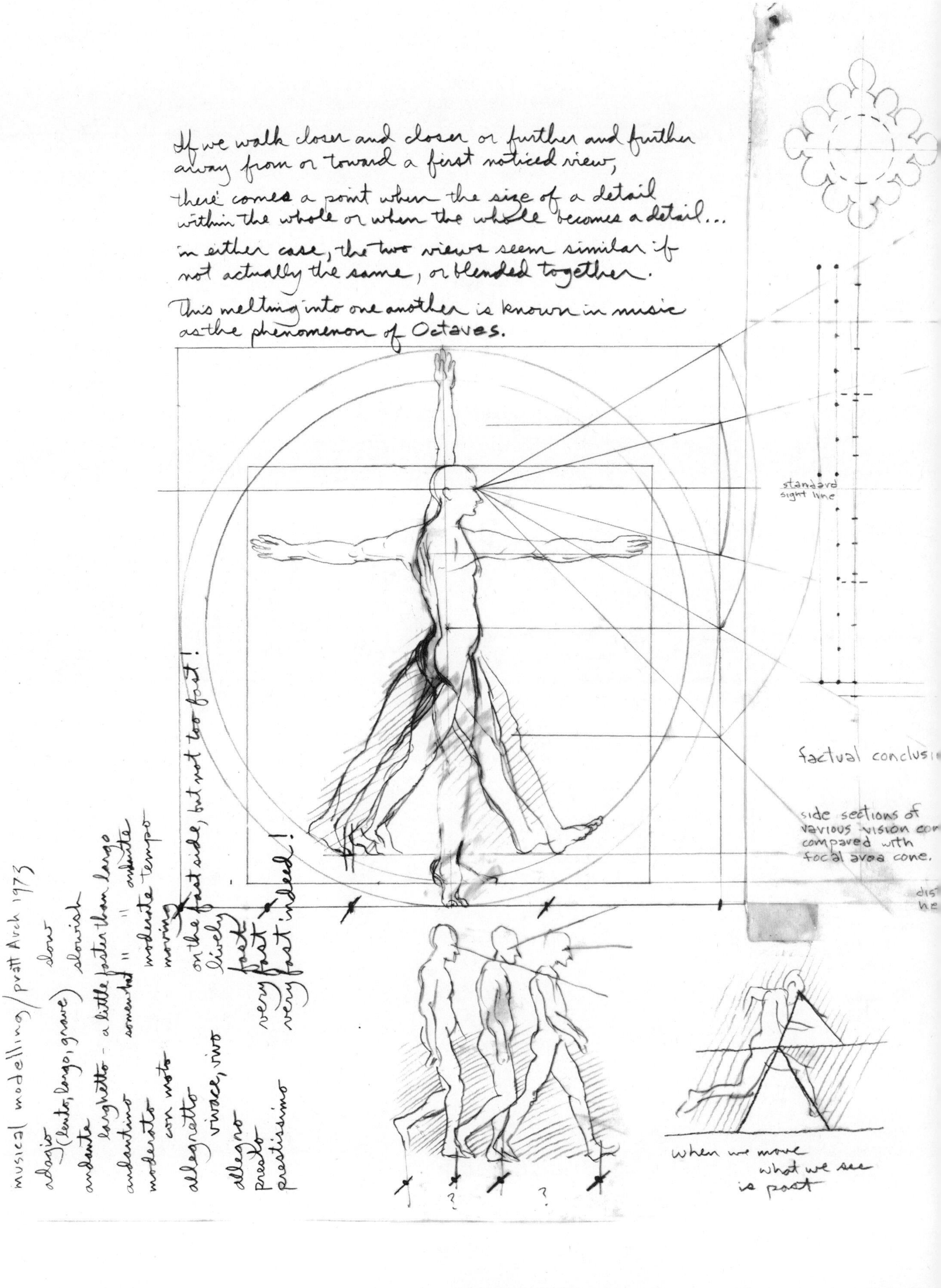
If we walk closer and closer or further and further away from or toward a first noticed view,
there comes a point when the size of a detail within the whole or when the whole becomes a detail...
in either case, the two views seem similar if not actually the same, or blended together.
This melting into one another is known in music as the phenomenon of Octaves.
standard sight line
factual conclusi
side sections of various vision con compared with focal area cone.
when we move what we see is past
musical modelling / pratt Arch 1973
adagio (lento, largo, grave) slow
andante slowish
larghetto - a little faster than largo
andantino somewhat " " andante
moderato moderate tempo
con moto moving
allegretto on the fast side, but not too fast!
vivace, vivo lively
allegro fast
presto very fast
prestissimo very fast indeed!

125. The visual octave is the harmonic combination of detail and visual field where the two blend as one.

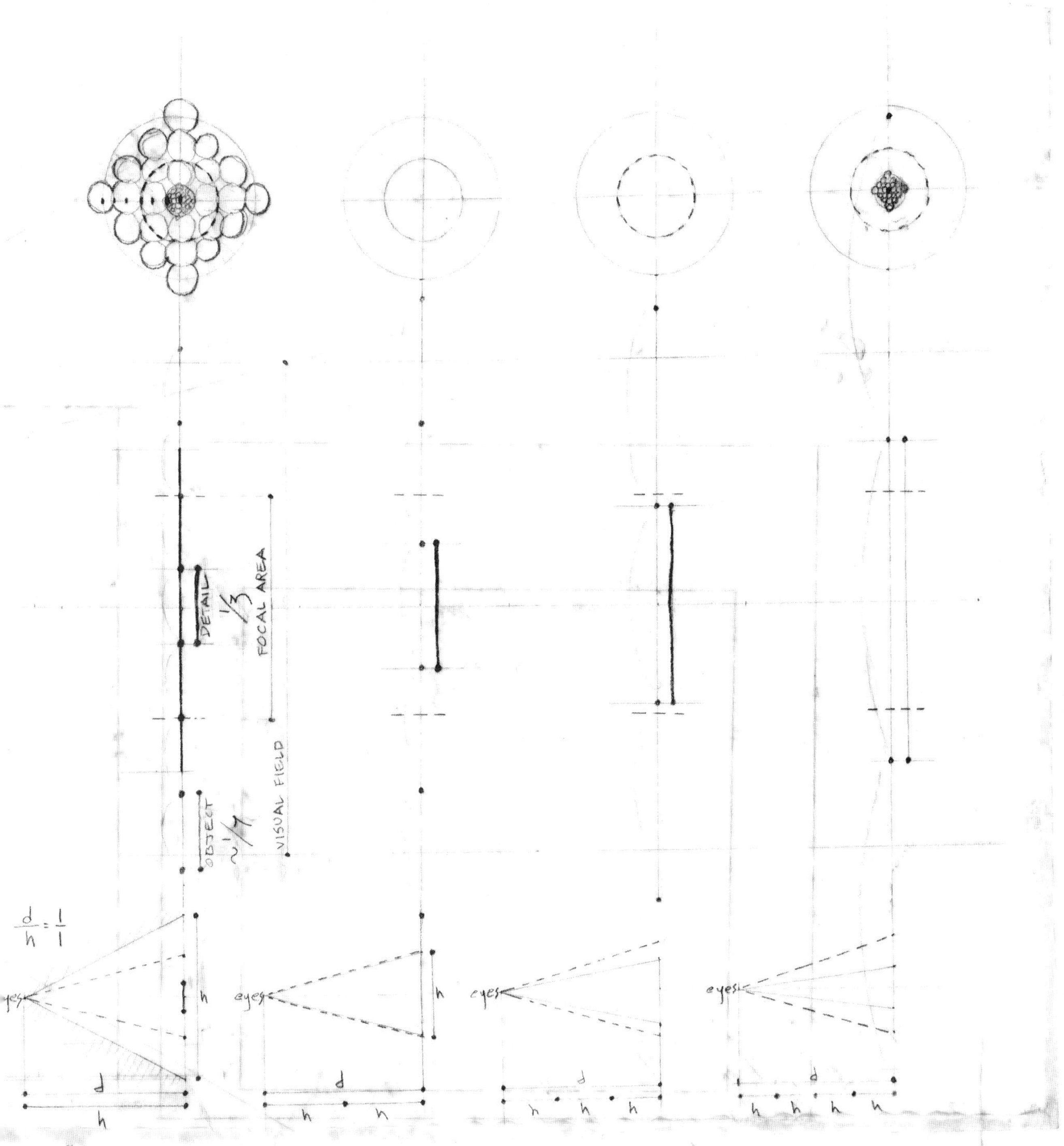

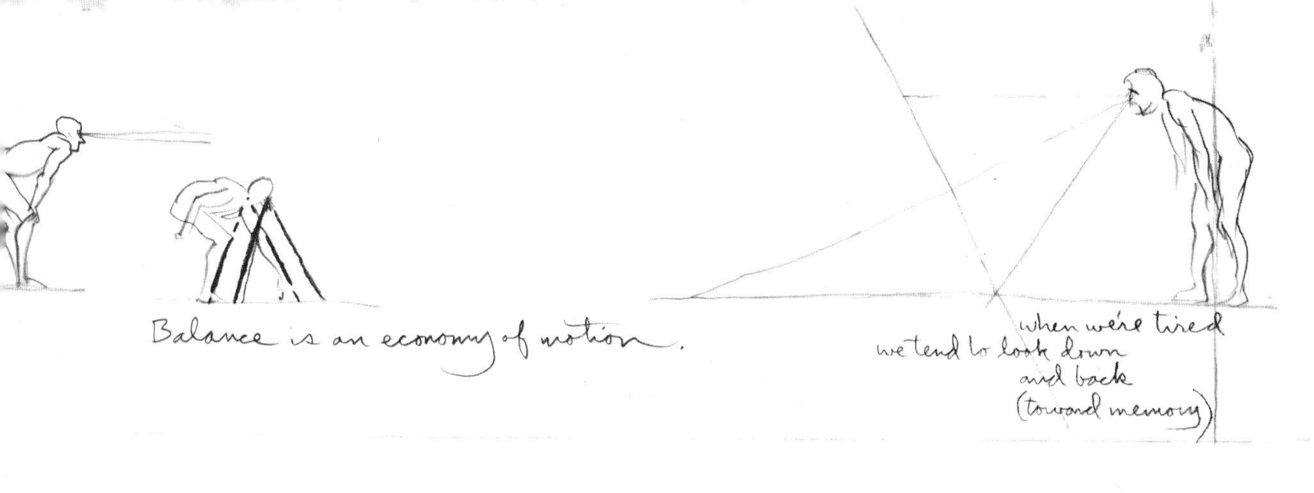

126. When the distance to an object from the eye equals the height of the object, the whole object is not seen, but rather a detail one third of the cone of vision is seen clearly without being lost in or dominating the surrounding object.

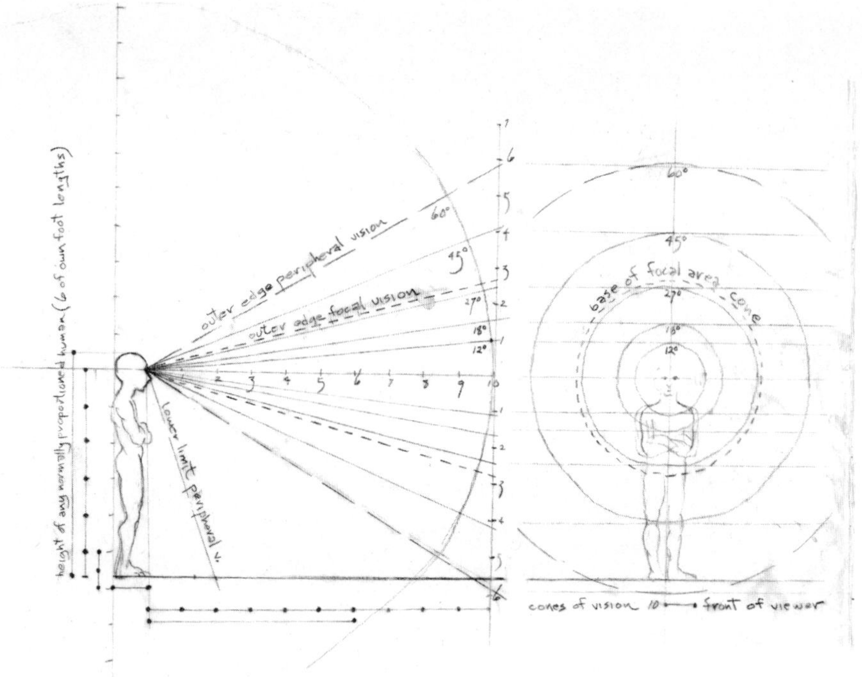

ANGLES OF SIGHT, PROXIMITY TO OBJECT, DETAILS, PARTS, & WHOLES

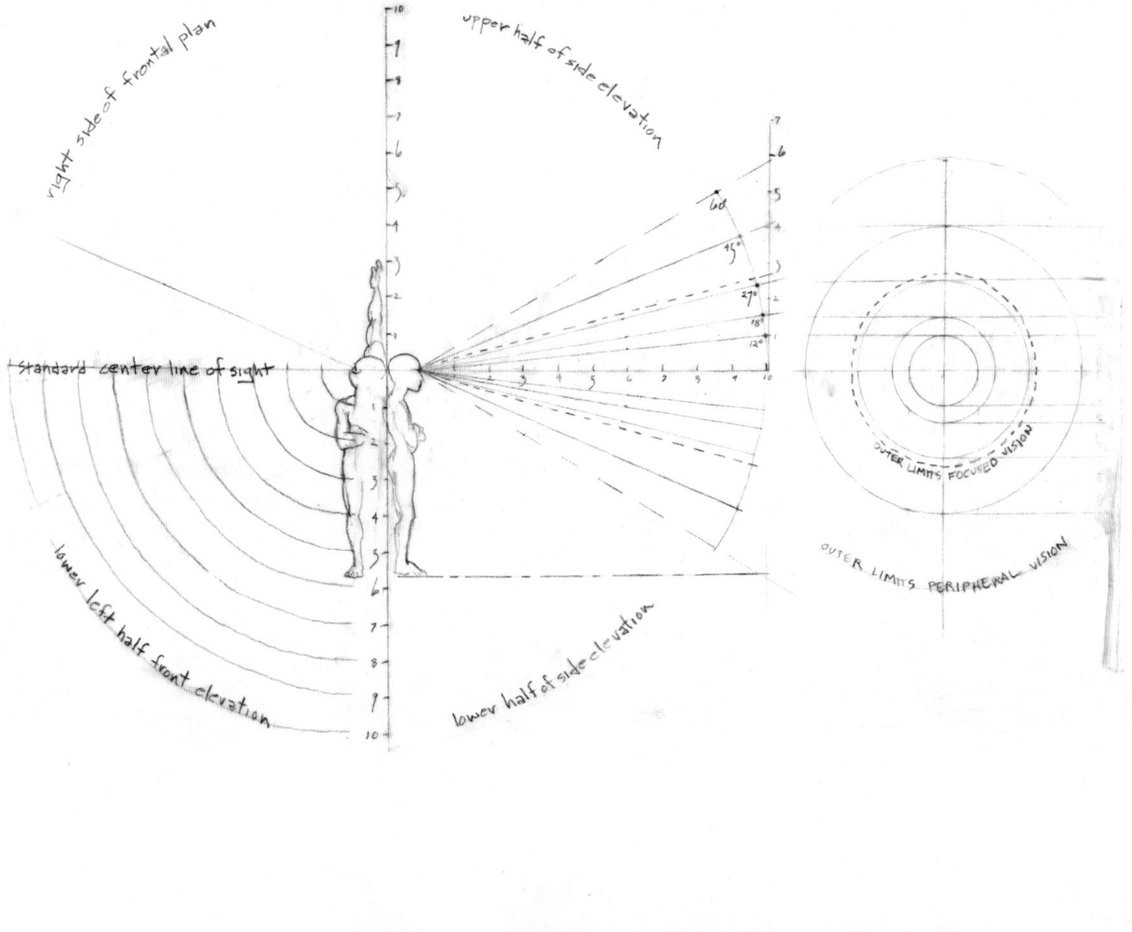

127. When the sight distance is twice the height of the object, the object is comfortably seen. When distance is three times height, the object dominates the visual field. When distance is four times height, the object becomes nondominant/not-lost detail.

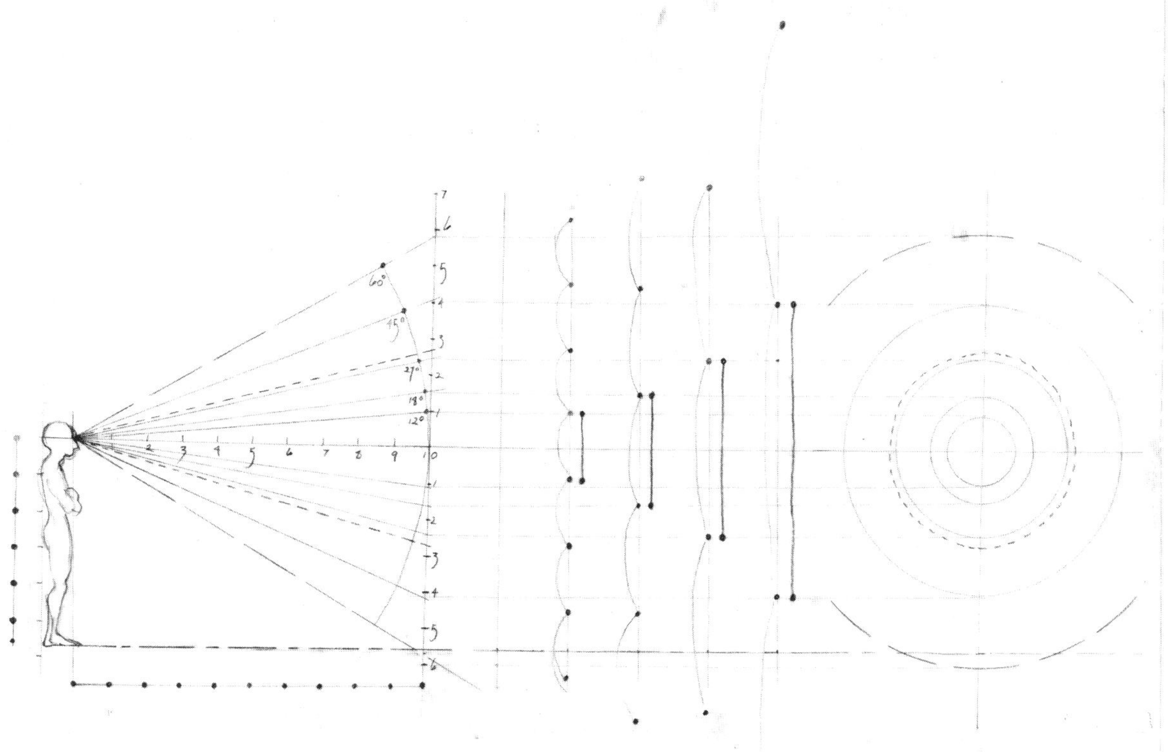

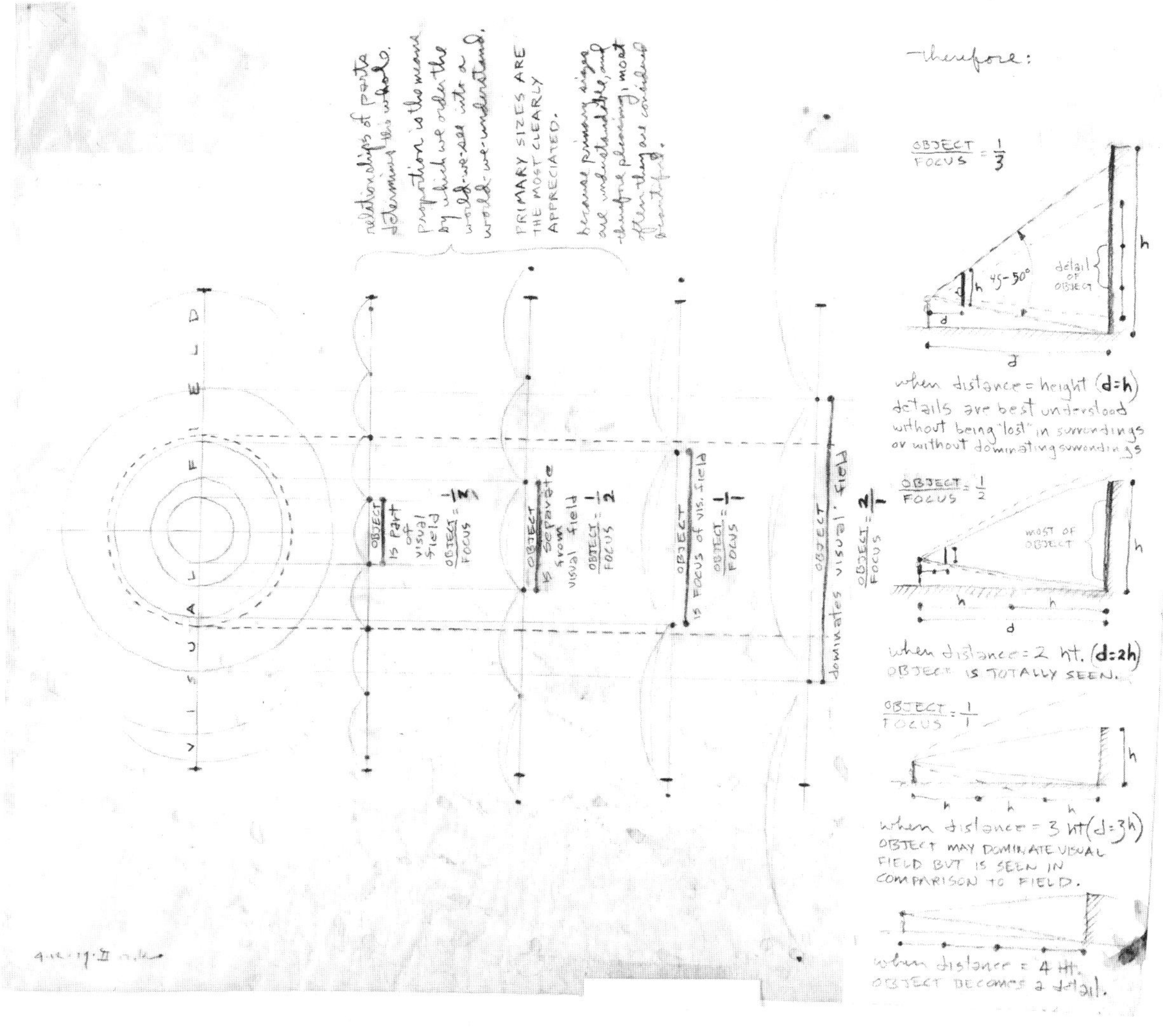
therefore:
OBJECT / FOCUS = 1/3
detail of object
when distance = height (d=h)
details are best understood without being "lost" in surroundings or without dominating surroundings
OBJECT / FOCUS = 1/2
MOST OF OBJECT
when distance = 2 ht. (d=2h)
OBJECT IS TOTALLY SEEN.
OBJECT / FOCUS = 1/1
when distance = 3 ht (d=3h)
OBJECT MAY DOMINATE VISUAL FIELD BUT IS SEEN IN COMPARISON TO FIELD.
when distance = 4 Ht.
OBJECT BECOMES a detail.
PRIMARY SIZES ARE THE MOST CLEARLY APPRECIATED.
VISUAL FIELD

128. If the width of a small room becomes more than the greater side of a golden rectangle to its height, and certainly if its width is more then twice its height, it will "feel" too low when crowded.

PEOPLE IN ONE SPACE

A room of simple symmetry focuses people together, unless the space is very large and even then the nature of sound will unite them.

A room of eccentric symmetries allows people to be together and also apart if they wish.

A room common to a group can be designed to allow conflicting interests to co-exist by appropriate space allocation and allow the group to feel related.

In a room crowded with people, two people can feel secluded and feel part of the group.

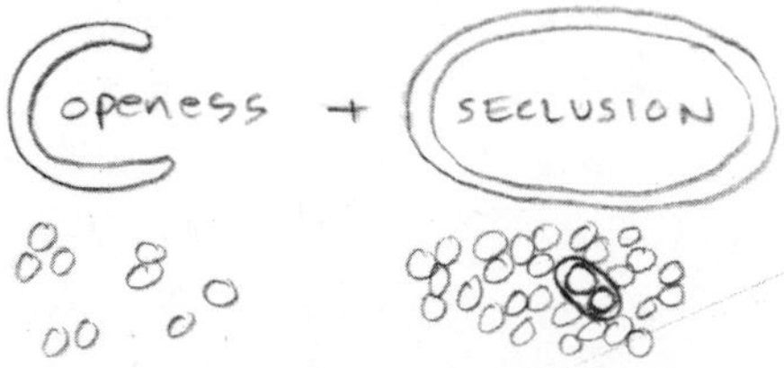
openess
+
SECLUSION

In a room loosely filled with family or other related group, duality of personal and group development can co-exist if both private and common spaces are available to each other.

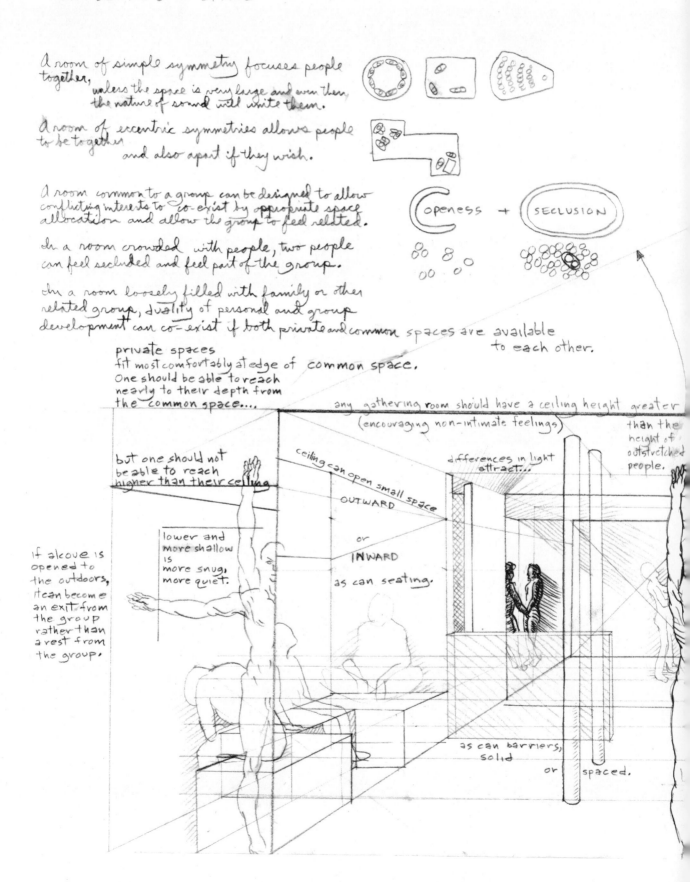

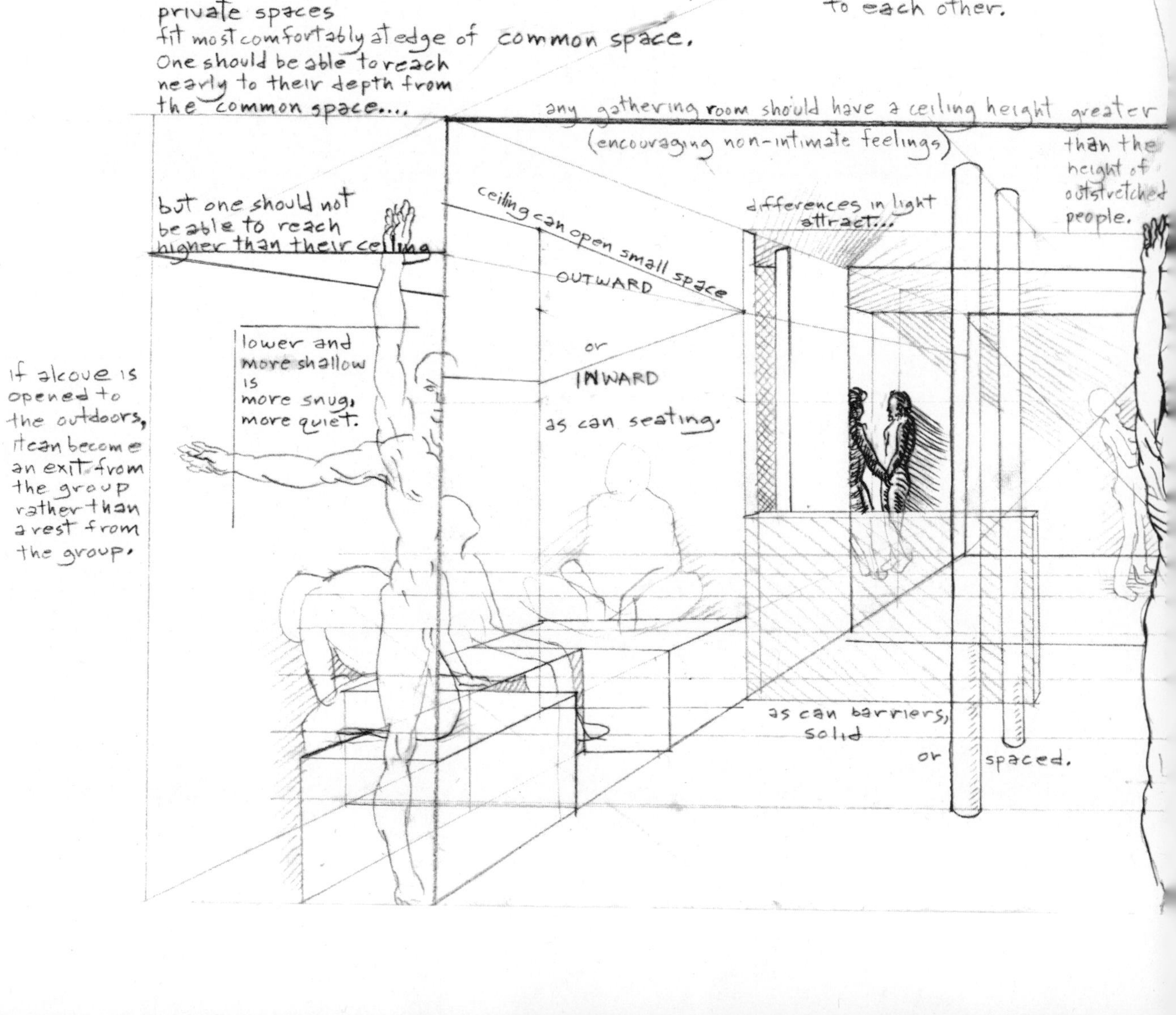
private spaces
fit most comfortably at edge of common space.
One should be able to reach nearly to their depth from the common space...
but one should not be able to reach higher than their ceiling
any gathering room should have a ceiling height greater than the height of outstretched people.
(encouraging non-intimate feelings)
ceiling can open small space
OUTWARD
or
INWARD
as can seating.
differences in light attract...
lower and more shallow is more snug, more quiet.
If alcove is opened to the outdoors, it can become an exit from the group rather than a rest from the group.
as can barriers, solid
or
spaced.

129. A room for gathering is a gathering of alcoves.

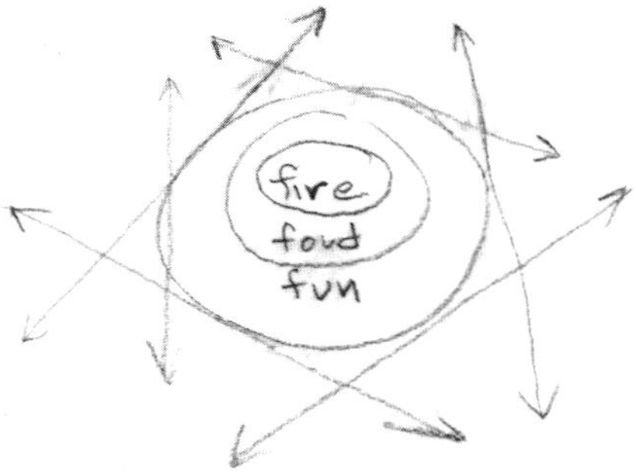
fire
food
fun

A COMMON ROOM is the main attraction for group life.
is not out of the way.
is on the way to more specific functions.
is interaction's center of gravity.

INTIMACY for a related group (family)
is provision for physical, emotional, and spiritual needs.
is sensitivity to group members needs.
is maintaining growing relationships.
is giving, accepting and being flexible.

Communal space is a gathering of alcoves.

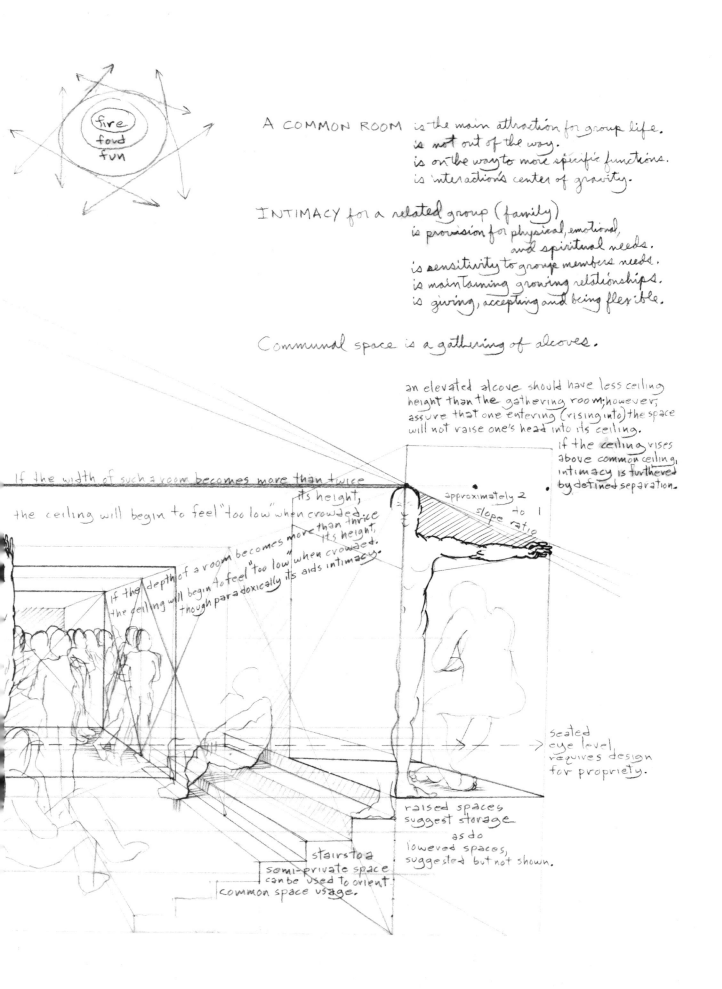

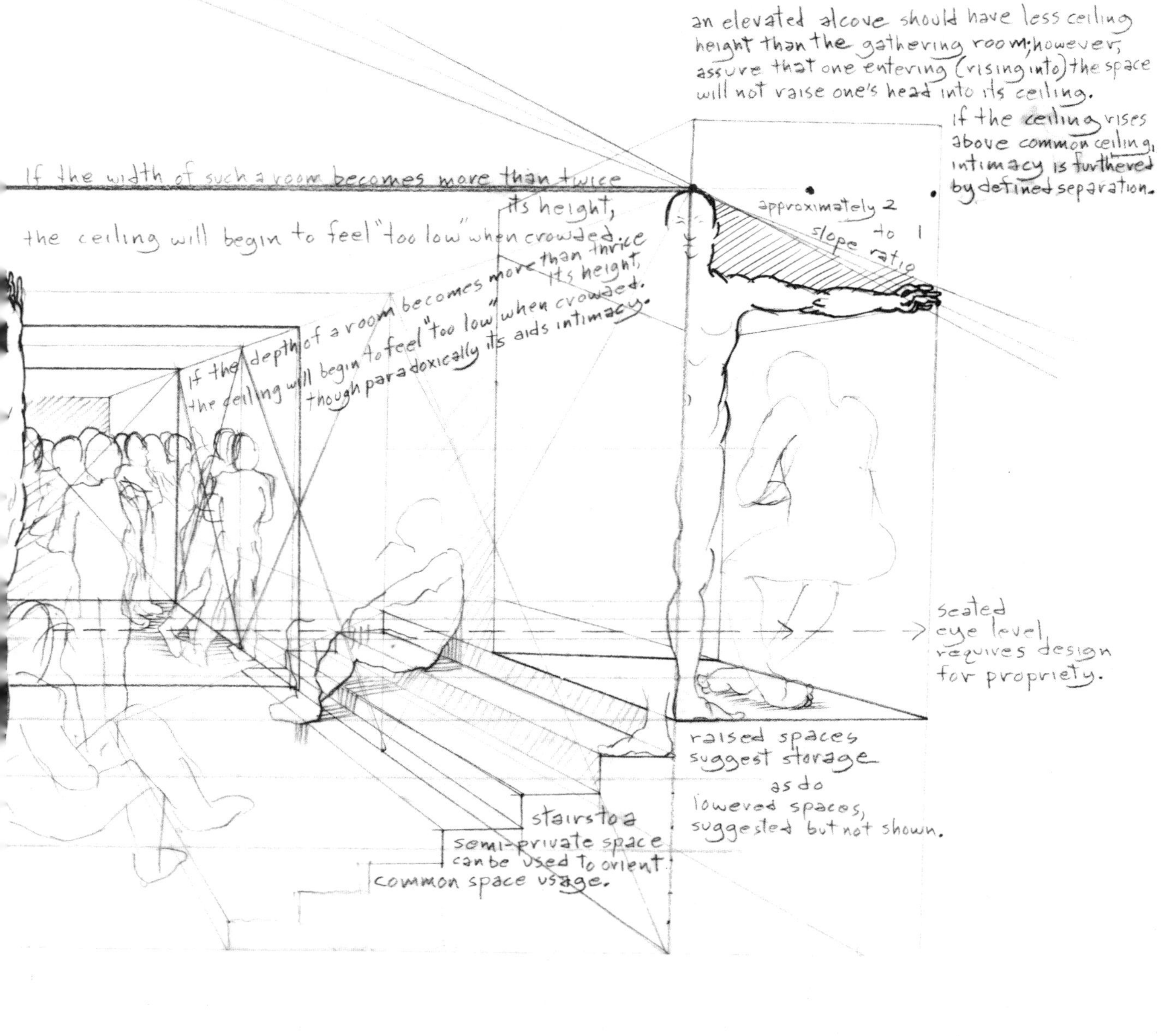
an elevated alcove should have less ceiling height than the gathering room; however, assure that one entering (rising into) the space will not raise one's head into its ceiling.
If the ceiling rises above common ceiling, intimacy is furthered by defined separation.
If the width of such a room becomes more than twice its height, the ceiling will begin to feel "too low" when crowded.
approximately 2 to 1 slope ratio
If the depth of a room becomes more than thrice its height, the ceiling will begin to feel "too low" when crowded, though paradoxically its aids intimacy.
seated eye level requires design for propriety.
raised spaces suggest storage as do lowered spaces, suggested but not shown.
stairs to a semi-private space can be used to orient common space usage.

130. When ascending and descending space, the step becomes the stair and the stair becomes the staircase, i.e., the human step dimensions the stair step, and the staircase is in the same proportions as its stair step.

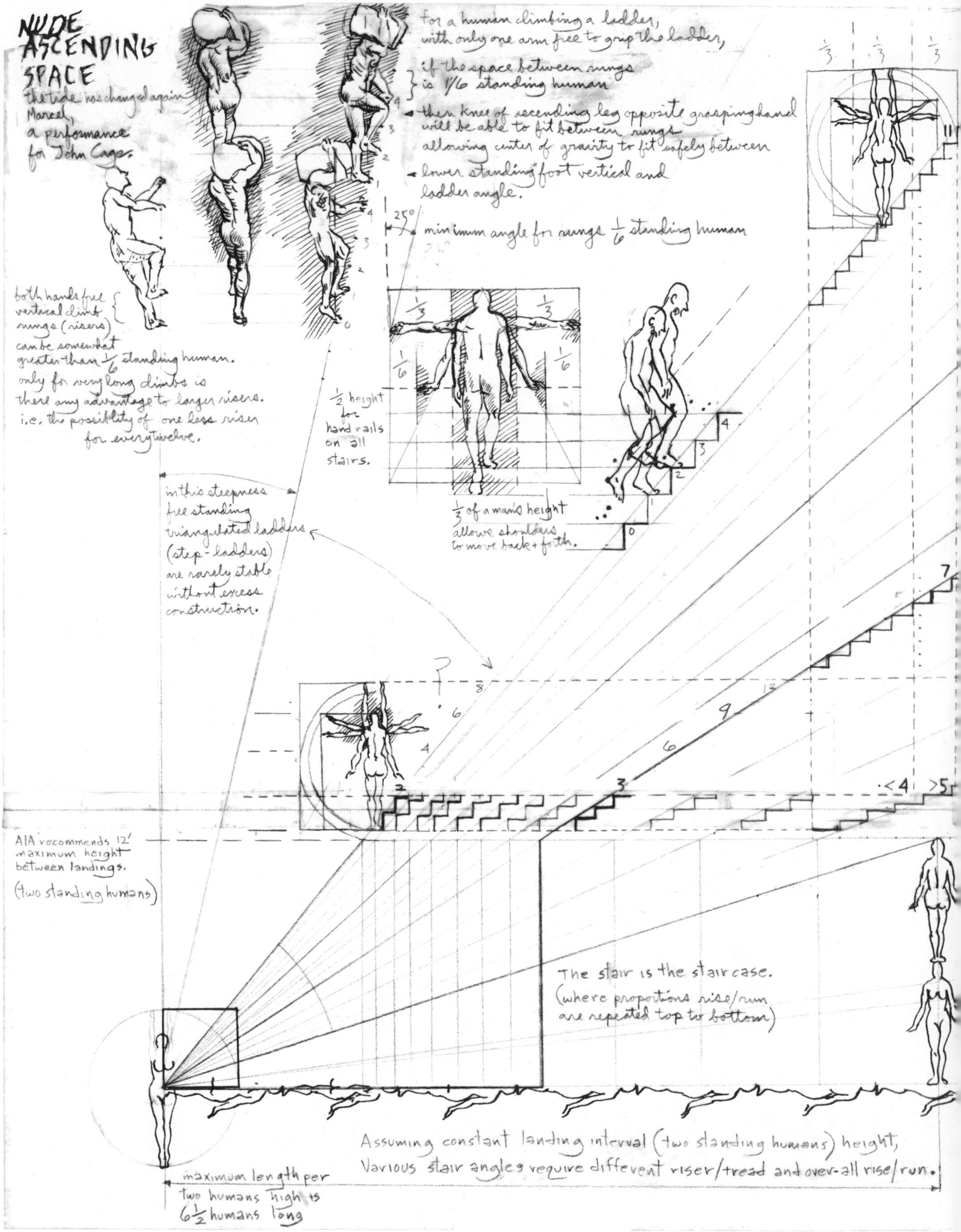
NUDE ASCENDING SPACE
the tide has changed again Marcel, a performance for John Cage.
for a human climbing a ladder, with only one arm free to grip the ladder,
if the space between rungs is 1/6 standing human
then knee of ascending leg opposite grasping hand will be able to fit between rungs allowing center of gravity to fit safely between
lower standing foot vertical and ladder angle.
minimum angle for rungs 1/6 standing human
both hands free vertical climb rungs (risers) can be somewhat greater than 1/6 standing human. only for very long climbs is there any advantage to larger risers. i.e. the possibility of one less riser for every twelve.
1/2 height for hand rails on all stairs.
1/3 of a man's height allows shoulders to move back + forth.
in this steepness free standing triangulated ladders (step-ladders) are rarely stable without excess construction.
AIA recommends 12' maximum height between landings. (two standing humans)
The stair is the staircase. (where proportions rise/run are repeated top to bottom)
Assuming constant landing interval (two standing humans) height, Various stair angles require different riser/tread and over-all rise/run.
maximum length per two humans high is 6 1/2 humans long

131. Most of our life we are on stairs, or in spaces affected by stair spaces or near walls hiding stair spaces; one flight is two stories.

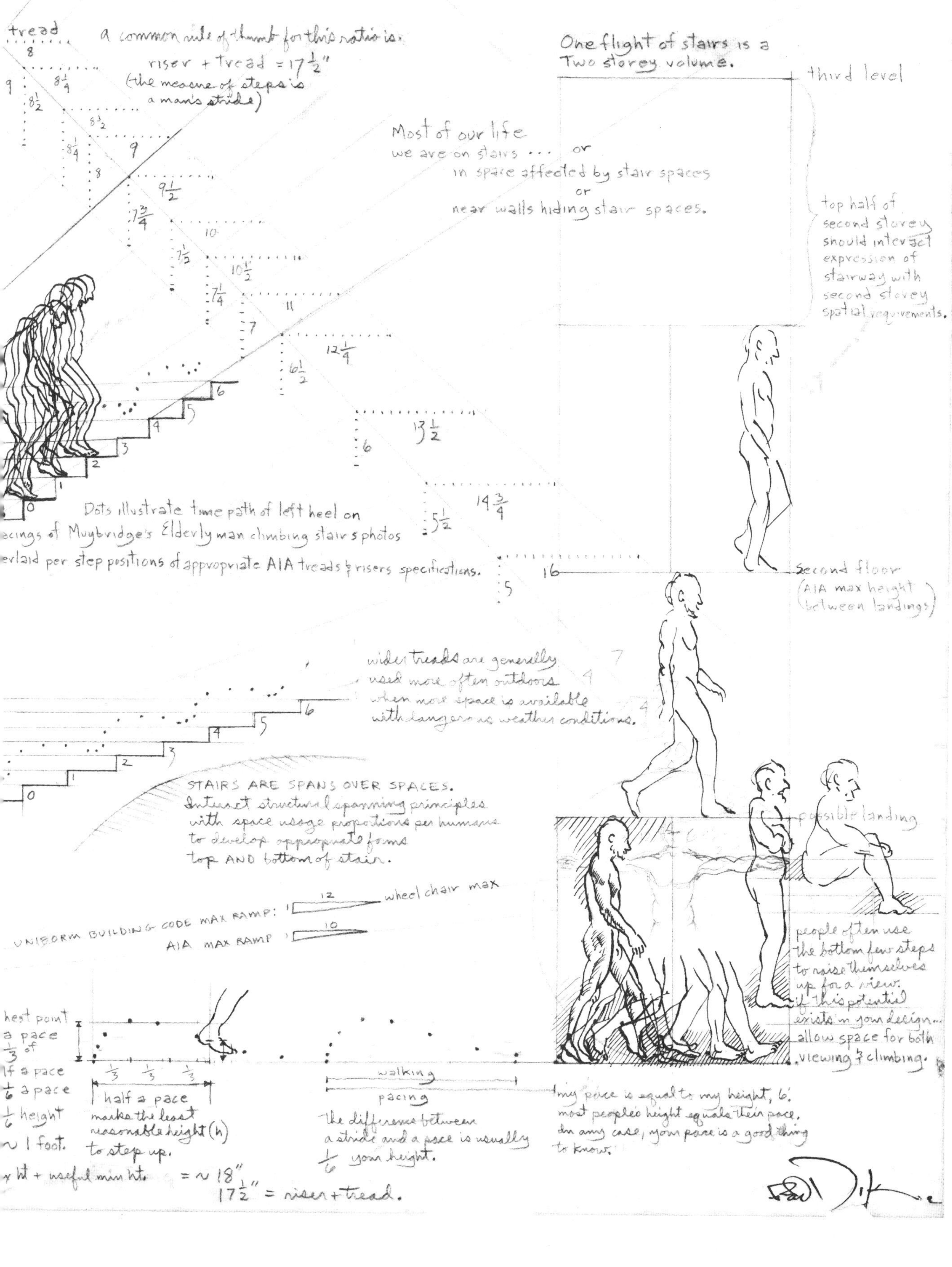

tread
a common rule of thumb for this ratio is:
riser + tread = 17 1/2"
(the measure of steps is a man's stride)
One flight of stairs is a Two storey volume.
third level
Most of our life we are on stairs ... or in space affected by stair spaces or near walls hiding stair spaces.
top half of second storey should interact expression of stairway with second storey spatial requirements.
Dots illustrate time path of left heel on tracings of Muybridge's Elderly man climbing stairs photos overlaid per step positions of appropriate AIA treads & risers specifications.
second floor (AIA max height between landings)
wider treads are generally used more often outdoors when more space is available with dangerous weather conditions.
STAIRS ARE SPANS OVER SPACES.
Interact structural spanning principles with space usage proportions per humans to develop appropriate forms top AND bottom of stair.
possible landing
wheel chair max
UNIFORM BUILDING CODE MAX RAMP:
AIA MAX RAMP
people often use the bottom few steps to raise themselves up for a view. if this potential exists in your design... allow space for both viewing & climbing.
half a pace marks the least reasonable height (h) to step up.
walking
pacing
The difference between a stride and a pace is usually 1/6 your height.
my pace is equal to my height, 6'. most people's height equals their pace. In any case, your pace is a good thing to know.
= ~ 18"
17 1/2" = riser + tread.

132. Our movement is structured by bones, and bones define the laws of structuring space.

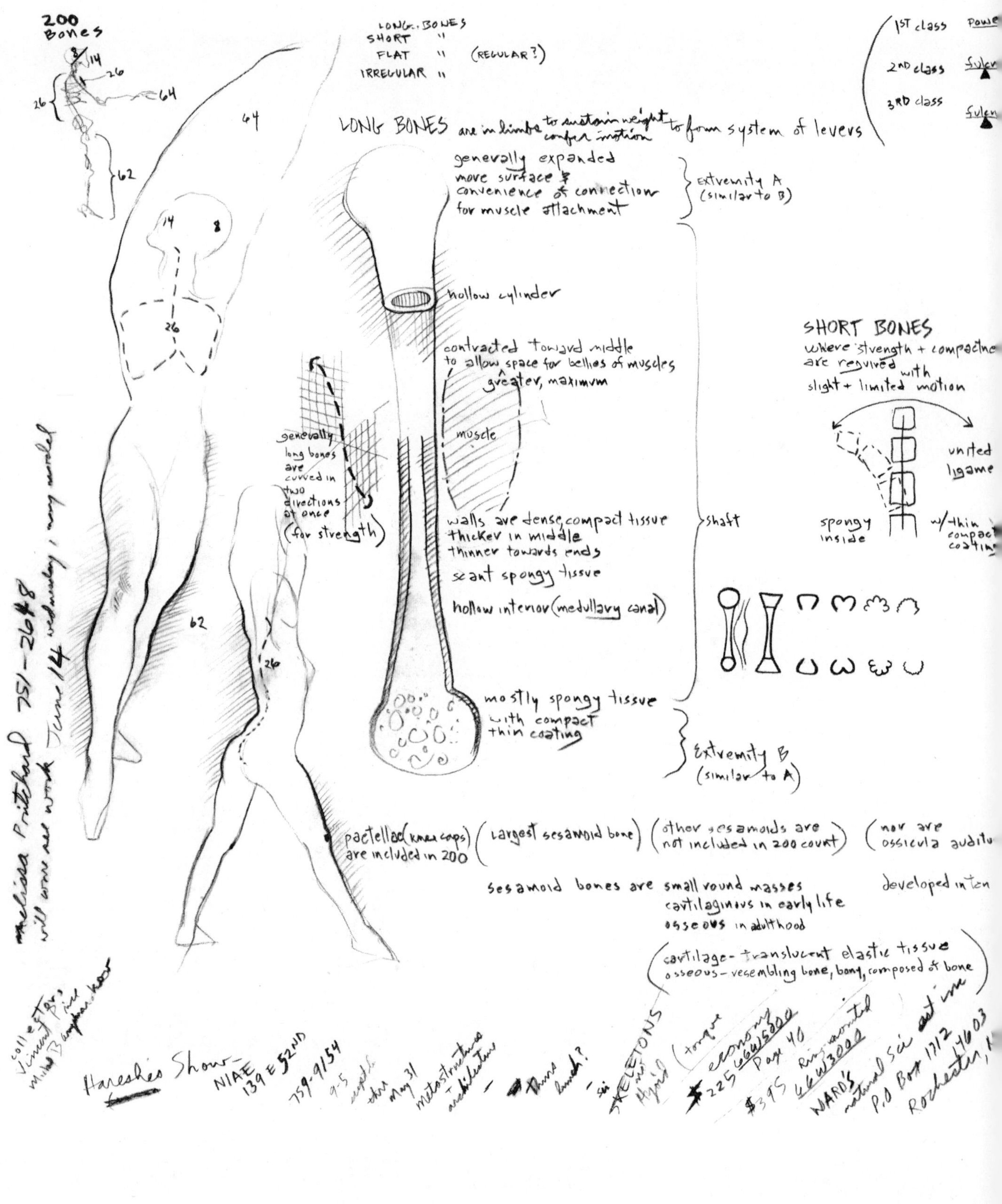

133. Bones offer logic for detailing and joining.

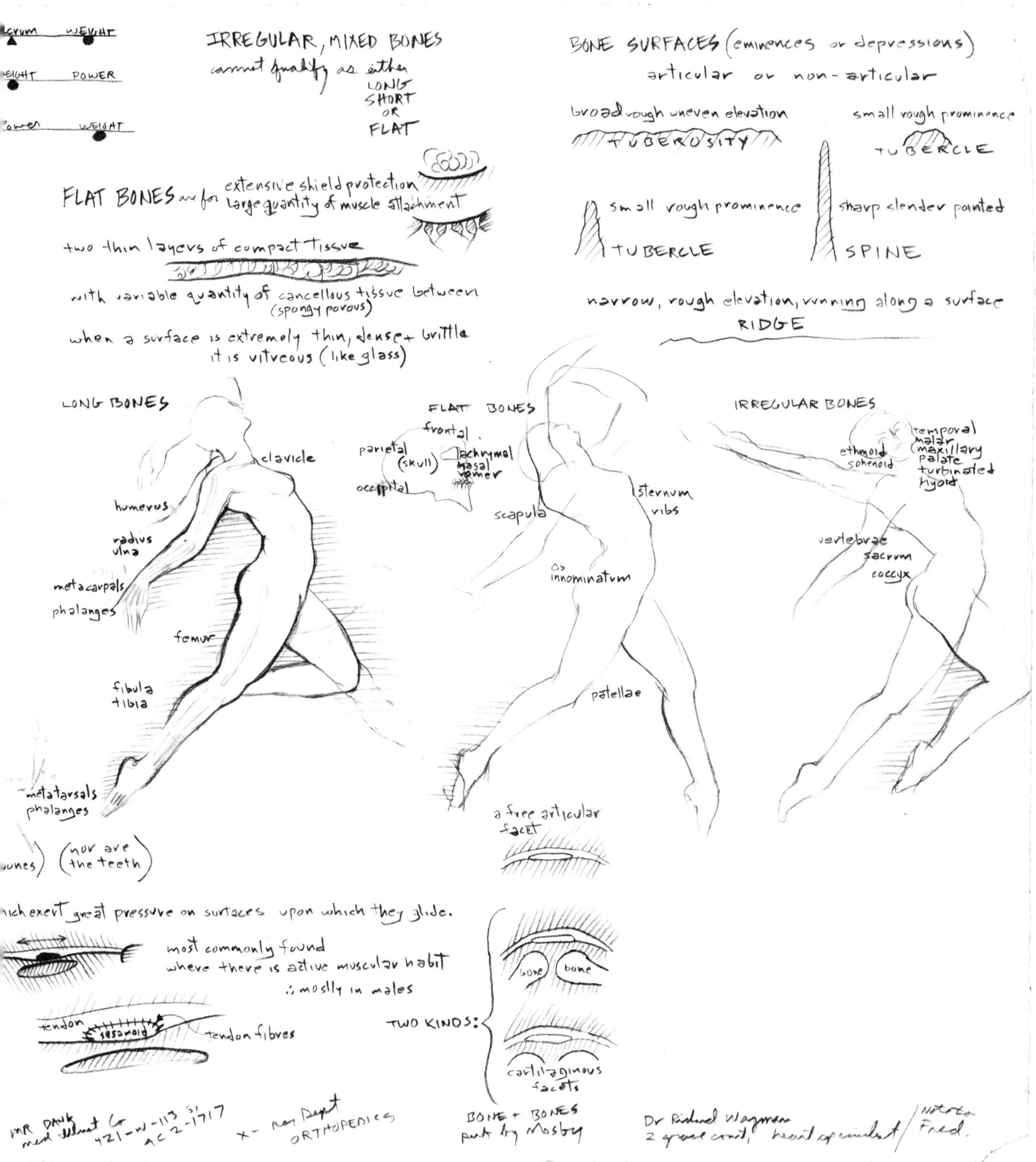

134. Our spine challenges us to design a flexible column, a responsive post.

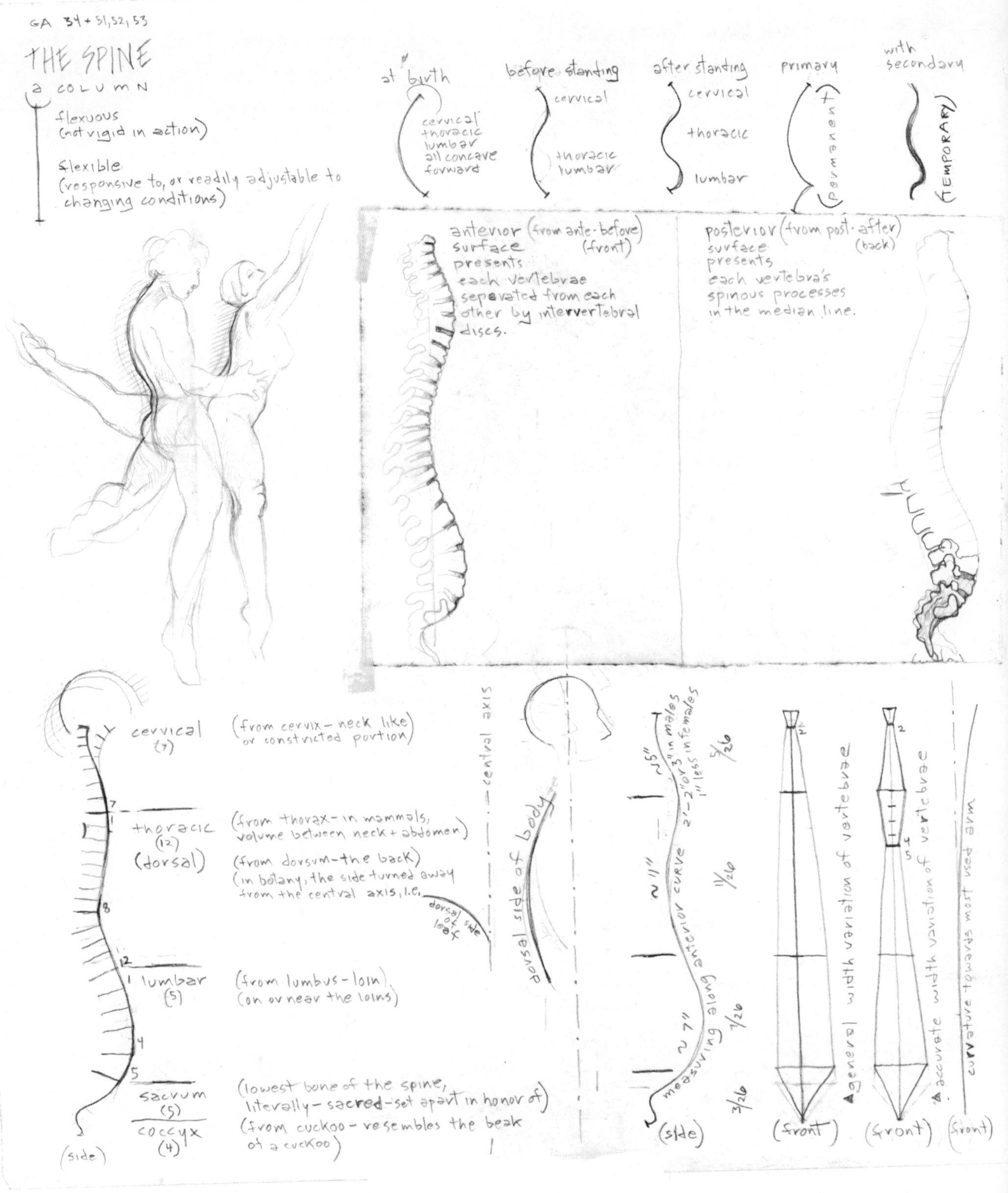

135. The interrelation of the skeleton and organs is an organizational message.

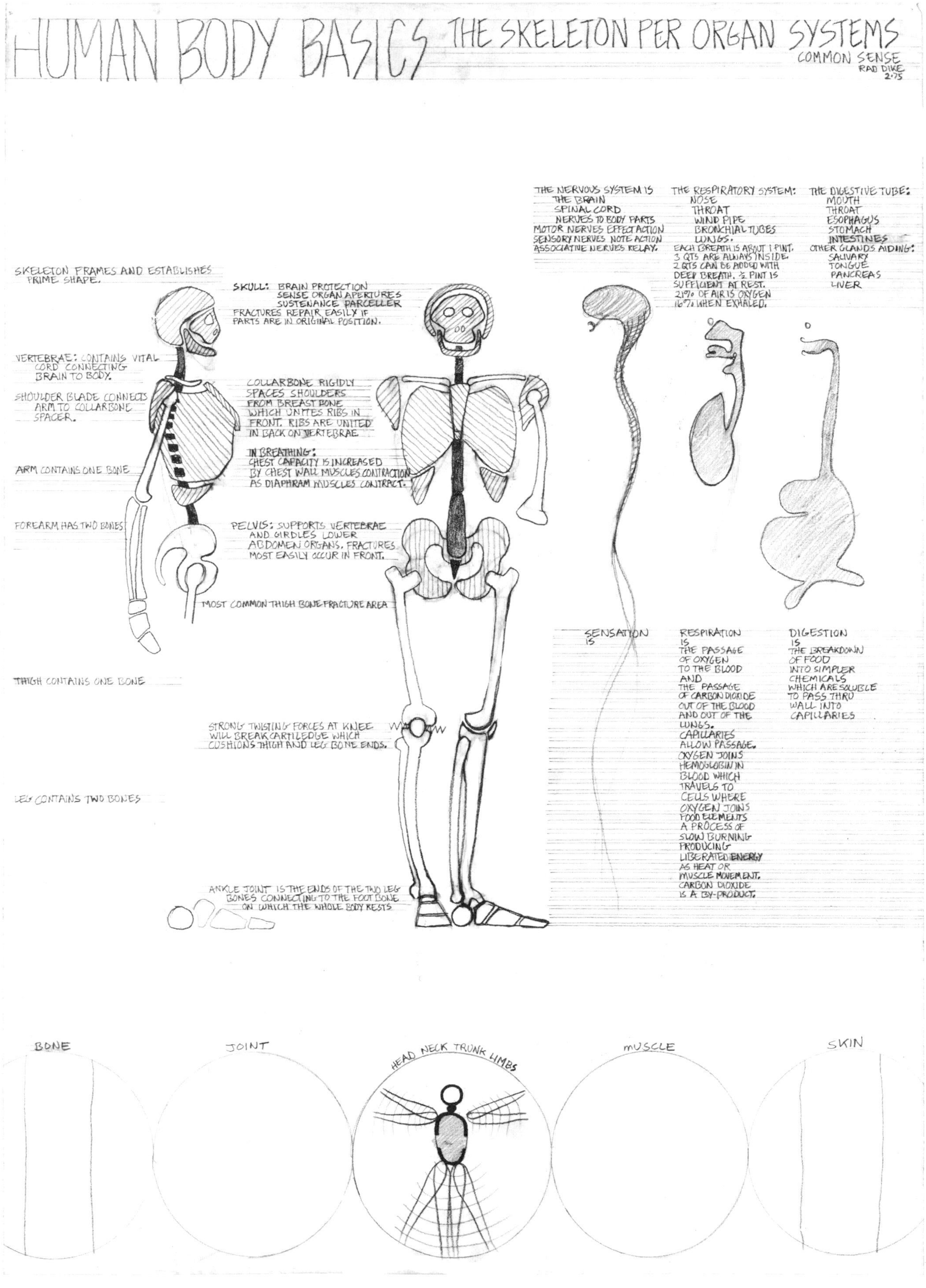

HUMAN BODY BASICS THE SKELETON PER ORGAN SYSTEMS
COMMON SENSE
RAD DIKE 2.75
THE NERVOUS SYSTEM IS THE BRAIN SPINAL CORD NERVES TO BODY PARTS MOTOR NERVES EFFECT ACTION SENSORY NERVES NOTE ACTION ASSOCIATIVE NERVES RELAY.
THE RESPIRATORY SYSTEM: NOSE THROAT WIND PIPE BRONCHIAL TUBES LUNGS.
EACH BREATH IS ABOUT 1 PINT. 3 QTS ARE ALWAYS INSIDE. 2 QTS CAN BE ADDED WITH DEEP BREATH. ½ PINT IS SUFFICIENT AT REST. 21% OF AIR IS OXYGEN 16% WHEN EXHALED.
THE DIGESTIVE TUBE: MOUTH THROAT ESOPHAGUS STOMACH INTESTINES
OTHER GLANDS AIDING: SALIVARY TONGUE PANCREAS LIVER
SKELETON FRAMES AND ESTABLISHES PRIME SHAPE.
SKULL: BRAIN PROTECTION SENSE ORGAN APERTURES SUSTENANCE PARCELLER
FRACTURES REPAIR EASILY IF PARTS ARE IN ORIGINAL POSITION.
VERTEBRAE: CONTAINS VITAL CORD CONNECTING BRAIN TO BODY.
SHOULDER BLADE CONNECTS ARM TO COLLARBONE SPACER.
COLLARBONE RIGIDLY SPACES SHOULDERS FROM BREAST BONE WHICH UNITES RIBS IN FRONT. RIBS ARE UNITED IN BACK ON VERTEBRAE
IN BREATHING: CHEST CAPACITY IS INCREASED BY CHEST WALL MUSCLES CONTRACTION AS DIAPHRAM MUSCLES CONTRACT.
ARM CONTAINS ONE BONE
FOREARM HAS TWO BONES
PELVIS: SUPPORTS VERTEBRAE AND GIRDLES LOWER ABDOMEN ORGANS. FRACTURES MOST EASILY OCCUR IN FRONT.
MOST COMMON THIGH BONE FRACTURE AREA
SENSATION IS
RESPIRATION IS THE PASSAGE OF OXYGEN TO THE BLOOD AND THE PASSAGE OF CARBON DIOXIDE OUT OF THE BLOOD AND OUT OF THE LUNGS. CAPILLARIES ALLOW PASSAGE. OXYGEN JOINS HEMOGLOBIN IN BLOOD WHICH TRAVELS TO CELLS WHERE OXYGEN JOINS FOOD ELEMENTS A PROCESS OF SLOW BURNING PRODUCING LIBERATED ENERGY AS HEAT OR MUSCLE MOVEMENT. CARBON DIOXIDE IS A BY-PRODUCT.
DIGESTION IS THE BREAKDOWN OF FOOD INTO SIMPLER CHEMICALS WHICH ARE SOLUBLE TO PASS THRU WALL INTO CAPILLARIES
THIGH CONTAINS ONE BONE
STRONG TWISTING FORCES AT KNEE WILL BREAK CARTILEDGE WHICH CUSHIONS THIGH AND LEG BONE ENDS.
LEG CONTAINS TWO BONES
ANKLE JOINT IS THE ENDS OF THE TWO LEG BONES CONNECTING TO THE FOOT BONE ON WHICH THE WHOLE BODY RESTS
BONE
JOINT
HEAD NECK TRUNK LIMBS
MUSCLE
SKIN

136. The posture of the spine is primary to the durability of the seating.

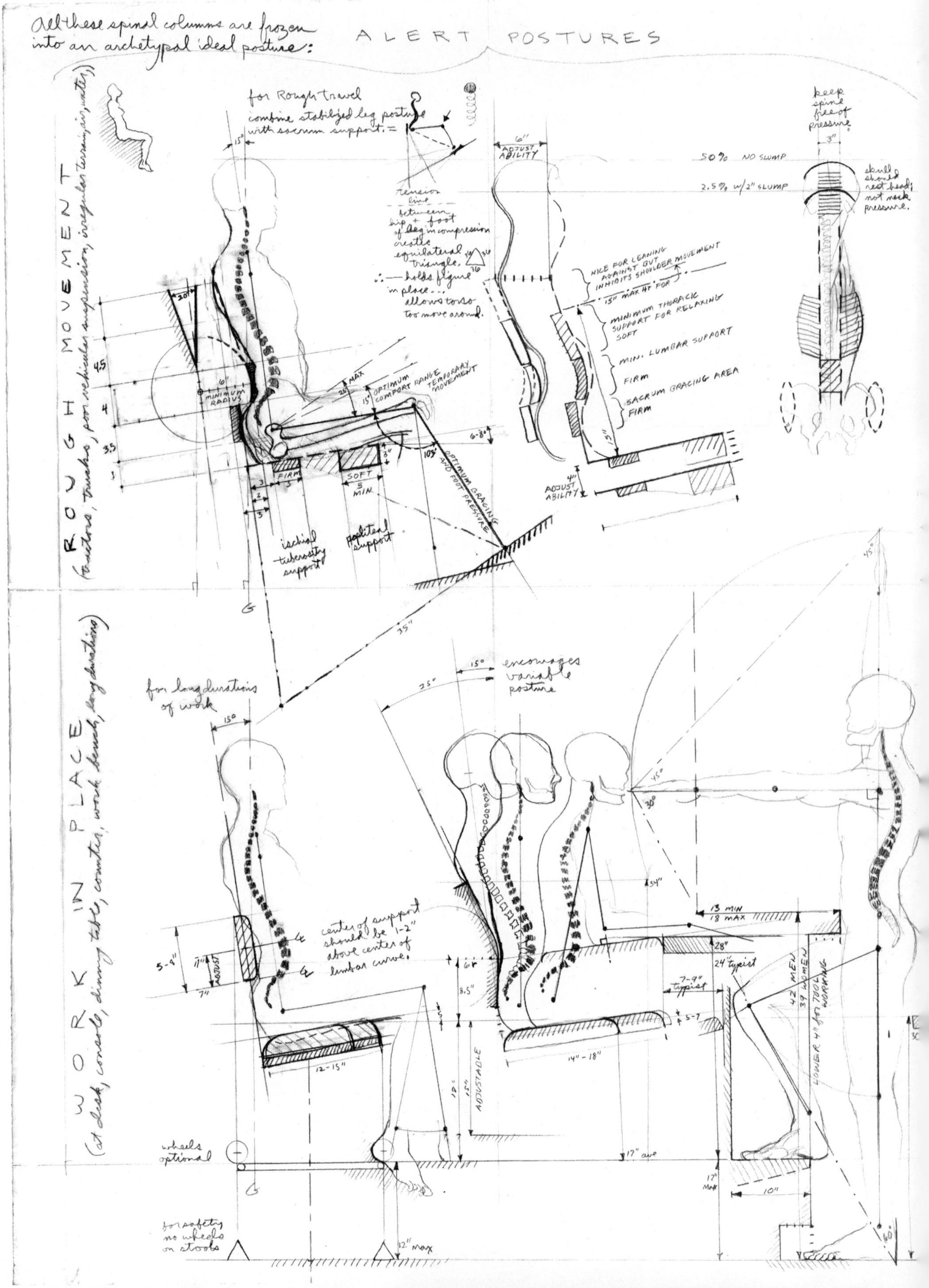

137. The length of the bones is primary to the convenience of the moving.

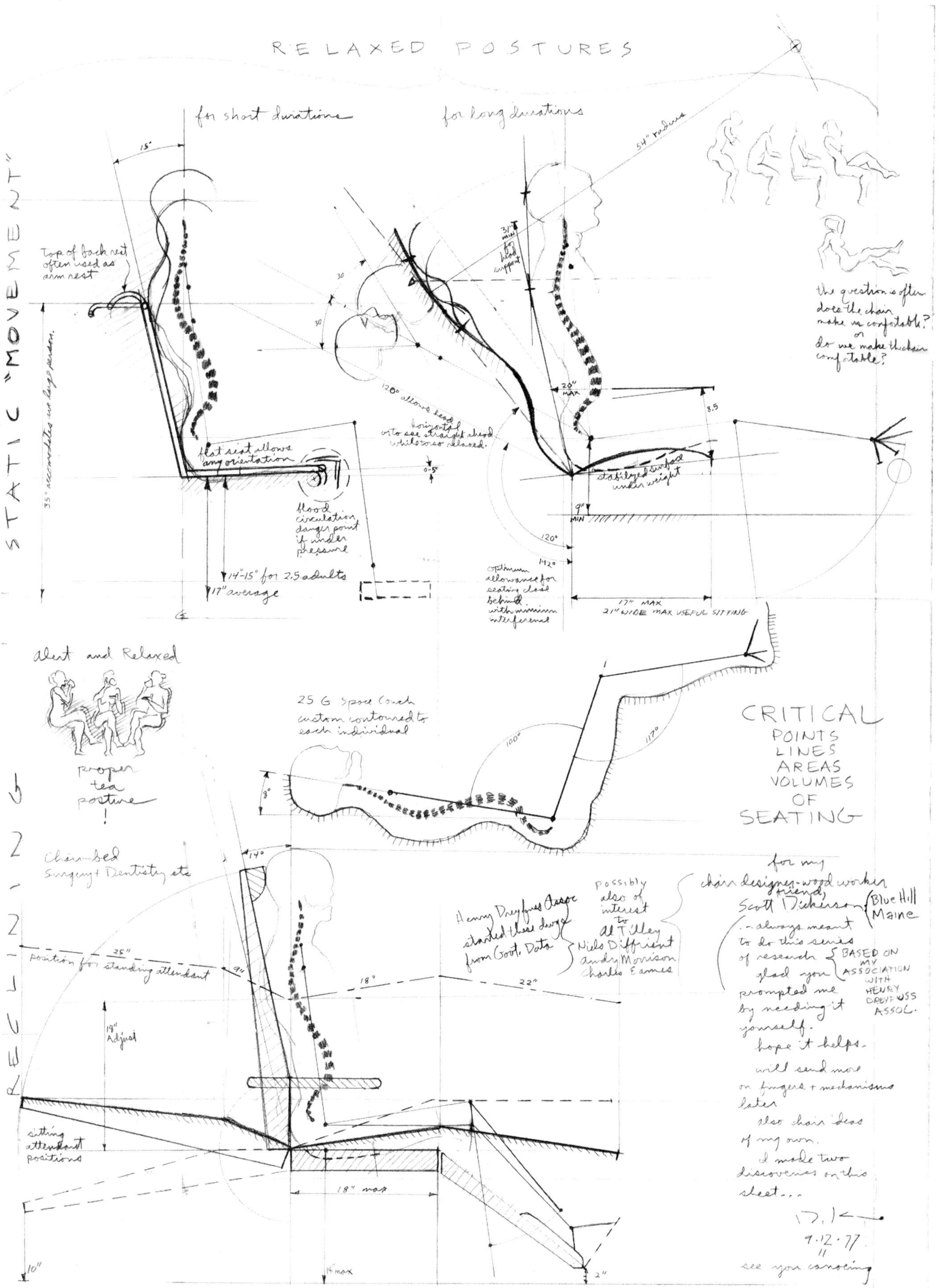

138. The flesh is primary to the surfaces and atmospheres, required for delight.

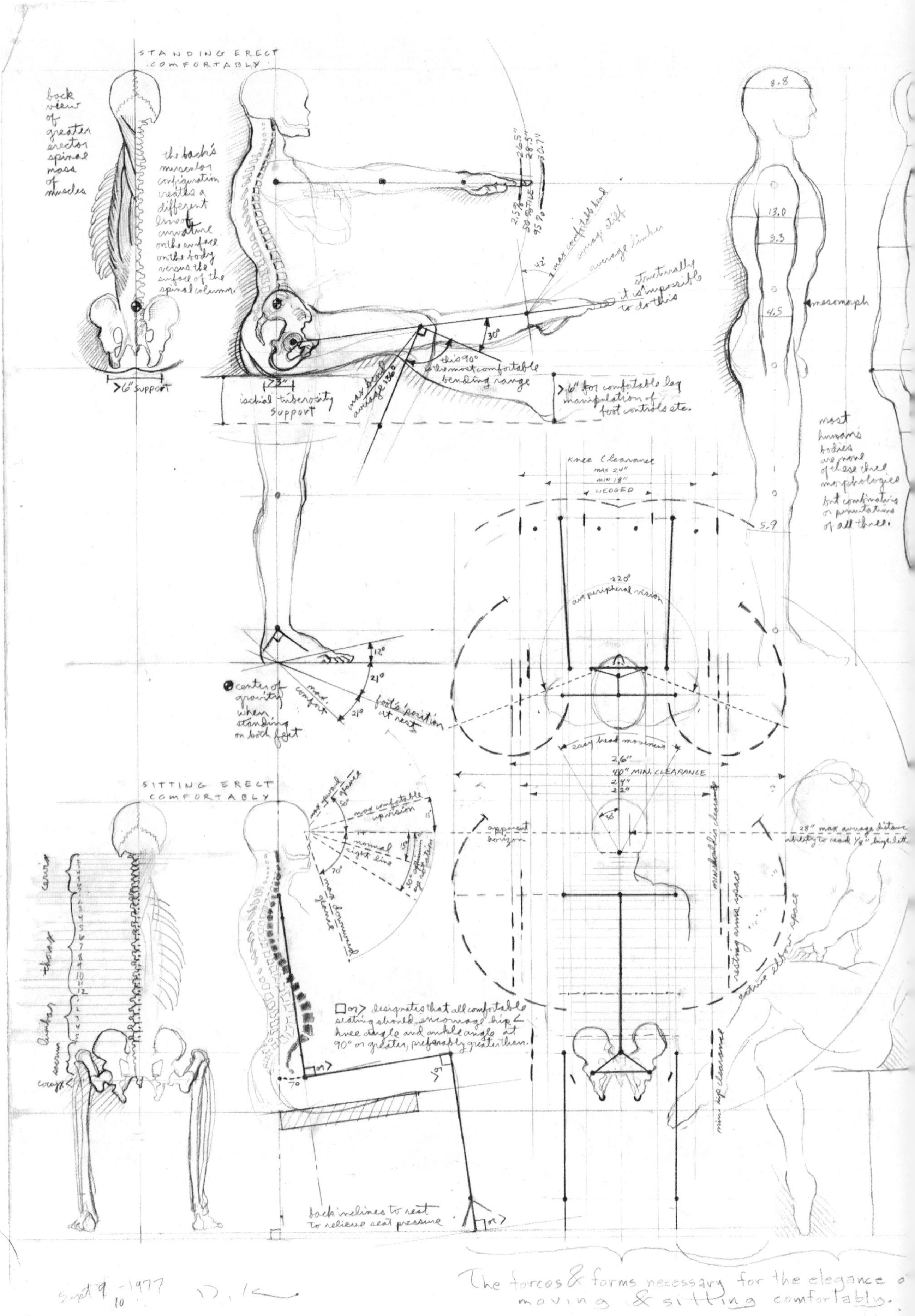

139. Architecture is built by numbers in the hopes of suiting our hopes and capacities, bound by our culture as it really is.

THE STRUCTURAL LOGIC OF STANDING ERECT
compression
tension
℄ Gravity running behind axis of rotation in pelvis and in front of knee making tension in those ligaments and compression in the bones.
compression
tension
The health of these muscles probably relates to frequency of back aches. (stomach holds backbone from sagging
SKIAGRAPHY is drawing in light + shade
(SKIA = SHADOW + GRAPHEIN = TO DELINEATE)
SKIASCOPE is a device for determining the refractive state of the eye by observing retinal lights + shadows.

140. The sins of architecture are not caused by the natural laws of proportion, but by their unnatural institutionalization.

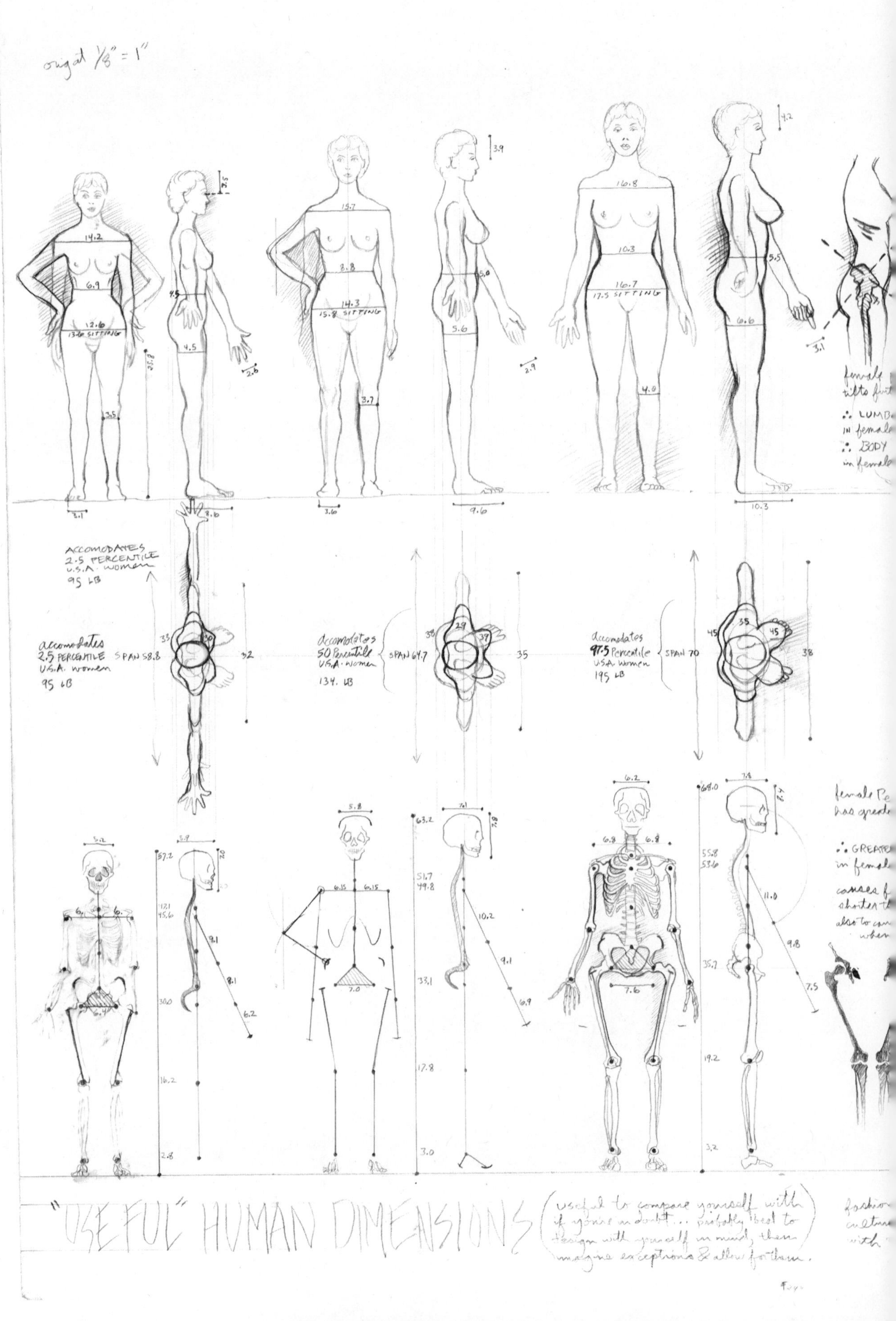

141. Though we are confined within an infinite variety of body form, we are apparently not limited to categories correlative with our possibilities.

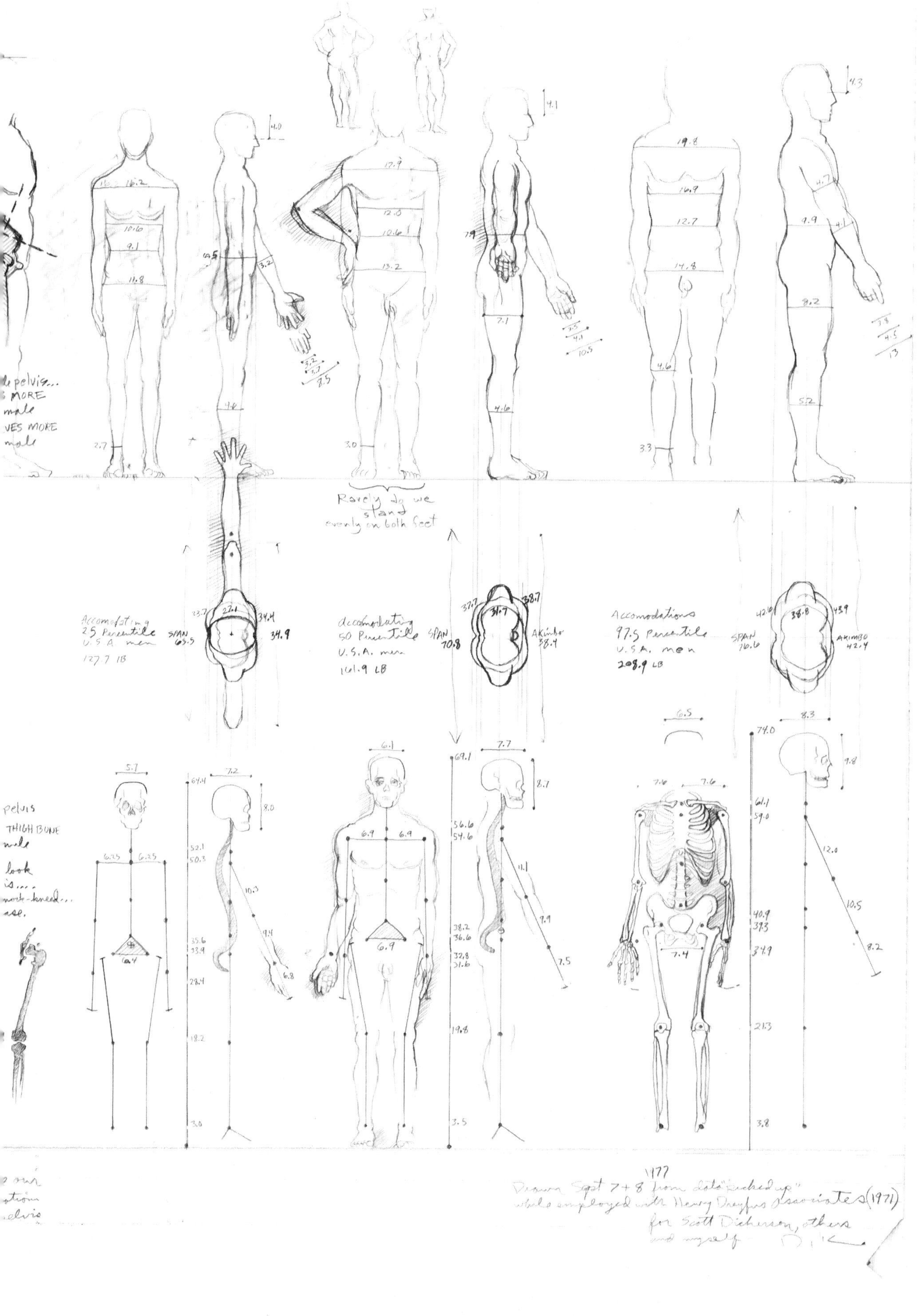

142. Clothing styles not only suit our culture; clothing is the first layer of shelter.

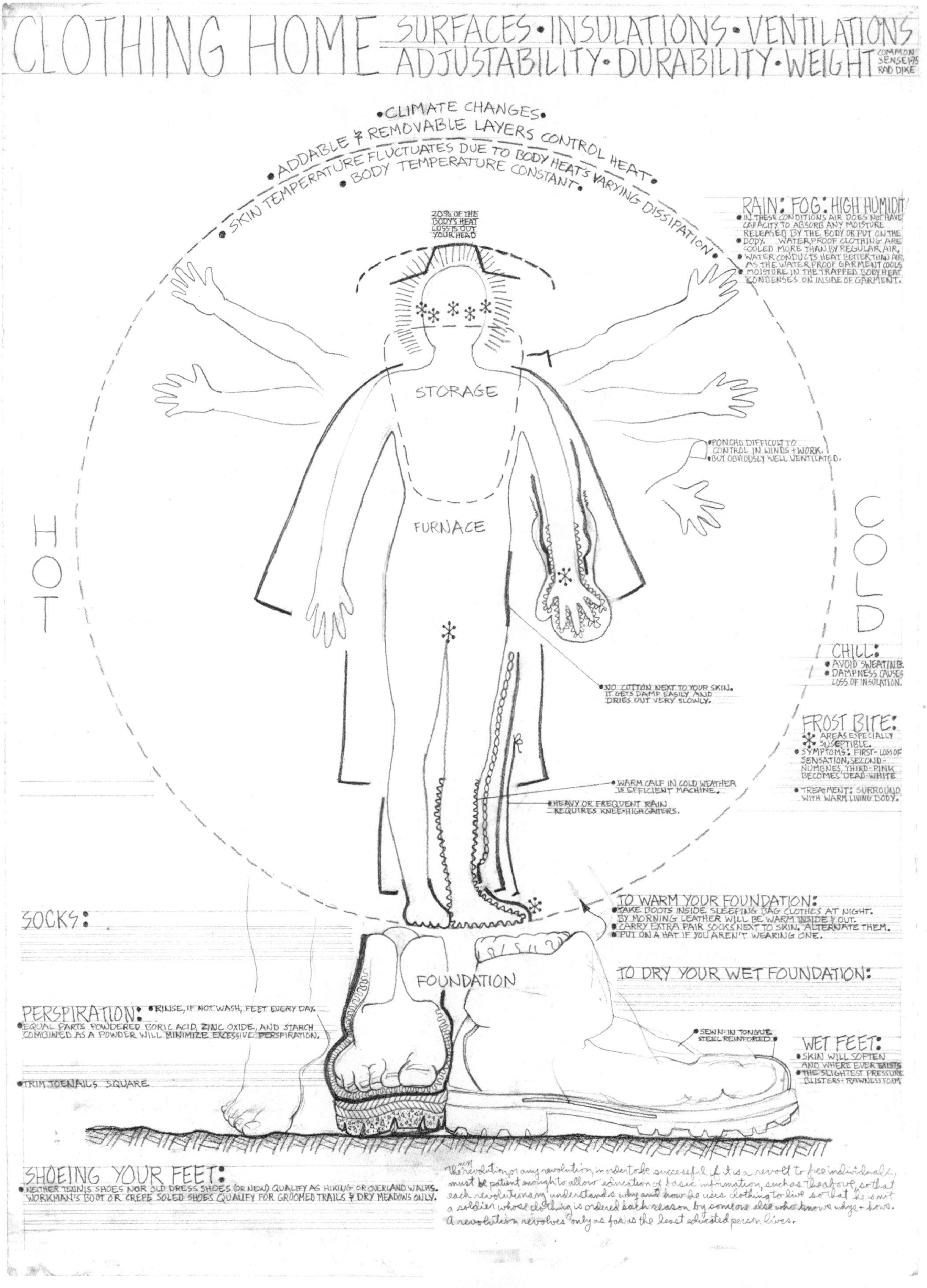

CLOTHING HOME
SURFACES • INSULATIONS • VENTILATIONS
ADJUSTABILITY • DURABILITY • WEIGHT
COMMON SENSE 1975 RAD DIKE
• CLIMATE CHANGES •
• ADDABLE & REMOVABLE LAYERS CONTROL HEAT •
• SKIN TEMPERATURE FLUCTUATES DUE TO BODY HEAT'S VARYING DISSIPATION •
• BODY TEMPERATURE CONSTANT •
20% OF THE BODY'S HEAT LOSS IS OUT YOUR HEAD
RAIN: FOG: HIGH HUMIDIT
• IN THESE CONDITIONS AIR DOES NOT HAVE CAPACITY TO ABSORB ANY MOISTURE RELEASED BY THE BODY OR PUT ON THE BODY.
• WATERPROOF CLOTHING ARE COOLED MORE THAN BY REGULAR AIR,
• WATER CONDUCTS HEAT BETTER THAN AIR. AS THE WATER PROOF GARMENT COOLS
• MOISTURE IN THE TRAPPED BODY HEAT CONDENSES ON INSIDE OF GARMENT.
STORAGE
FURNACE
• PONCHO DIFFICULT TO CONTROL IN WINDS + WORK.
• BUT OBVIOUSLY WELL VENTILATED.
HOT
COLD
CHILL:
• AVOID SWEATING.
• DAMPNESS CAUSES LOSS OF INSULATION.
• NO COTTON NEXT TO YOUR SKIN. IT GETS DAMP EASILY AND DRIES OUT VERY SLOWLY.
FROST BITE:
✻ AREAS ESPECIALLY SUSEPTIBLE.
• SYMPTOMS: FIRST - LOSS OF SENSATION, SECOND - NUMBNES, THIRD - PINK BECOMES DEAD-WHITE
• TREATMENT: SURROUND WITH WARM LIVING BODY.
• WARM CALF IN COLD WEATHER IS EFFICIENT MACHINE.
• HEAVY OR FREQUENT RAIN REQUIRES KNEE-HIGH GAITERS.
SOCKS:
TO WARM YOUR FOUNDATION:
• TAKE BOOTS INSIDE SLEEPING BAG CLOTHES AT NIGHT. BY MORNING LEATHER WILL BE WARM INSIDE & OUT.
• CARRY EXTRA PAIR SOCKS NEXT TO SKIN. ALTERNATE THEM.
• PUT ON A HAT IF YOU AREN'T WEARING ONE.
FOUNDATION
TO DRY YOUR WET FOUNDATION:
PERSPIRATION:
• RINSE, IF NOT WASH, FEET EVERY DAY.
• EQUAL PARTS POWDERED BORIC ACID, ZINC OXIDE, AND STARCH COMBINED AS A POWDER WILL MINIMIZE EXCESSIVE PERSPIRATION.
• TRIM TOENAILS SQUARE
• SEWN-IN TONGUE
STEEL REINFORCED •
WET FEET:
• SKIN WILL SOFTEN AND WHERE EVER EXISTS
• THE SLIGHTEST PRESSURE BLISTERS + RAWNESS FORM
SHOEING YOUR FEET:
• NEITHER TENNIS SHOES NOR OLD DRESS SHOES (OR NEW) QUALIFY AS HIKING OR OVERLAND WALKS. WORKMAN'S BOOT OR CREPE SOLED SHOES QUALIFY FOR GROOMED TRAILS & DRY MEADOWS ONLY.
The next revolution, or any revolution, in order to be successful if it is a revolt to free individuals, must be patient enough to allow education of basic information, such as the above, so that each revolutionary understands why and how he uses clothing to live so that he is not a soldier whose clothing is ordered each season by someone else who knows whys + hows.
A revolution revolves only as far as the least educated person lives.

143. When architecture requires bandaging, structure responsive to sun, site, and self is the only acceptable aesthetic retrofit.

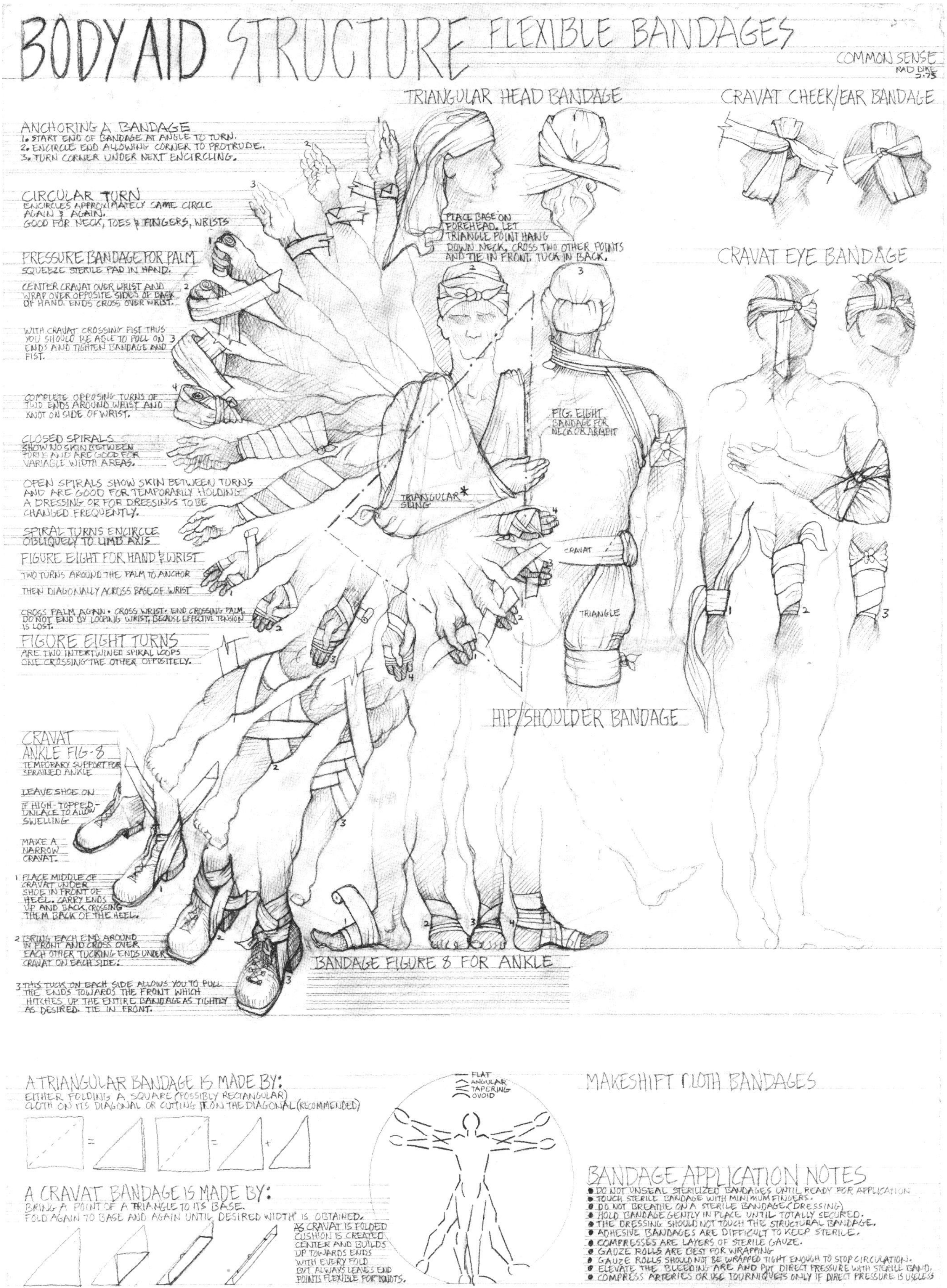
BODY AID STRUCTURE FLEXIBLE BANDAGES
COMMON SENSE
RAD DIKE
TRIANGULAR HEAD BANDAGE
CRAVAT CHEEK/EAR BANDAGE
ANCHORING A BANDAGE
1. START END OF BANDAGE AT ANGLE TO TURN.
2. ENCIRCLE END ALLOWING CORNER TO PROTRUDE.
3. TURN CORNER UNDER NEXT ENCIRCLING.
CIRCULAR TURN
ENCIRCLES APPROXIMATELY SAME CIRCLE AGAIN & AGAIN.
GOOD FOR NECK, TOES & FINGERS, WRISTS
PLACE BASE ON FOREHEAD. LET TRIANGLE POINT HANG DOWN NECK. CROSS TWO OTHER POINTS AND TIE IN FRONT. TUCK IN BACK.
PRESSURE BANDAGE FOR PALM
SQUEEZE STERILE PAD IN HAND.
CENTER CRAVAT OVER WRIST AND WRAP OVER OPPOSITE SIDES OF BACK OF HAND. ENDS CROSS OVER WRIST.
WITH CRAVAT CROSSING FIST THUS YOU SHOULD BE ABLE TO PULL ON ENDS AND TIGHTEN BANDAGE AND FIST.
COMPLETE OPPOSING TURNS OF TWO ENDS AROUND WRIST AND KNOT ON SIDE OF WRIST.
CRAVAT EYE BANDAGE
FIG. EIGHT BANDAGE FOR NECK OR ARMPIT
CLOSED SPIRALS
SHOW NO SKIN BETWEEN TURNS AND ARE GOOD FOR VARIABLE WIDTH AREAS.
OPEN SPIRALS SHOW SKIN BETWEEN TURNS AND ARE GOOD FOR TEMPORARILY HOLDING A DRESSING OR FOR DRESSINGS TO BE CHANGED FREQUENTLY.
TRIANGULAR SLING*
SPIRAL TURNS ENCIRCLE OBLIQUELY TO LIMB AXIS
CRAVAT
FIGURE EIGHT FOR HAND & WRIST
TWO TURNS AROUND THE PALM TO ANCHOR
THEN DIAGONALLY ACROSS BASE OF WRIST
CROSS PALM AGAIN. CROSS WRIST. END CROSSING PALM. DO NOT END BY LOOPING WRIST, BECAUSE EFFECTIVE TENSION IS LOST.
TRIANGLE
FIGURE EIGHT TURNS
ARE TWO INTERTWINED SPIRAL LOOPS ONE CROSSING THE OTHER OPPOSITELY.
HIP/SHOULDER BANDAGE
CRAVAT ANKLE FIG-8
TEMPORARY SUPPORT FOR SPRAINED ANKLE
LEAVE SHOE ON
IF HIGH-TOPPED-UNLACE TO ALLOW SWELLING
MAKE A NARROW CRAVAT.
1 PLACE MIDDLE OF CRAVAT UNDER SHOE IN FRONT OF HEEL. CARRY ENDS UP AND BACK, CROSSING THEM BACK OF THE HEEL.
2 BRING EACH END AROUND IN FRONT AND CROSS OVER EACH OTHER TUCKING ENDS UNDER CRAVAT ON EACH SIDE.
3 THIS TUCK ON EACH SIDE ALLOWS YOU TO PULL THE ENDS TOWARDS THE FRONT WHICH HITCHES UP THE ENTIRE BANDAGE AS TIGHTLY AS DESIRED. TIE IN FRONT.
BANDAGE FIGURE 8 FOR ANKLE
A TRIANGULAR BANDAGE IS MADE BY:
EITHER FOLDING A SQUARE (POSSIBLY RECTANGULAR) CLOTH ON ITS DIAGONAL OR CUTTING IT ON THE DIAGONAL (RECOMMENDED)
FLAT
ANGULAR
TAPERING
OVOID
MAKESHIFT CLOTH BANDAGES
A CRAVAT BANDAGE IS MADE BY:
BRING A POINT OF A TRIANGLE TO ITS BASE.
FOLD AGAIN TO BASE AND AGAIN UNTIL DESIRED WIDTH IS OBTAINED.
AS CRAVAT IS FOLDED CUSHION IS CREATED CENTER AND BUILDS UP TOWARDS ENDS WITH EVERY FOLD BUT ALWAYS LEAVES END POINTS FLEXIBLE FOR KNOTS.
BANDAGE APPLICATION NOTES
• DO NOT UNSEAL STERILIZED BANDAGES UNTIL READY FOR APPLICATION
• TOUCH STERILE BANDAGE WITH MINIMUM FINGERS.
• DO NOT BREATHE ON A STERILE BANDAGE (DRESSING)
• HOLD BANDAGE GENTLY IN PLACE UNTIL TOTALLY SECURED.
• THE DRESSING SHOULD NOT TOUCH THE STRUCTURAL BANDAGE.
• ADHESIVE BANDAGES ARE DIFFICULT TO KEEP STERILE.
• COMPRESSES ARE LAYERS OF STERILE GAUZE.
• GAUZE ROLLS ARE BEST FOR WRAPPING
• GAUZE ROLLS SHOULD NOT BE WRAPPED TIGHT ENOUGH TO STOP CIRCULATION.
• ELEVATE THE BLEEDING ARE AND PUT DIRECT PRESSURE WITH STERILE BAND.
• COMPRESS ARTERIES OR USE TOURNIQUETS ONLY IF DIRECT PRESSURE IS USELESS

144. Stand Da Vinci's man on his toes and upstretch his hands, then you will have begun to visualize the reasonable extremities of man.

Anatomical ratios and proportions are of course not universal.

For an artist, this knowledge underlies his ability to understand and demonstrate differences in human nature.

For an architect, this knowledge of general ratios & proportions underlies his ability to establish useful spaces for human nature.

Vitruvius informed da Vinci of the proportions of the human body which da Vinci drew and upon which he elaborated.

Da Vinci drew his canon of proportions man in relation to the inner circle and inner square of the drawing below.

Nowhere did either Vitruvius or da Vinci relate the human body to a related circle and square (the squared circle.).

the navel is the cons
in the 3 diagrams bel

	da vinci dp	me
head top		
belly button	25	28
bot feet	64	44
total ht	89	72"

(39/100)

6 39 72 78 273 100) 28.08 28.1

~1/14 = 5/72

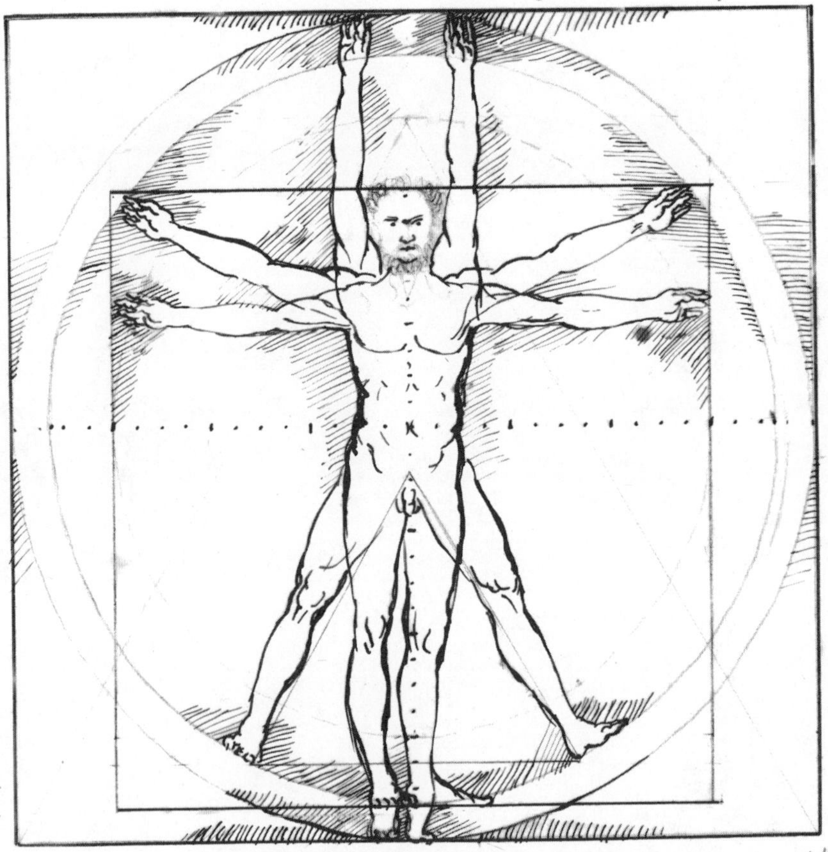

da Vinci's square & circle & equilateral triangle

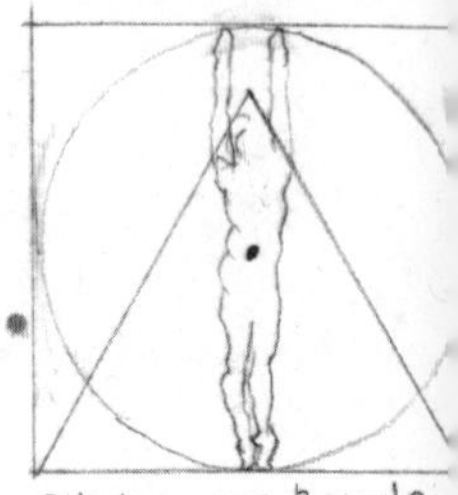

Dike's square & circle & equilateral triangle

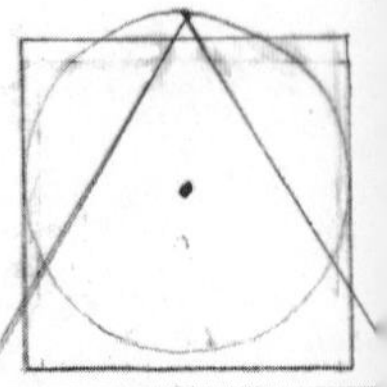

da Vinci's square
Dike's equilateral triang
and the circle linking
the two proportional c

INNER SQUARE: "The length of a man's outspread arms equals his height."

INNER CIRCLE: "If you open your legs so much as to decrease your height 1/14 and spread and raise your arms til your middle fingers touch the level of the top of your head you must know that the centre of the outspread limbs will be in the navel..."

INNER EQUILATERAL TRIANGLE: "... and the space between the legs" (opened to decrease ht. 1/14) "will be an equilateral triangle."

OUTER SQUARE: The length of a man's outstretched standing body is twice the distance from his navel to either extremity...

OUTER CIRCLE: ... which is a radius encircling his navel at a diameter 1/3 greater than his height (man 6' = circle 8')

OUTER EQUILATERAL TRIANGLE: The distance from the peak of an equilateral triangle to the navel is a radius of a circle squared by da Vinci's square.

145. The regular divisions of man are halves or doubles of each other.

These proportions are relevant to your measurement and development of space no matter what your overall height or difference in build. At least they will aid you in discovering your own.

1/2 an outstretched body is the radius of a circle touched by toe and finger tips

1/2 a standing body from bottom of foot to top of skull is center of genitals

1/3 standing body height equals increase from standing to outstretched body height.

1/4 an outstretched body is from sole of toes to knee cap.

1/4 a standing body is from sole of foot to bottom of kneecap.

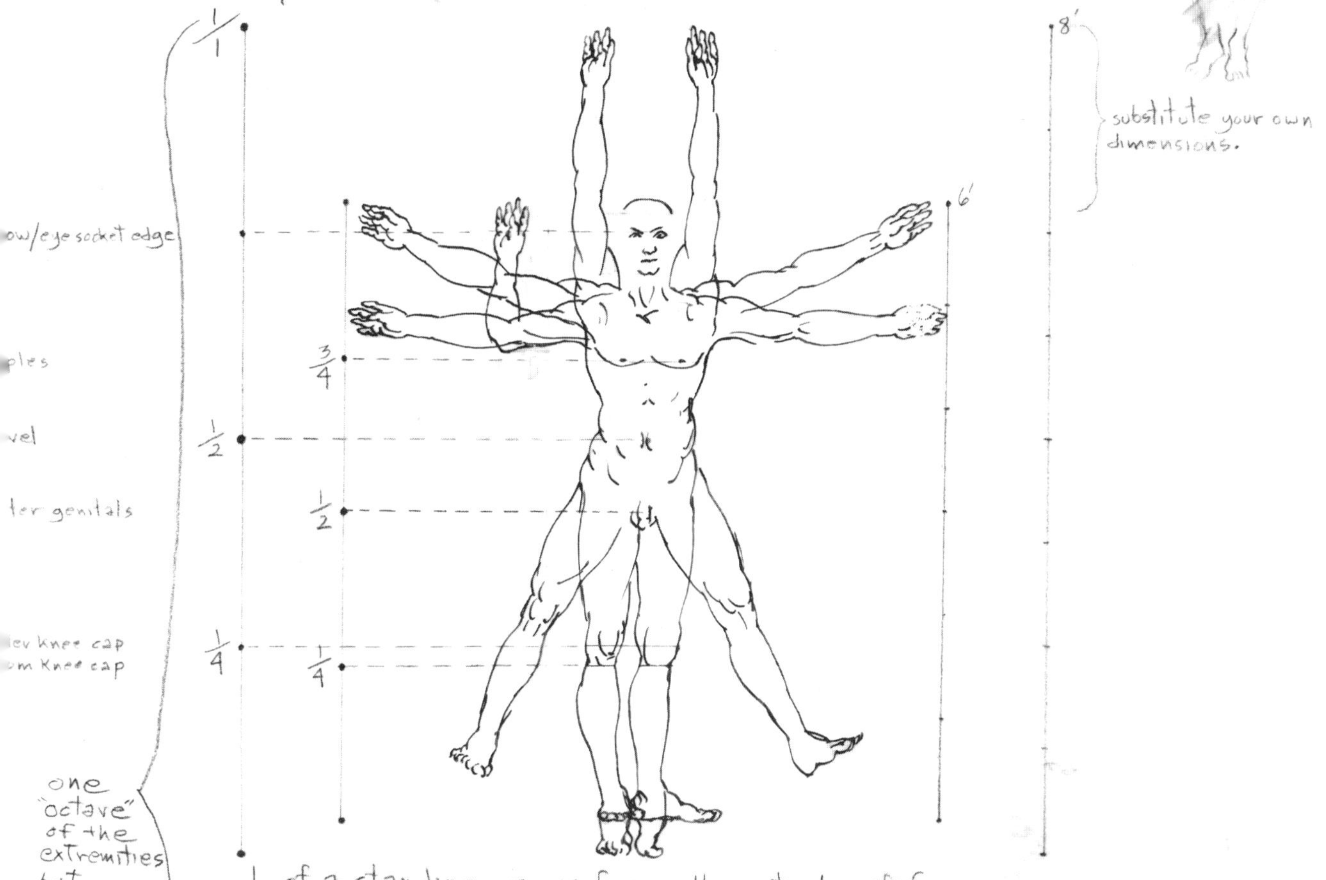

1/5 of a standing man is from elbow to tip of fingers.

1/6 of a standing man is from top of breast to top of head.

1/7 of a standing man is from top of breast to hairline.

1/8 of a standing body is from elbow to armpit.

1/9

1/10 of a standing man is the length of his hand.

one "octave" of the extremities but not similar to a musical octave. see Golden Visual Octave.

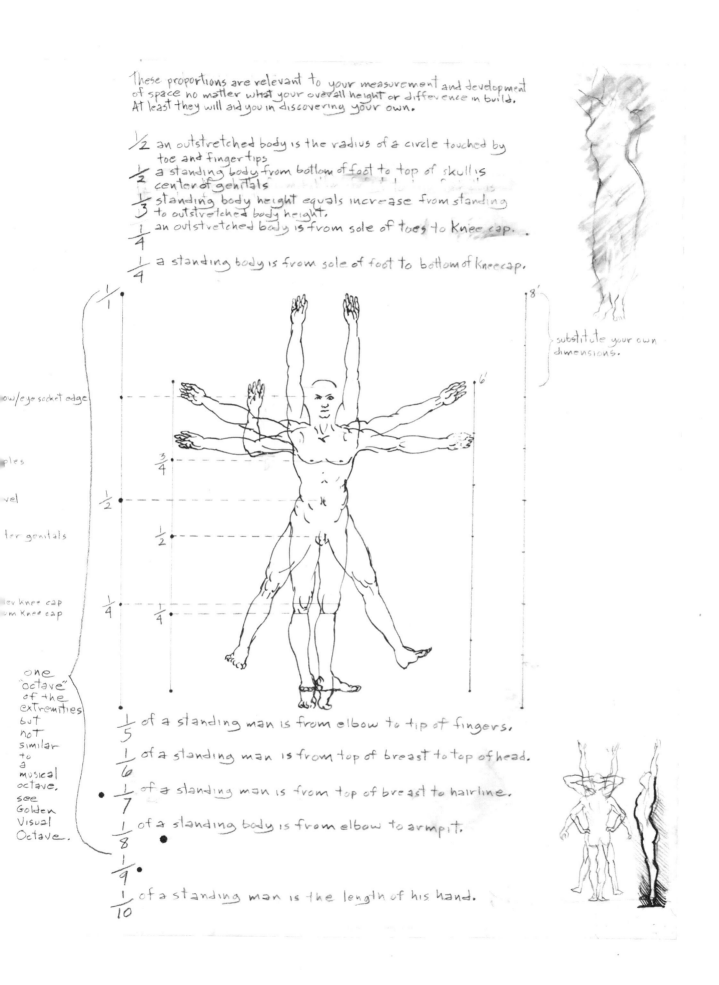

146. The extremities of man can be shown as a cube and sphere about the navel in top, front, and side views.

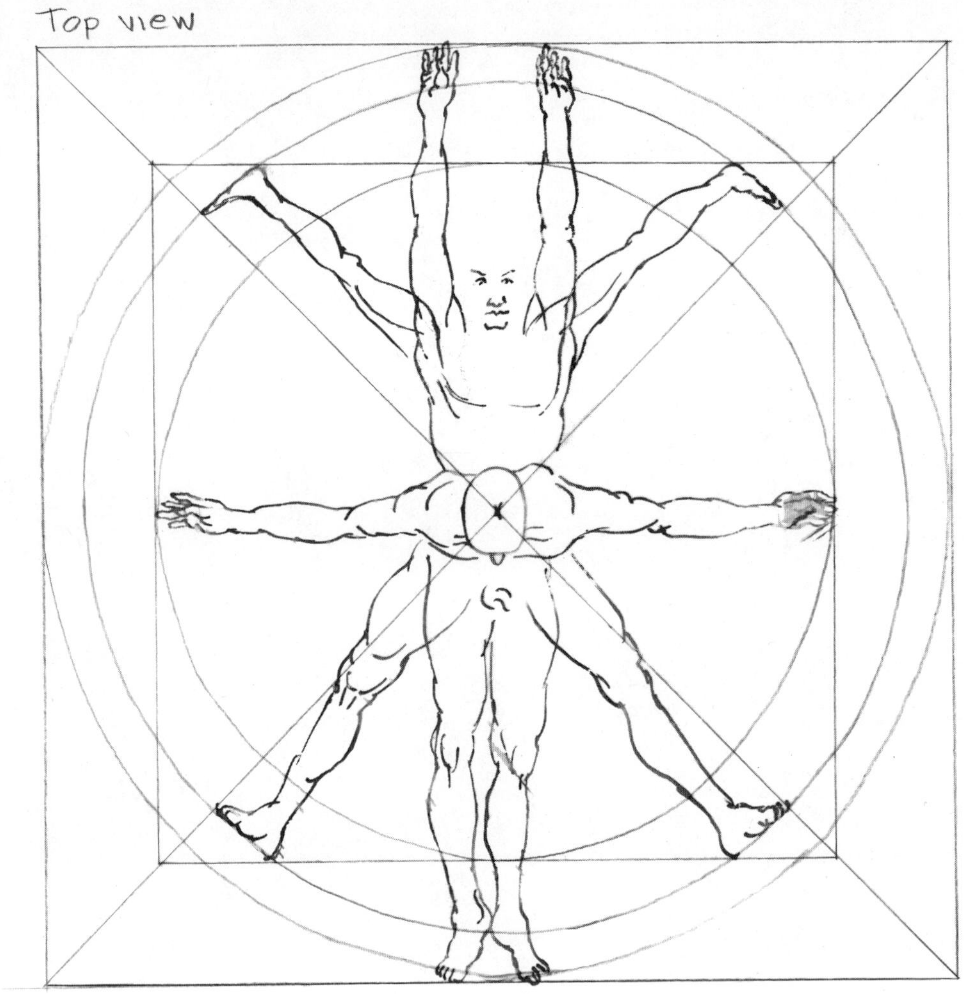

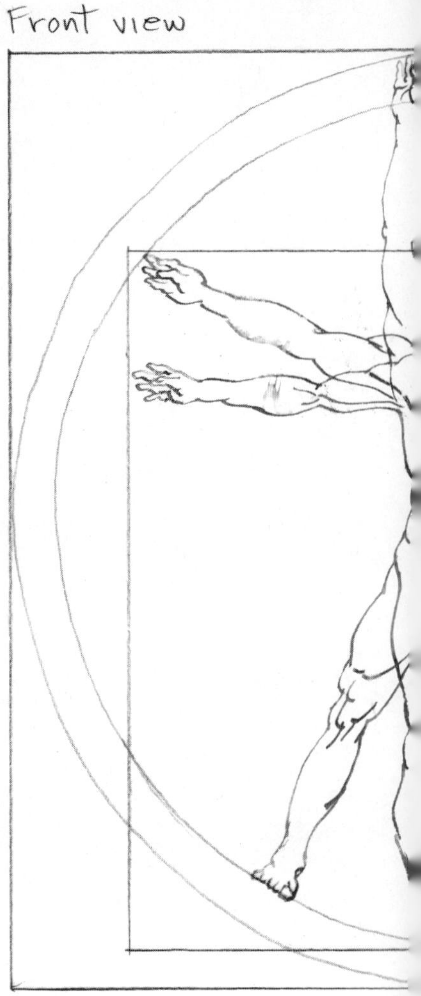

147. Great horizons may be seen in the splatter of plaster, but there is no scale to them without human proportions.

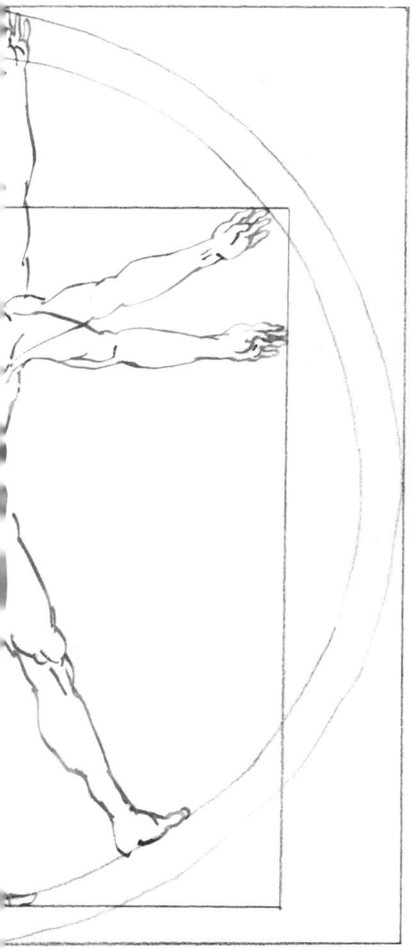

Side view

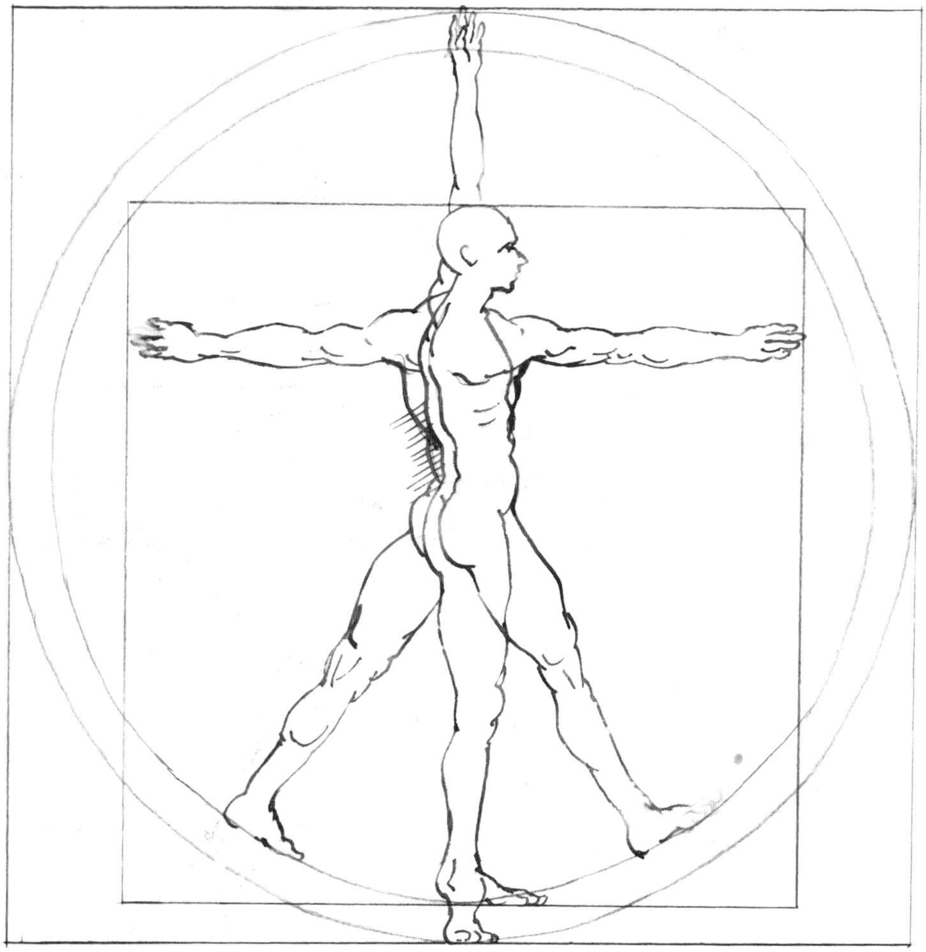

148. Rules of human proportion are not necessary when we see and remember our basic behavior.

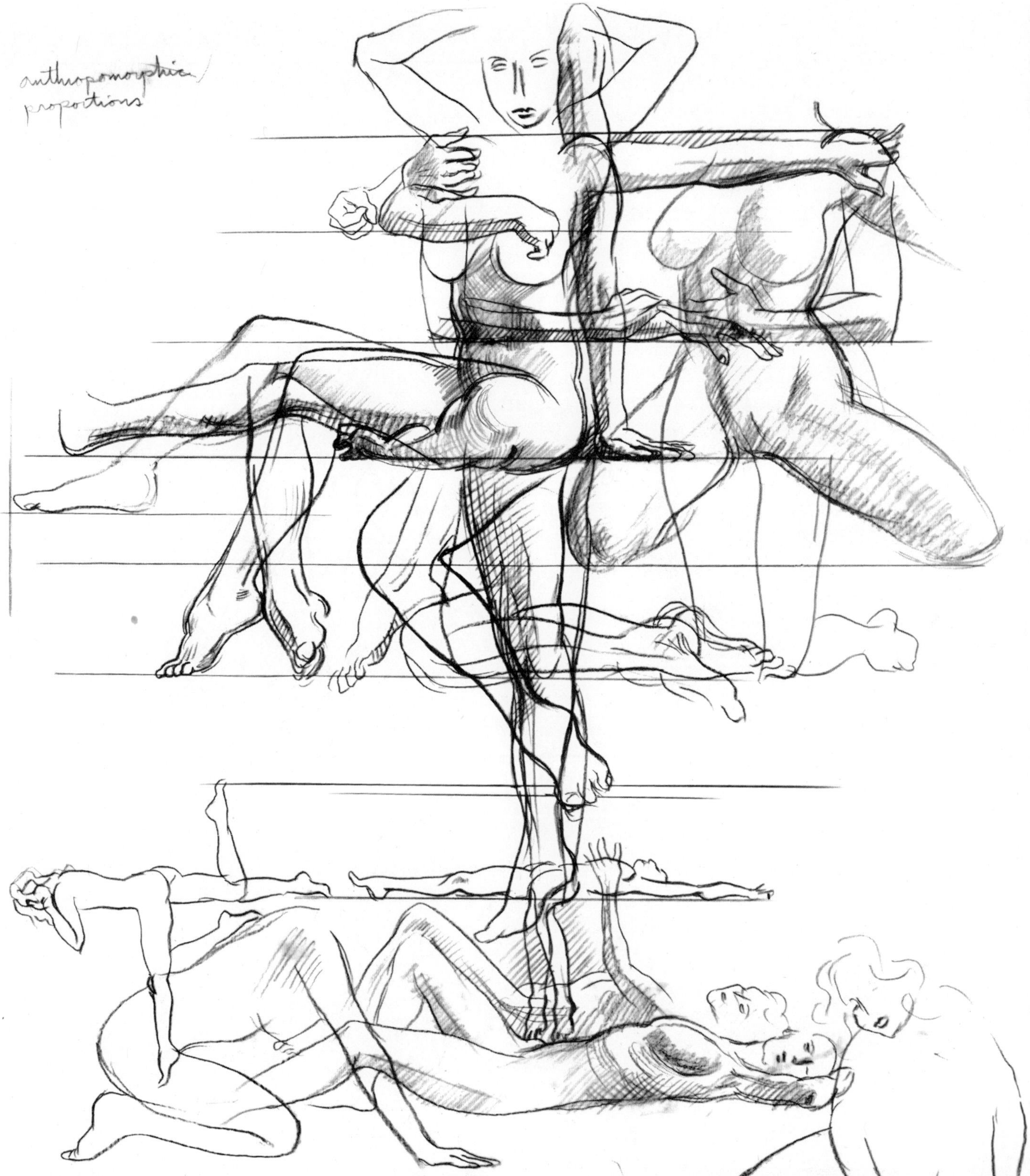

149. However, elegance is the dignity and grace which restrains the great wealth of our behavior.

150. Expression of the logic symbolized by woman is one of my basic intentions.

when squaring and simultaneously circling the navel,
The height of the head doesn't change,
standing on tiptoe and stretching upwards
raises the navel above the equilateral triangle
which violates da Vinci's canon
of spreading the legs to lower the height by 1/14.
da Vinci's Circle
Dike's Circle
1/14 ht.

151. The irrational proportions of the pentagon's harmony (golden ratio, divine proportion, etc.) are expressed by a woman giving birth.

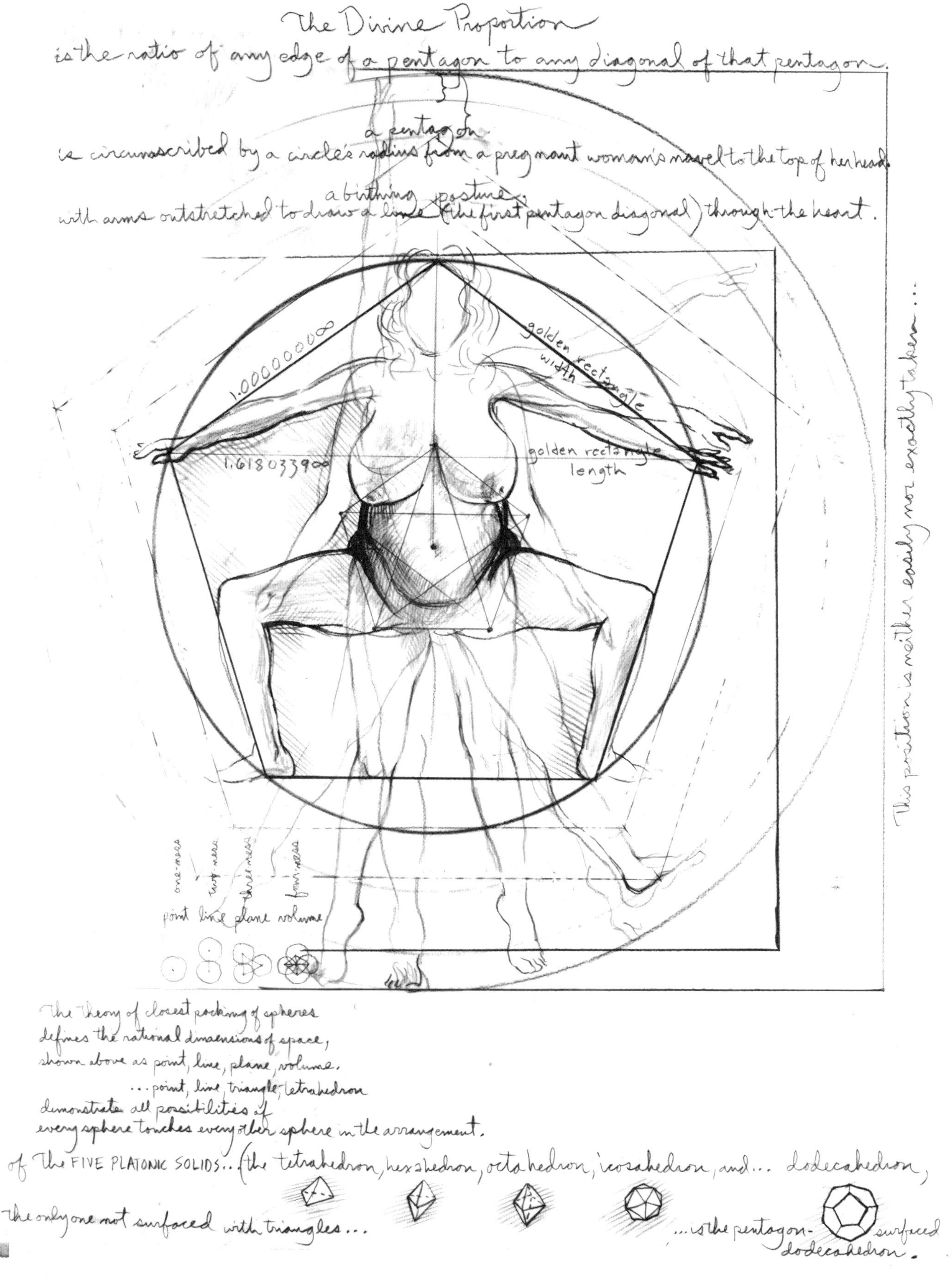
The Divine Proportion
is the ratio of any edge of a pentagon to any diagonal of that pentagon.
a pentagon
is circumscribed by a circle's radius from a pregnant woman's navel to the top of her head
a birthing posture
with arms outstretched to draw a line (the first pentagon diagonal) through the heart.
1.000000000
golden rectangle width
golden rectangle length
1.618033900
This position is neither easily nor exactly taken...
one-ness
two-ness
three-ness
four-ness
point line plane volume
The theory of closest packing of spheres
defines the rational dimensions of space,
shown above as point, line, plane, volume.
...point, line, triangle, tetrahedron
demonstrate all possibilities of
every sphere touches every other sphere in the arrangement.
of The FIVE PLATONIC SOLIDS... (the tetrahedron, hexahedron, octahedron, icosahedron, and... dodecahedron,
the only one not surfaced with triangles...
...is the pentagon-surfaced dodecahedron.

152. Le Corbusier's relaxed modular man was drawn to fit divine proportions rather than found in the proportions of actual man.

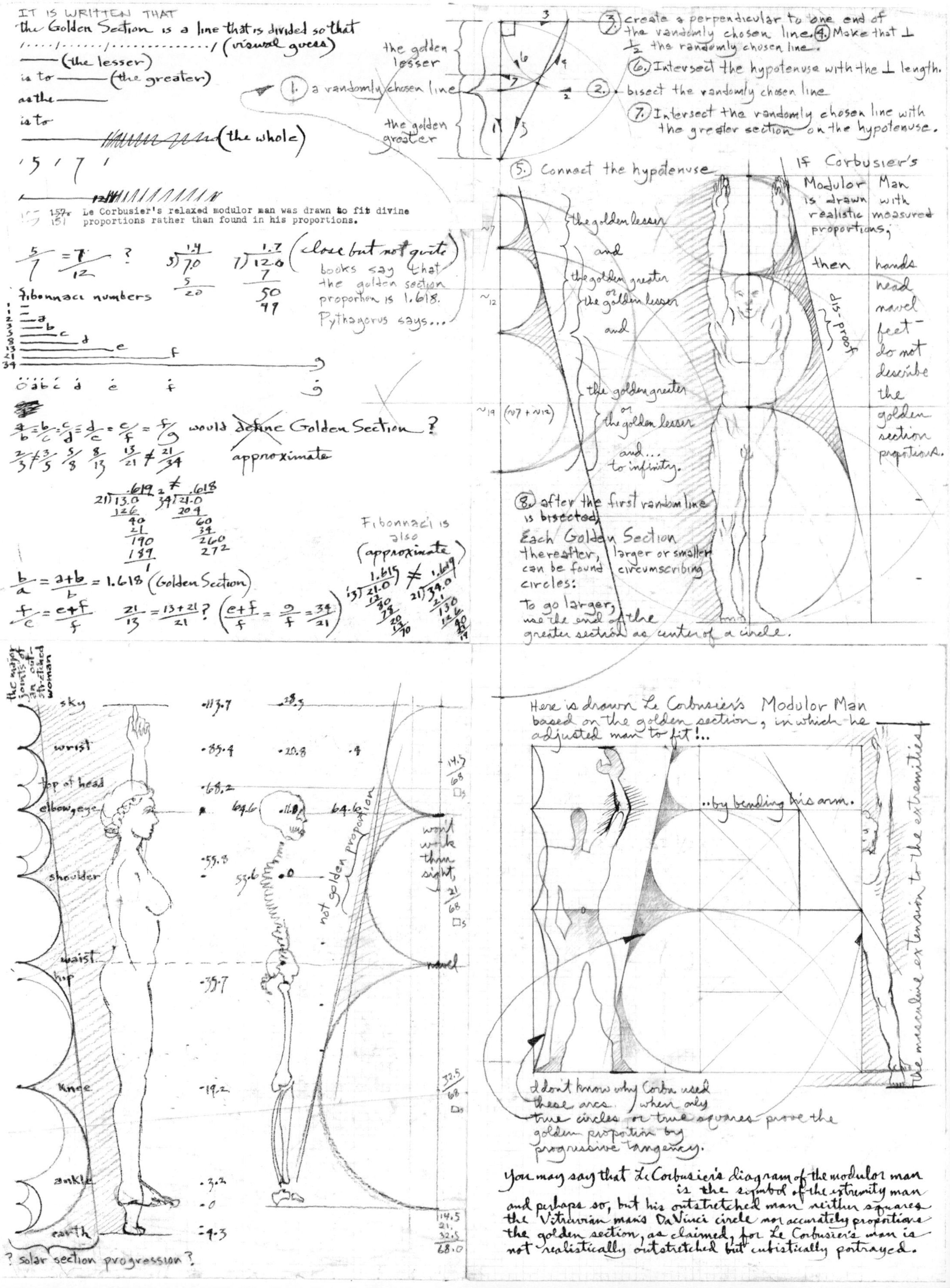

IT IS WRITTEN THAT
the Golden Section is a line that is divided so that
(visual guess)
(the lesser)
is to (the greater)
as the
is to
(the whole)
157. Le Corbusier's relaxed modulor man was drawn to fit divine proportions rather than found in his proportions.
(close but not quite)
books say that the golden section proportion is 1.618.
Pythagorus says...
Fibonnaci numbers
would define Golden Section?
approximate
Fibonnaci is also (approximate)
= 1.618 (Golden Section)
the golden lesser
the golden greater
1. a randomly chosen line
2. bisect the randomly chosen line
3. create a perpendicular to one end of the randomly chosen line. 4. Make that ⊥ ½ the randomly chosen line.
6. Intersect the hypotenuse with the ⊥ length.
7. Intersect the randomly chosen line with the greater section on the hypotenuse.
5. Connect the hypotenuse
the golden lesser
and
the golden greater or the golden lesser
and
the golden greater or the golden lesser
and... to infinity.
8. after the first random line is bisected, Each Golden Section thereafter, larger or smaller can be found circumscribing circles:
To go larger, use the end of the greater section as center of a circle.
If Corbusier's Modulor Man is drawn with realistic measured proportions; then hands head navel feet do not describe the golden section proportions.
dis-proof
the major joints of an out-stretched woman
sky
wrist
top of head
elbow, eye
shoulder
waist
hip
knee
ankle
earth
? solar section progression ?
not golden proportion
won't work thru sight,
navel
Here is drawn Le Corbusier's Modulor Man based on the golden section, in which he adjusted man to fit!..
..by bending his arm.
the masculine extension to the extremities
I don't know why Corbu used these arcs when only true circles or true squares prove the golden proportion by progressive tangency.
You may say that Le Corbusier's diagram of the modulor man is the symbol of the extremity man and perhaps so, but his outstretched man neither squares the Vitruvian man's DaVinci circle nor accurately proportions the golden section, as claimed, for Le Corbusier's man is not realistically outstretched but cubistically portrayed.

153. One day I found that a drawing of woman's extremities fit the divine proportions.

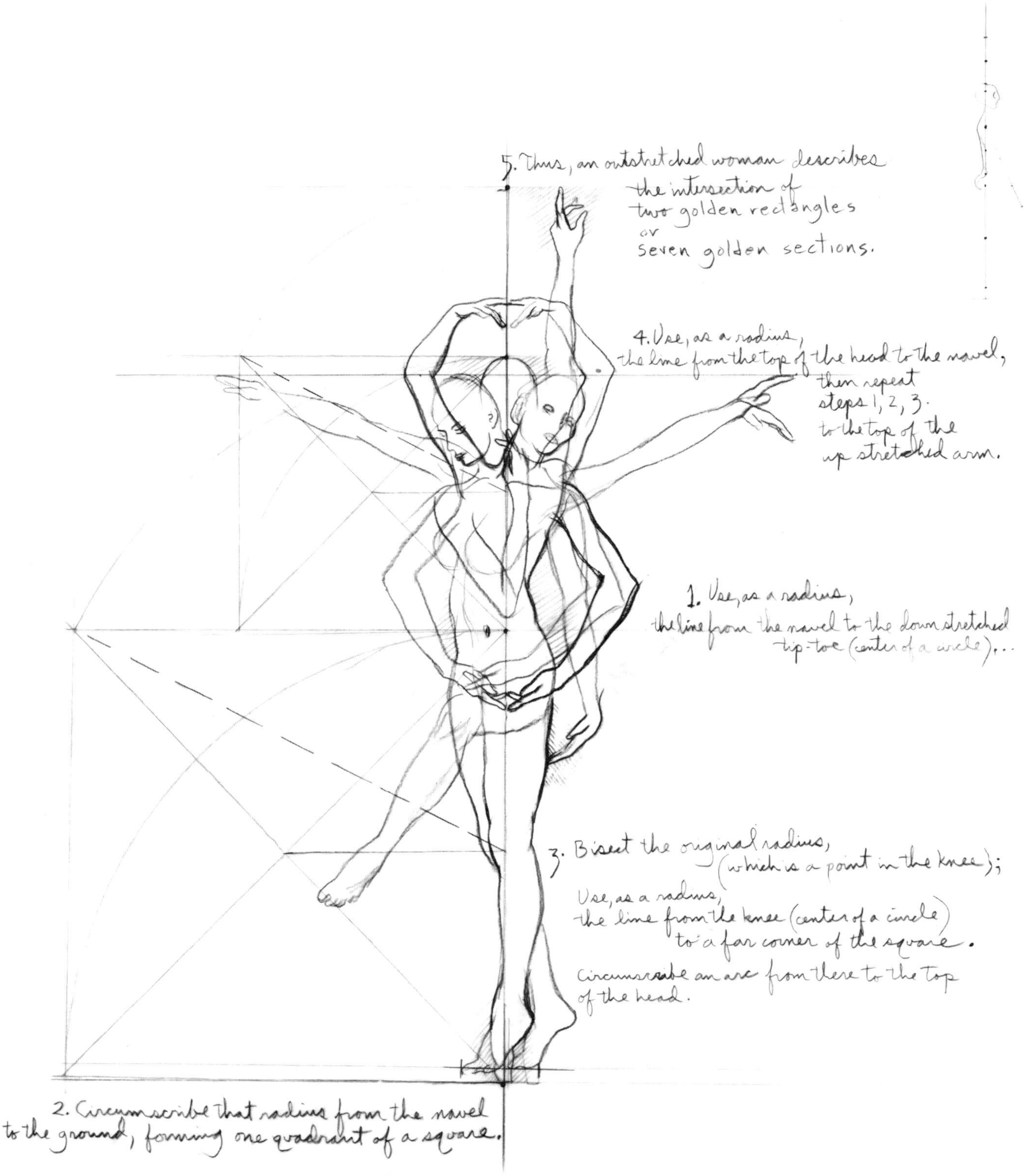

5. Thus, an outstretched woman describes the intersection of two golden rectangles or seven golden sections.
4. Use, as a radius, the line from the top of the head to the navel, then repeat steps 1, 2, 3. to the top of the up stretched arm.
1. Use, as a radius, the line from the navel to the down stretched tip-toe (center of a circle)...
3. Bisect the original radius, (which is a point in the knee); Use, as a radius, the line from the knee (center of a circle) to a far corner of the square. Circumscribe an arc from there to the top of the head.
2. Circumscribe that radius from the navel to the ground, forming one quadrant of a square.

154. This discovery only proved that my ideal fit the divine proportions.

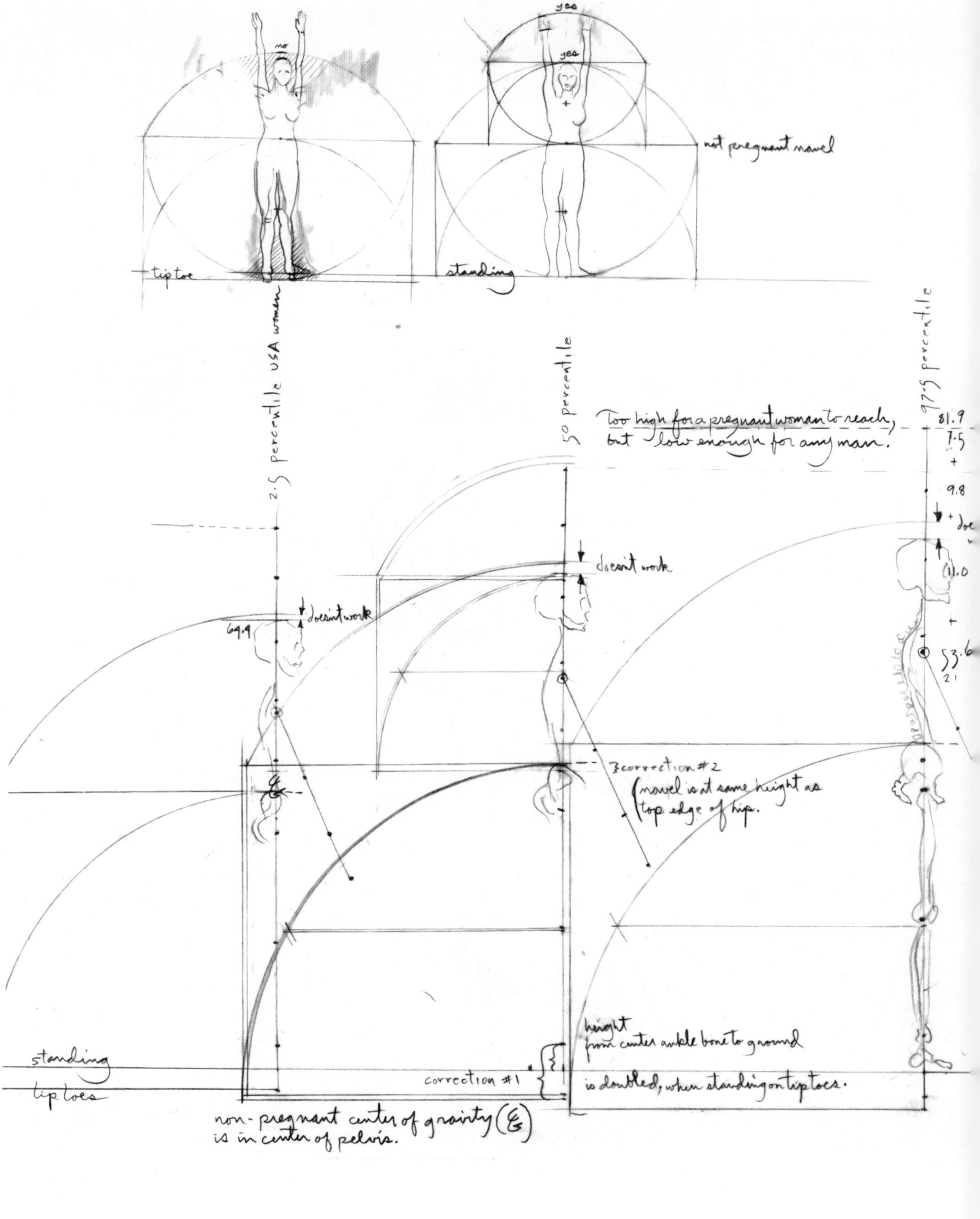

155. An anthropometric study of Americans was used to show that neither our male nor female extremities fit the divien proportion.

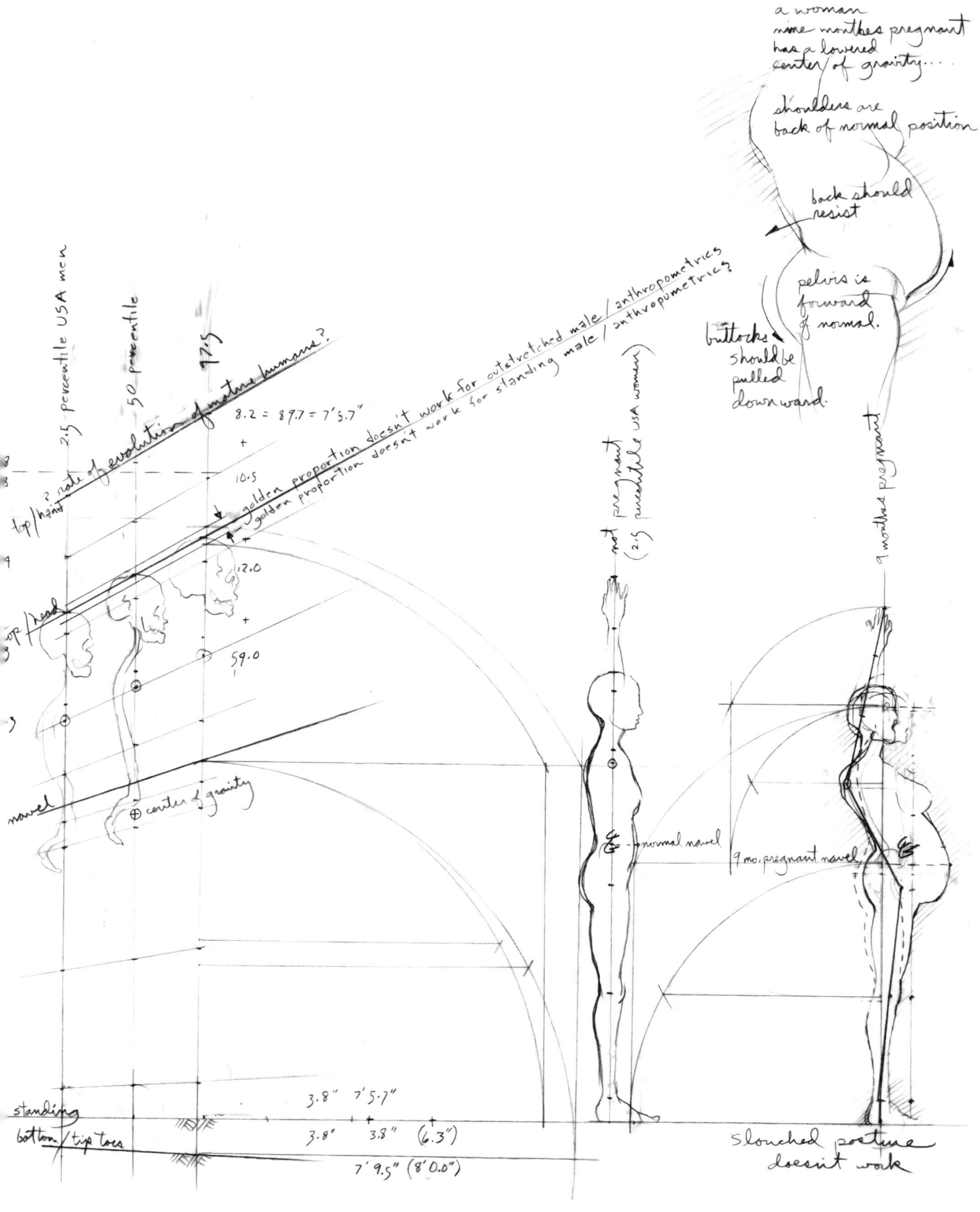

156. It then seemed that only pregnancy, which changes the position of the navel, might allow the divine proportion to be found in anthropometrics.

normal navel	Sag	Up	lower navel	Sag	up
flat feet / top head	no	no	flat feet / top head	n	no
" " / eye level	no	no	←	n	no
forehead	no	no	←	n	no
nose	no	n	←	n	no
tip toes / top head	no	n		n	no
eye level	no	n		n	no
f...	no	n		n	no

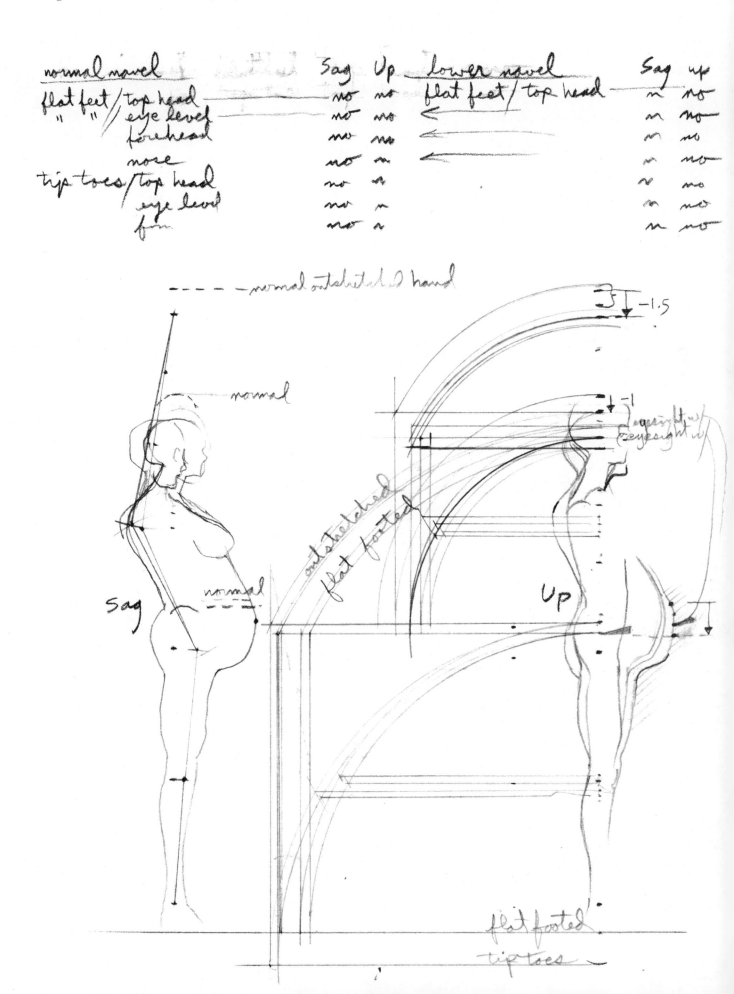

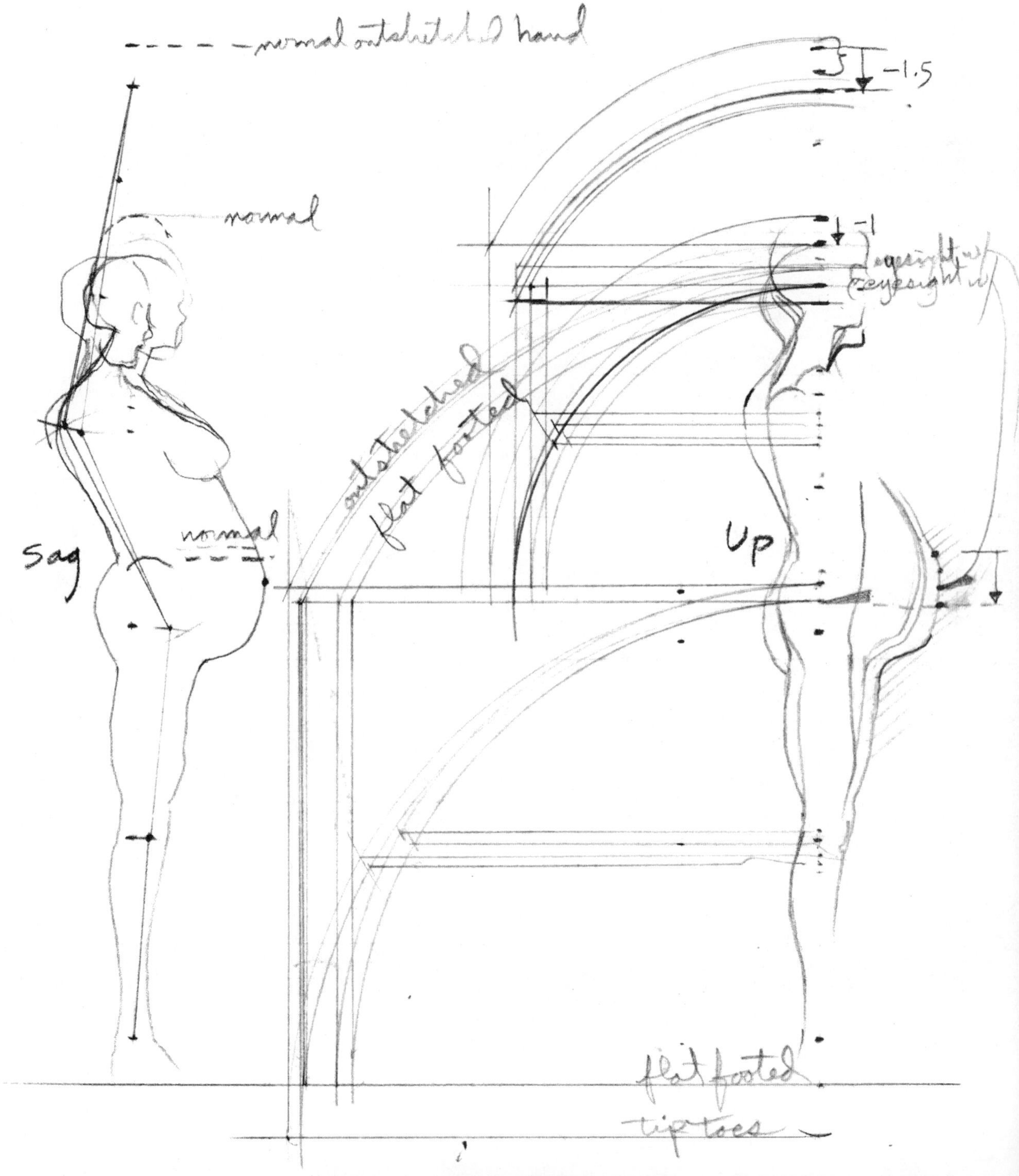

157. A lower than normal navel is bad posture and doesn't allow the divine proportion; good pregnant posture required new control of the natural postural reflex.

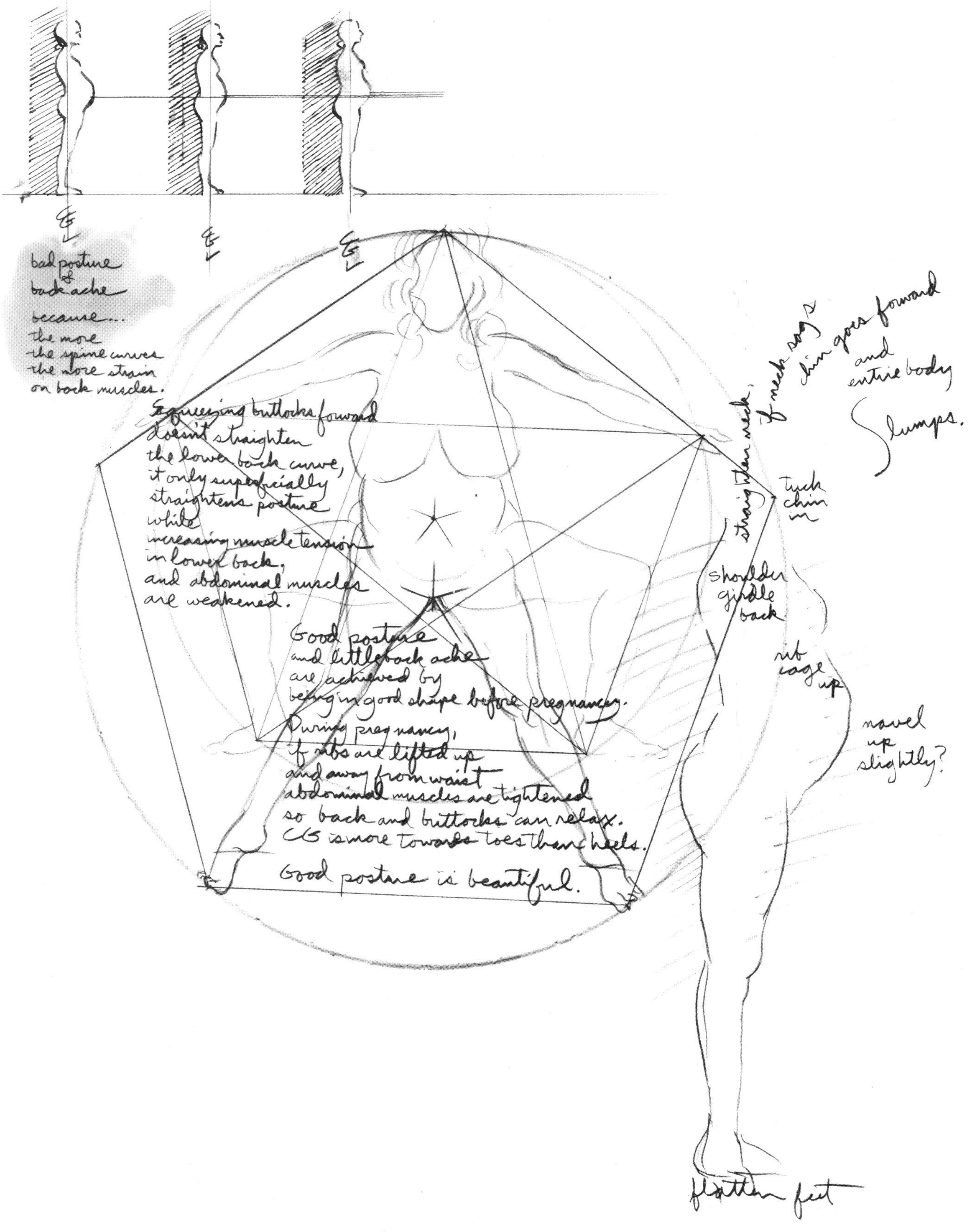

bad posture & back ache because... the more the spine curves the more strain on back muscles.
Squeezing buttocks forward doesn't straighten the lower back curve, it only superficially straightens posture while increasing muscle tension in lower back, and abdominal muscles are weakened.
Good posture and little back ache are achieved by being in good shape before pregnancy.
During pregnancy, if ribs are lifted up and away from waist abdominal muscles are tightened so back and buttocks can relax.
CG is more towards toes than heels.
Good posture is beautiful.
straighten neck, if neck sags chin goes forward and entire body Slumps.
tuck chin in
shoulder girdle back
rib cage up
navel up slightly?
flatten feet

158. The healthy posture of the fully pregnant estate of woman does find the divine proportion.

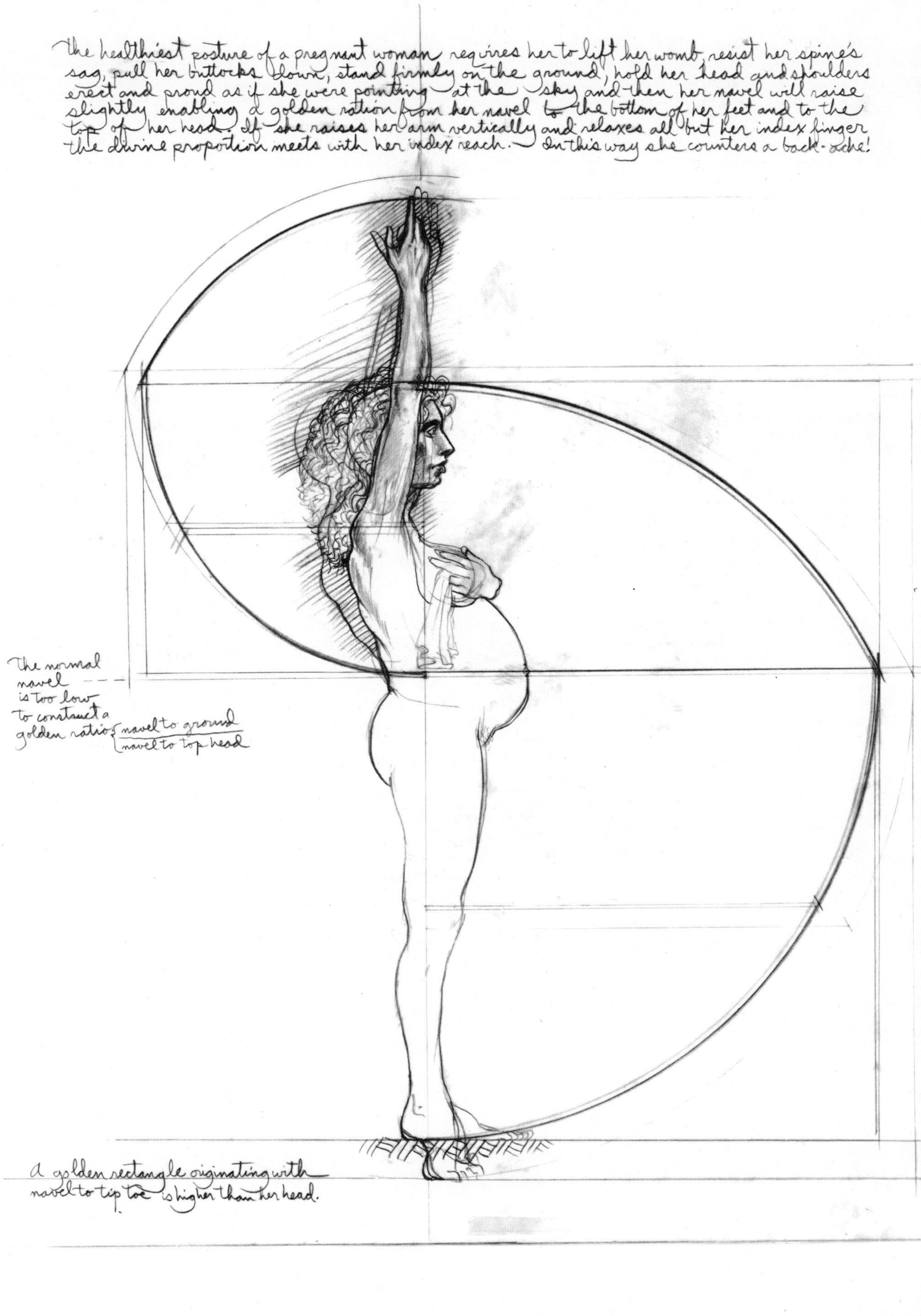
the healthiest posture of a pregnant woman requires her to lift her womb, resist her spine's sag, pull her buttocks down, stand firmly on the ground, hold her head and shoulders erect and proud as if she were pointing at the sky and then her navel will raise slightly enabling a golden ration from her navel to the bottom of her feet and to the top of her head. If she raises her arm vertically and relaxes all but her index finger the divine proportion meets with her index reach. In this way she counters a back-ache!
The normal navel is too low to construct a golden ratio: navel to ground / navel to top head
A golden rectangle originating with navel to tip toe is higher than her head.

159. This moment is the full term symbol. This full term moment is the symbol of the closest unity of man, woman, and child.

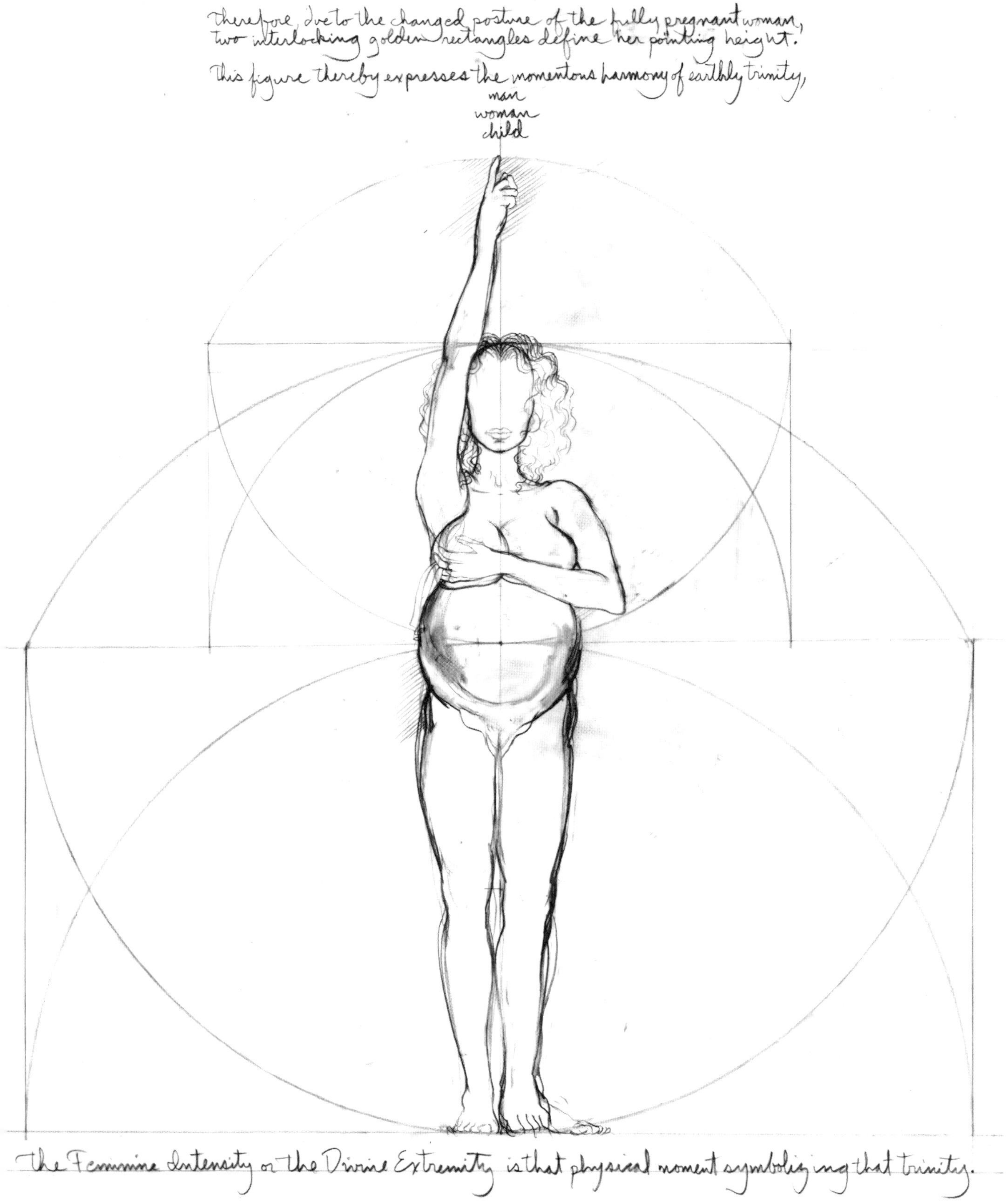

Therefore, due to the changed posture of the fully pregnant woman,
two interlocking golden rectangles define her pointing height.
This figure thereby expresses the momentous harmony of earthly trinity,
man
woman
child
The Feminine Intensity or the Divine Extremity is that physical moment symbolizing that trinity.

CONCLUSION

160. Significant relations of the minimums and maximums of different invisible architectural parts can be symbolized by finding their divine proportions.

Henry F. Arnold, in his book, "Trees in urban design" (Van Nostrand) p.128, states: for a large tree, 70' height, allow 4' soil depth, of a volume 21.3 cubic yards for the roots. (paraphrased from diagram 6-13)

The diagram at right shows a tree 70' height with 4' depth of soil.

$21.3 \times 27 = 1491 + 446 = 595.1$

$595.1 \div 36 = 16$ (36, 235, 216, 19)

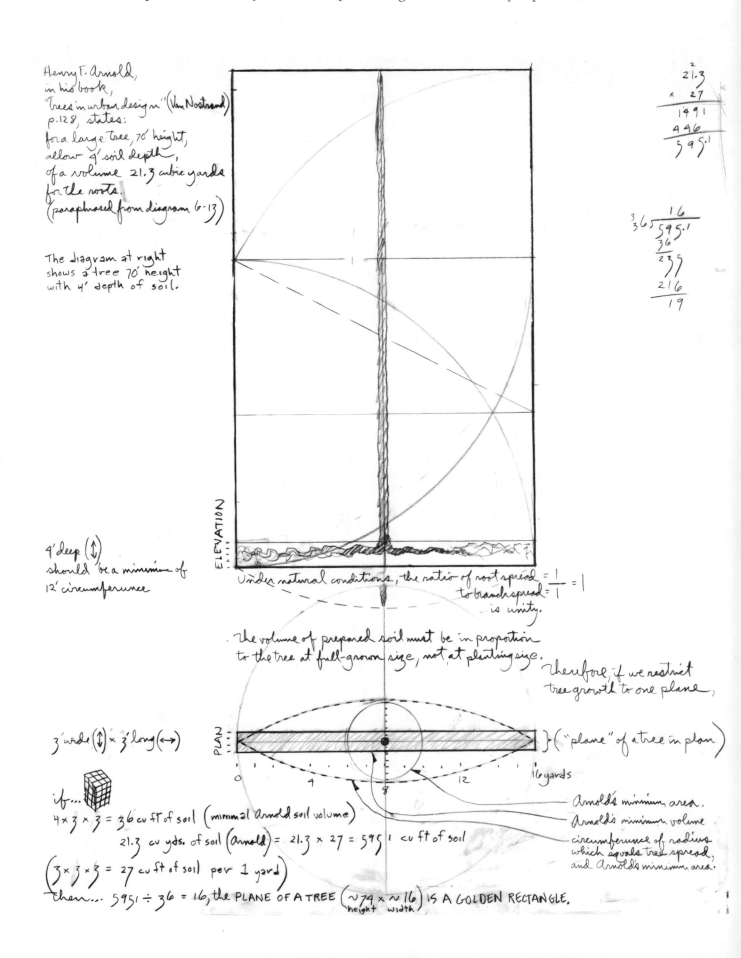

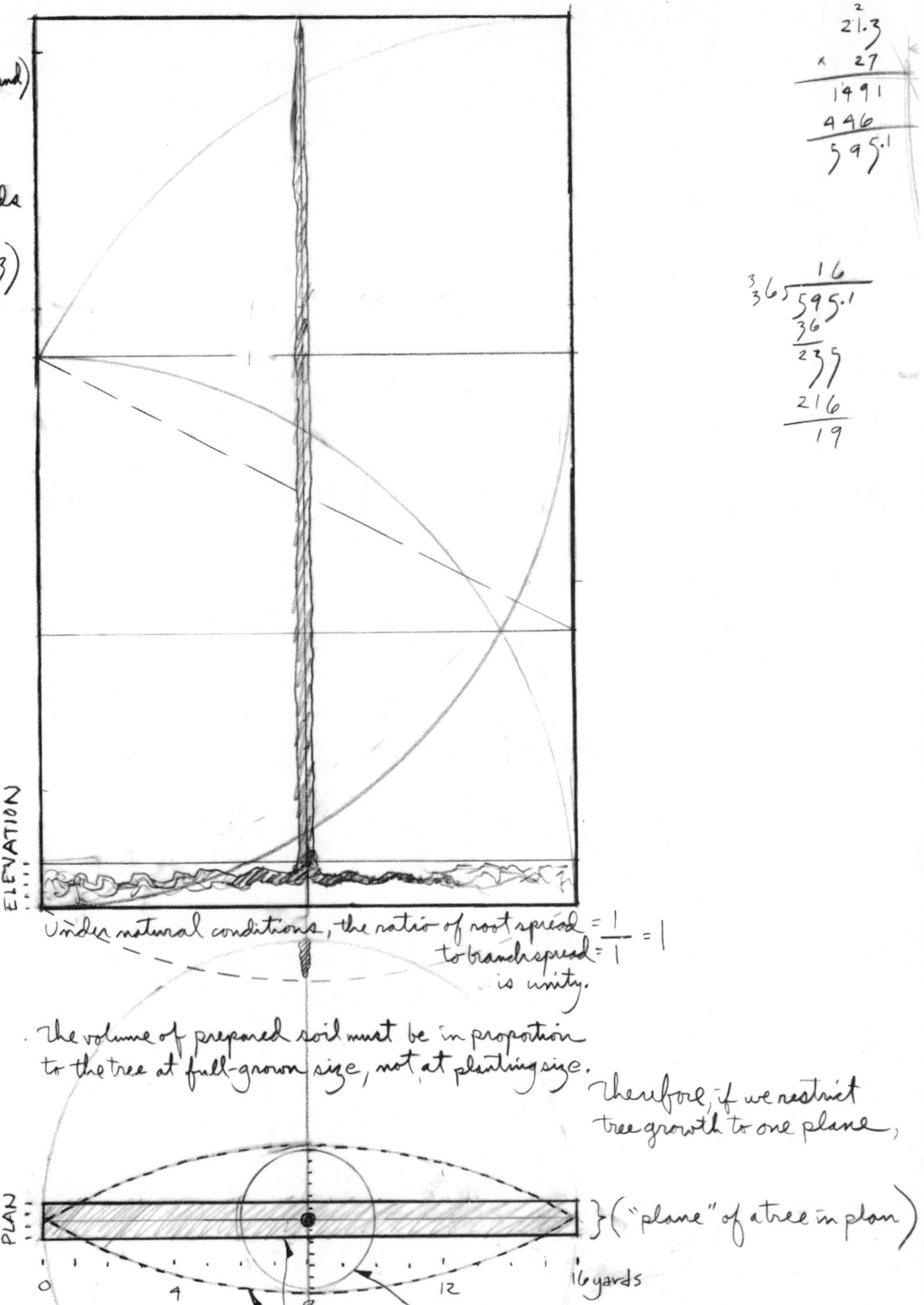

ELEVATION
Under natural conditions, the ratio of root spread to branch spread is unity. = 1/1 = 1
The volume of prepared soil must be in proportion to the tree at full-grown size, not at planting size.
Therefore, if we restrict tree growth to one plane,
PLAN
("plane" of a tree in plan)
0
4
8
12
16 yards
Arnold's minimum area.
Arnold's minimum volume
circumference of radius which equals tree spread, and Arnold's minimum area.

4' deep (↕) should be a minimum of 12' circumference

3' wide (↕) × 3' long (↔)

if...

$4 \times 3 \times 3 = 36$ cu ft of soil (minimal Arnold soil volume)

21.3 cu yds. of soil (Arnold) $= 21.3 \times 27 = 595.1$ cu ft of soil

$(3 \times 3 \times 3 = 27$ cu ft of soil per 1 yard$)$

then... $595.1 \div 36 = 16$, the PLANE OF A TREE ($\sim 74 \times \sim 16$ height width) IS A GOLDEN RECTANGLE.

161. For instance, a tree's height is in golden ratio with its roots' spread if the soil volume is arranged as the thickness of the tree's golden rectangle.

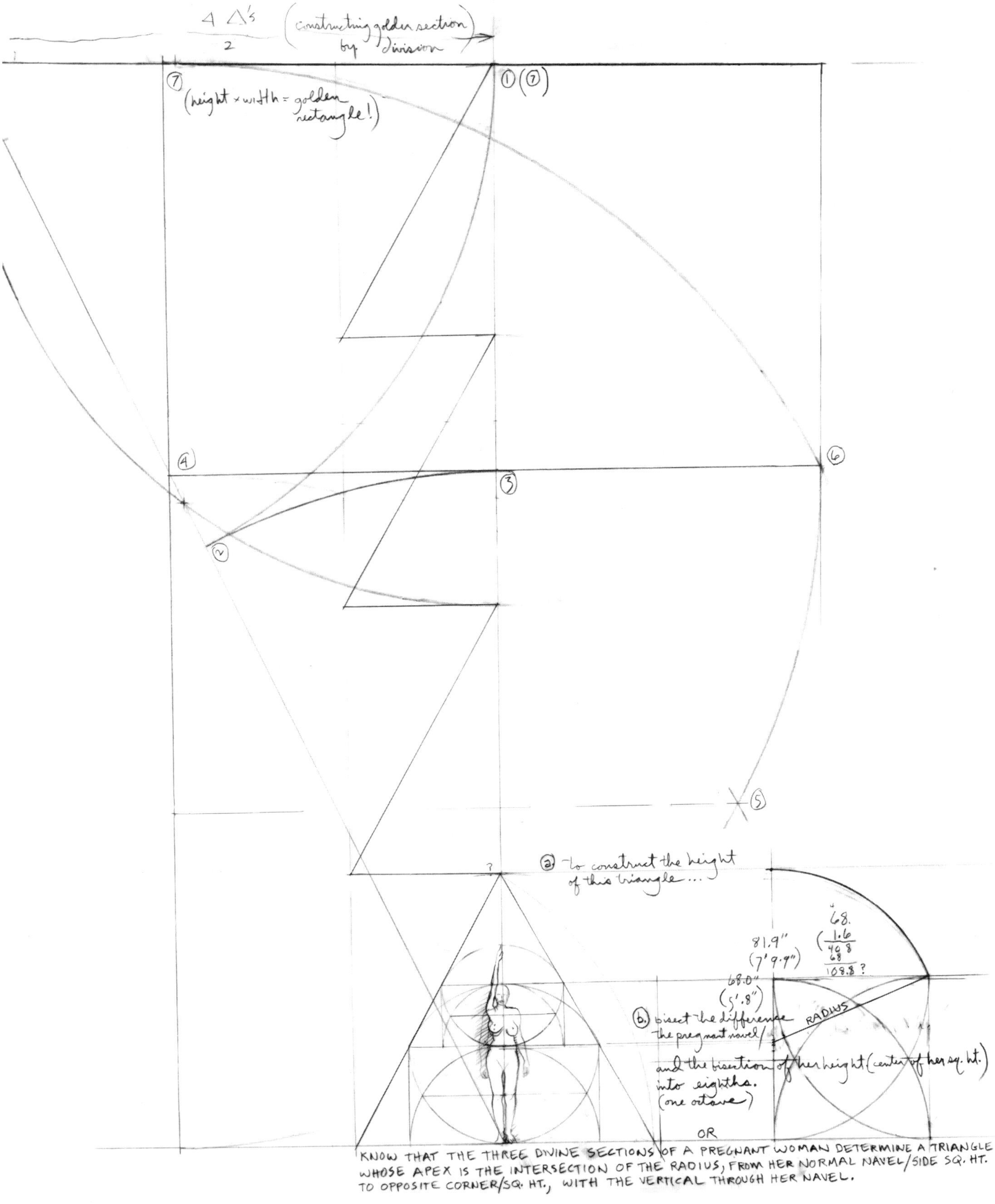

4 Δ's / 2
(constructing golden section by division)
①(⑦)
⑦ (height × width = golden rectangle!)
④
③
⑥
②
⑤
ⓐ to construct the height of this triangle...
81.9" (7'9.9")
68.0" (5'.8")
RADIUS
ⓑ bisect the difference the pregnant navel/ and the bisection of her height (center of her sq. ht.) into eighths. (one octave)
OR
KNOW THAT THE THREE DIVINE SECTIONS OF A PREGNANT WOMAN DETERMINE A TRIANGLE WHOSE APEX IS THE INTERSECTION OF THE RADIUS, FROM HER NORMAL NAVEL/SIDE SQ. HT. TO OPPOSITE CORNER/SQ. HT., WITH THE VERTICAL THROUGH HER NAVEL.

162. The average tree's height is four stories; in human extremity terms, four stories is four triangles high, each formed about the four golden rectangles about the human navel.

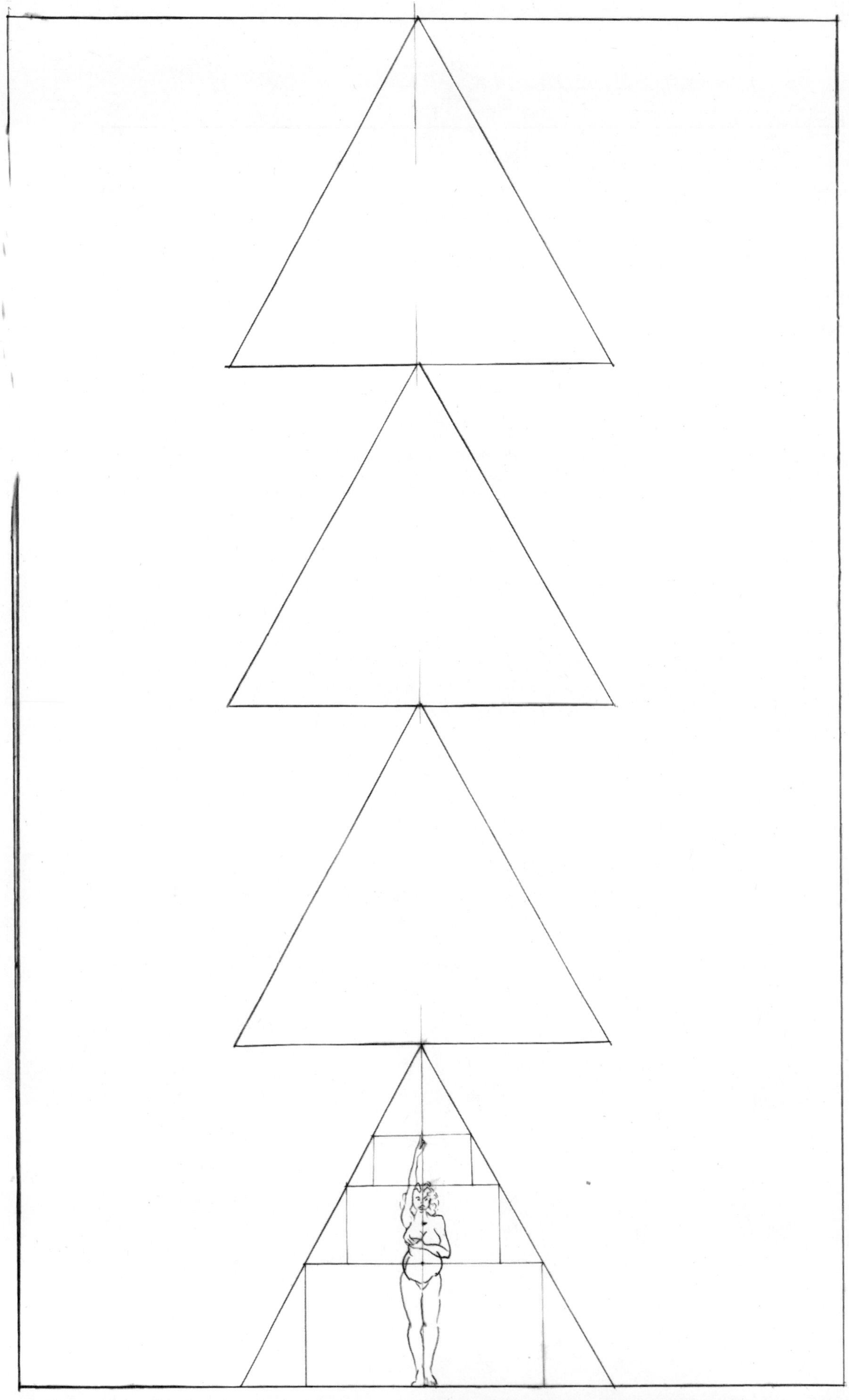

163. A medium sized tree's golden rectangle is based on the triangle base formed by the human's golden rectangles. A small sized tree's golden rectangle is based on the triangle base formed by the squares originating the human rectangles.

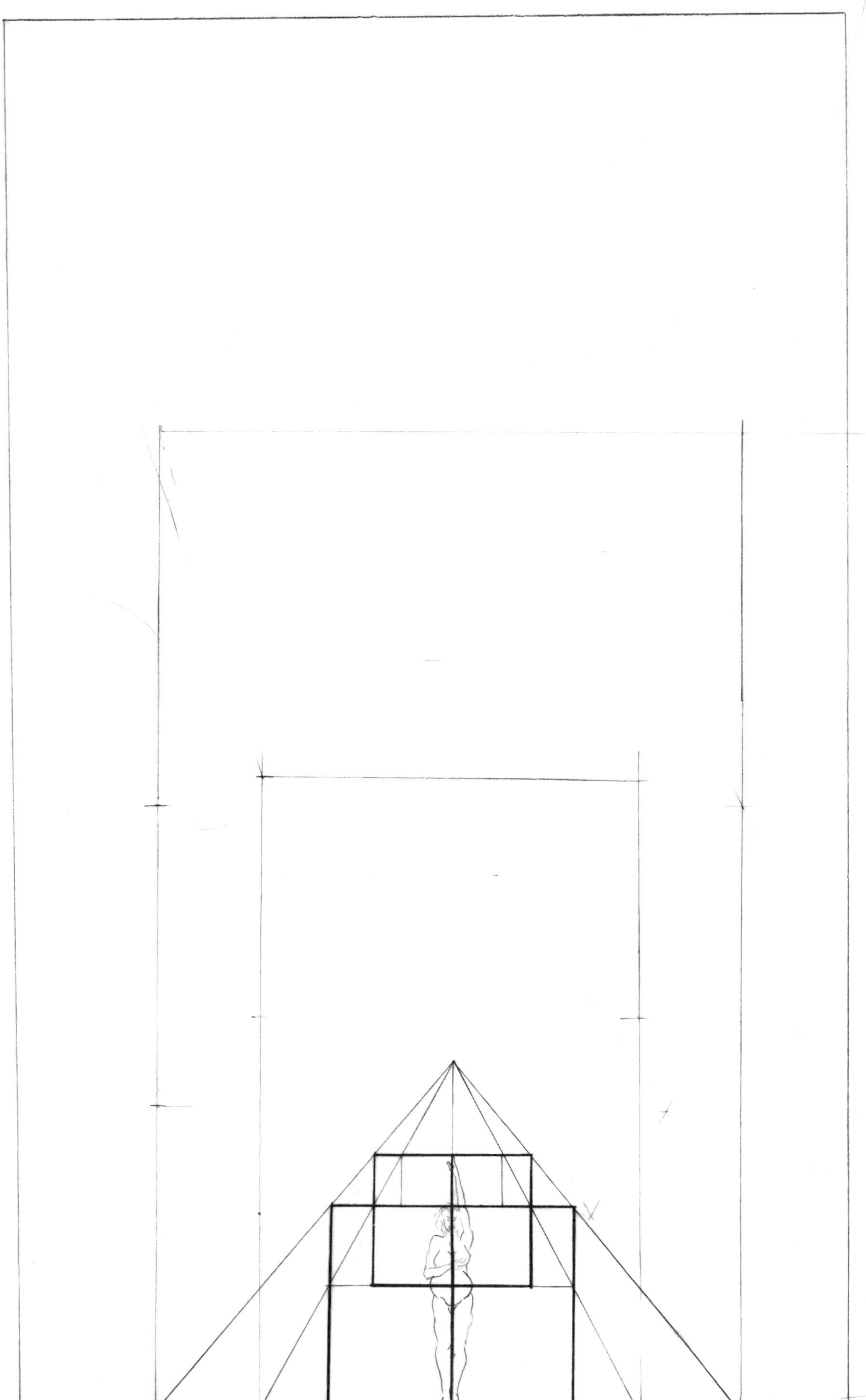

164. One eighth of the visual octave is determined by a golden rectangle the height of the viewer and the length of the intersection of the peripheral vision cone with the ground. One quarter of the visual octave is confirmed by the focal cone's intersection with the ground. One half is twice one fourth, etc.

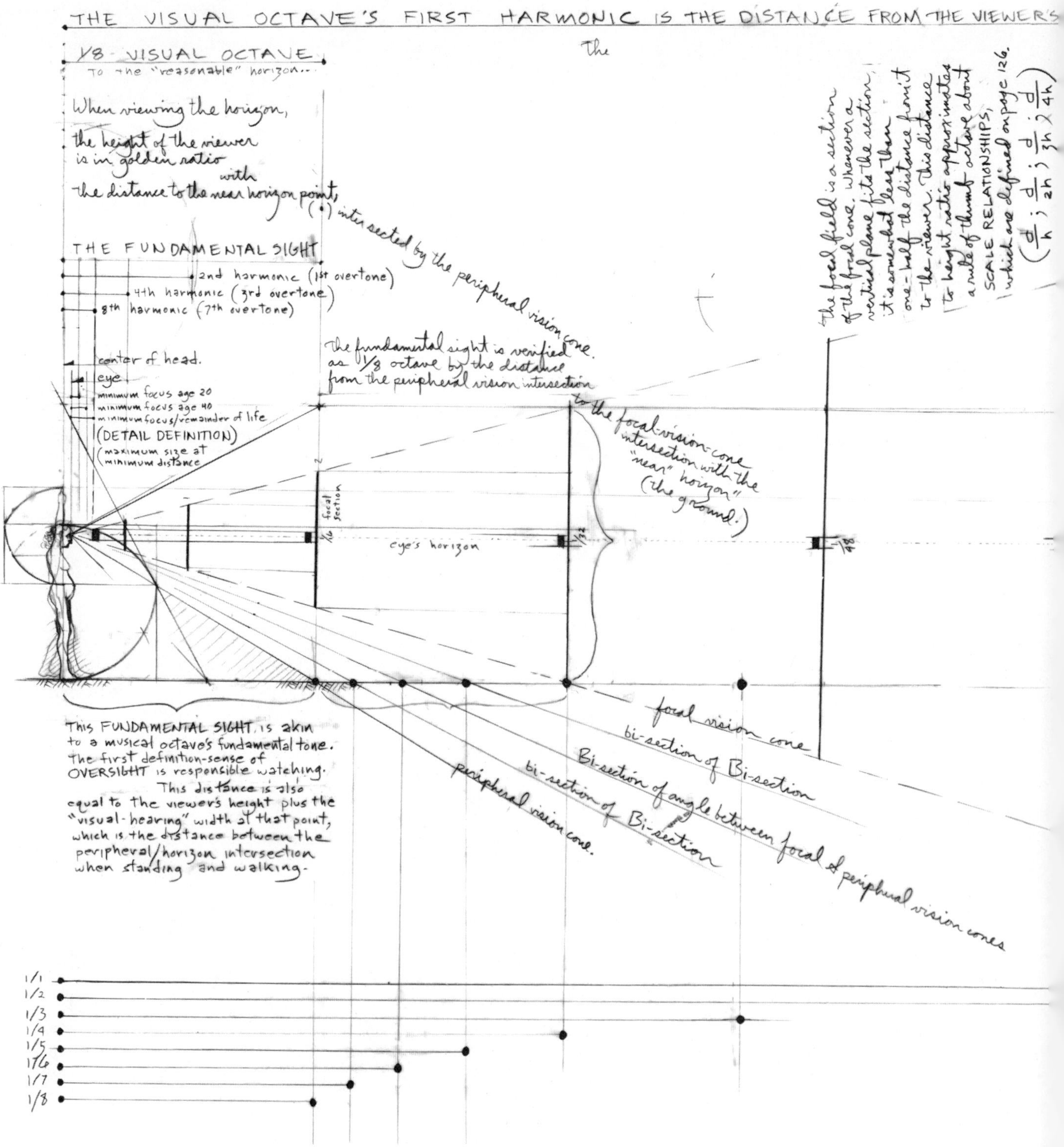

THE VISUAL OCTAVE'S FIRST HARMONIC IS THE DISTANCE FROM THE VIEWER'S
The
1/8 VISUAL OCTAVE
To the "reasonable" horizon..
When viewing the horizon, the height of the viewer is in golden ratio with the distance to the near horizon point, (•) intersected by the peripheral vision cone.
THE FUNDAMENTAL SIGHT
2nd harmonic (1st overtone)
4th harmonic (3rd overtone)
8th harmonic (7th overtone)
center of head.
eye
minimum focus age 20
minimum focus age 40
minimum focus/remainder of life
(DETAIL DEFINITION)
(maximum size at minimum distance
The fundamental sight is verified as 1/8 octave by the distance from the peripheral vision intersection to the focal-vision-cone intersection with the "near" horizon (the ground.)
focal section
eye's horizon
The focal field is a section of the focal cone whenever a vertical plane fits the section; it is somewhat less than one-half the distance from it to the viewer. This distance to height ratio approximates a rule of thumb octave about SCALE RELATIONSHIPS, which are defined on page 126.
(d/h ; d/2h ; d/3h ; d/4h)
This FUNDAMENTAL SIGHT is akin to a musical octave's fundamental tone. The first definition-sense of OVERSIGHT is responsible watching. This distance is also equal to the viewer's height plus the "visual-hearing" width at that point, which is the distance between the peripheral/horizon intersection when standing and walking.
focal vision cone
bi-section of Bi-section
Bi-section of angle between focal & peripheral vision cones
bi-section of Bi-section
peripheral vision cone.
1/1
1/2
1/3
1/4
1/5
1/6
1/7
1/8

165. One sixth of the visual octave is the bisection of the angle between the peripheral and the focal cones with the ground. One fifth and one seventh are the bisections of the previous bisection with the ground.

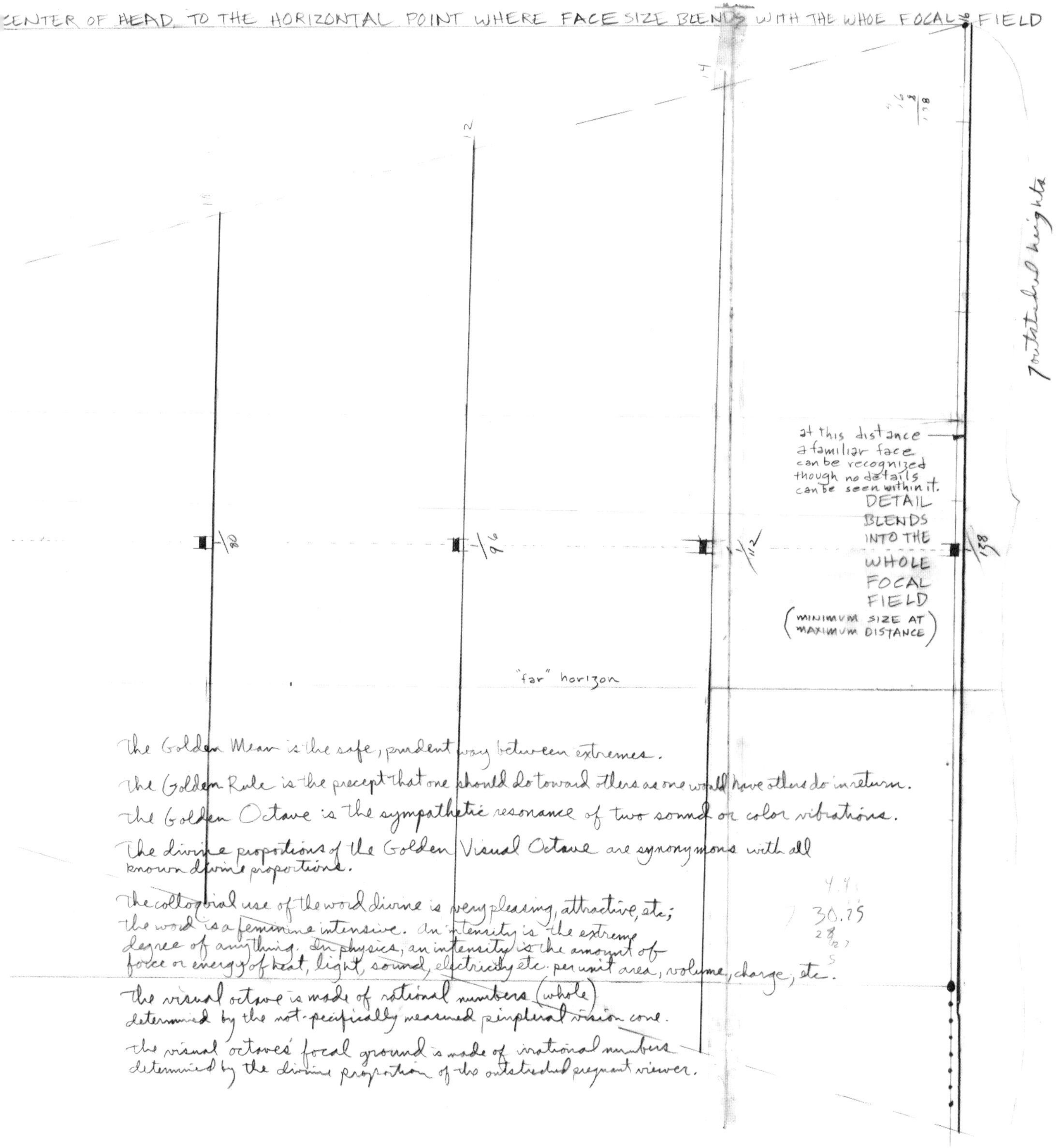
CENTER OF HEAD TO THE HORIZONTAL POINT WHERE FACE SIZE BLENDS WITH THE WHOE FOCAL FIELD
at this distance a familiar face can be recognized though no details can be seen within it.
DETAIL BLENDS INTO THE WHOLE FOCAL FIELD
(MINIMUM SIZE AT MAXIMUM DISTANCE)
7 outstretched heights
"far" horizon
The Golden Mean is the safe, prudent way between extremes.
The Golden Rule is the precept that one should do toward others as one would have others do in return.
the Golden Octave is the sympathetic resonance of two sound or color vibrations.
The divine proportions of the Golden Visual Octave are synonymous with all known divine proportions.
The colloquial use of the word divine is very pleasing, attractive, etc;
the word is a feminine intensive. An intensity is the extreme degree of anything. In physics, an intensity is the amount of force or energy of heat, light, sound, electricity etc. per unit area, volume, charge, etc.
the visual octave is made of rational numbers (whole) determined by the not-specifically measured peripheral vision cone.
the visual octaves' focal ground is made of irrational numbers determined by the divine proportion of the outstretched pregnant viewer.

166. The Golden Visual Octave's fundamental tone is an octave itself, ranging from 1/8 of the closest focal distance to 1/8 of the distance where that detail blends with the whole focal field, whose diameter is four human triangles high, the height of a mature tree's golden rectangle and a fair approximation of the average townhouse.

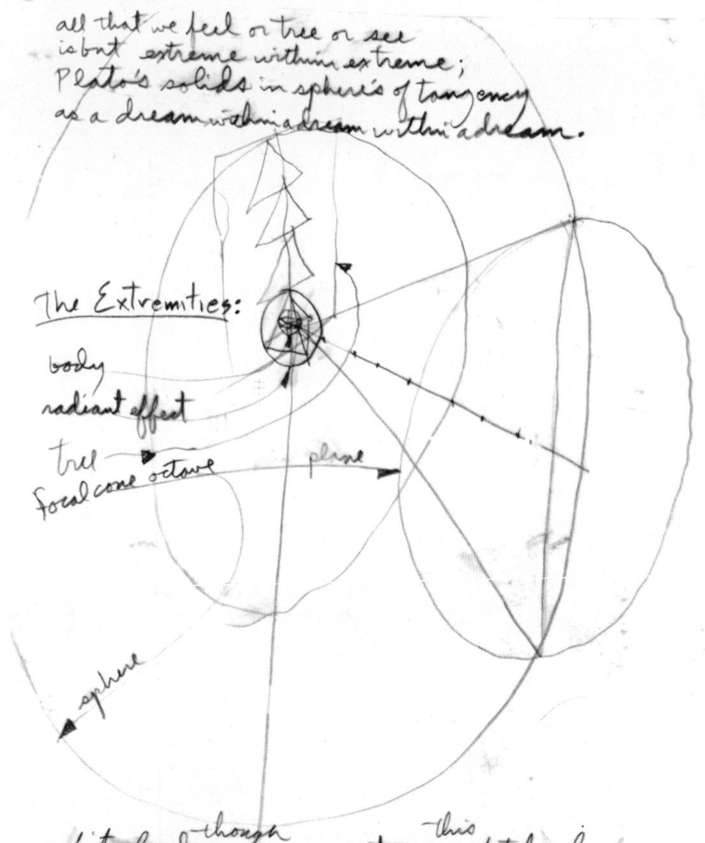

all that we feel or tree or see
is but extreme within extreme;
Plato's solids in sphere's of tangency
as a dream within a dream within a dream.
The Extremities:
body
radiant effect
tree
focal cone octave
plane
sphere

TO CONSTRUCT
THE GOLDEN VISUAL OCTAVE

1. first draw a profile of the viewer, (the Feminine Extremity, the Pregnant Intensity, man, woman, child Unity, two golden rectangles, p. 159)
2. Construct a square into the viewer's view, using the viewer's head height as one side.
3. Construct a golden rectangle further into the viewer's view, extending the square so that its height is the golden rectangle's length (into the view). This is the first golden rectangle and first interval of the visual octave. (1/8)
4. Scribe the second golden rectangle's length upon the ground. This is where the focal cone intersects the ground (1/4)
5. Bisect the angle between the peripheral cone and the focal cone. (1/6 ∴ 1/3)
6. Bisect that bisection on either side and you will approximate 1/5 and 1/7
7. Construct a golden rectangle at the 8th interval:

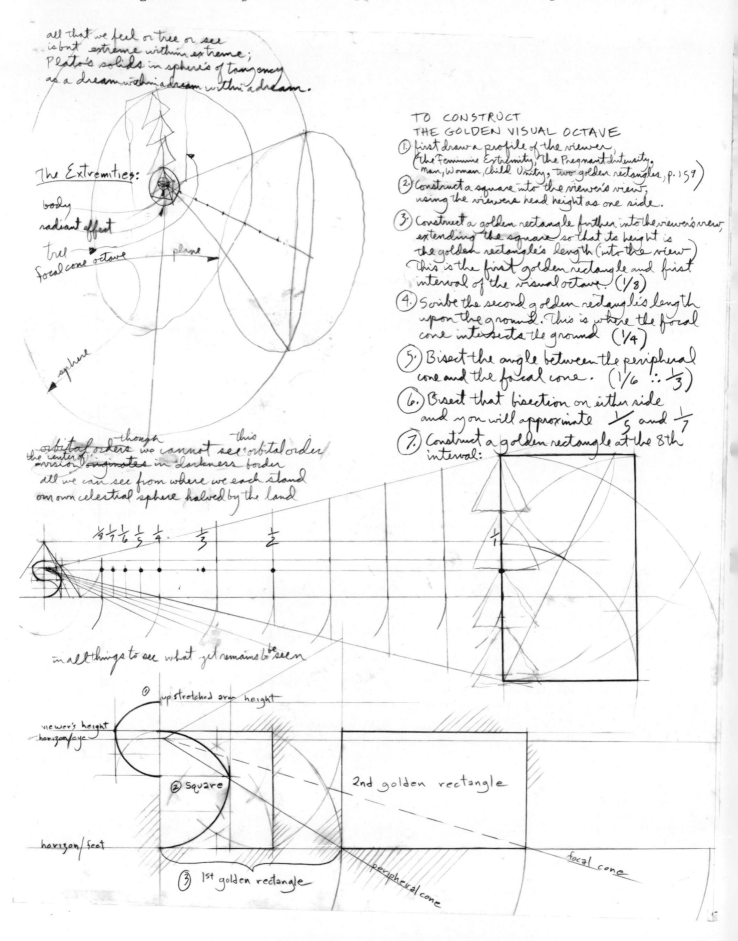

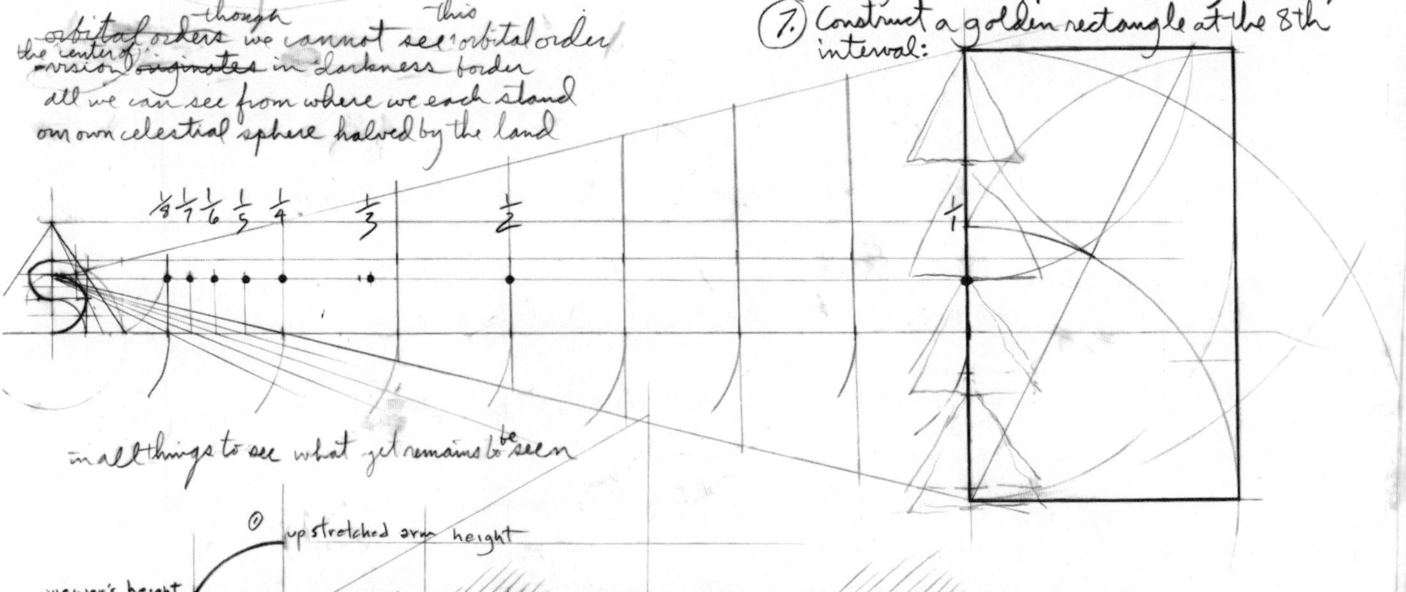

orbital orders though we cannot see this orbital order
the center of vision originates in darkness border
all we can see from where we each stand
our own celestial sphere halved by the land
1/8 1/7 1/6 1/5 1/4
1/3
1/2
1
in all things to see what yet remains to be seen

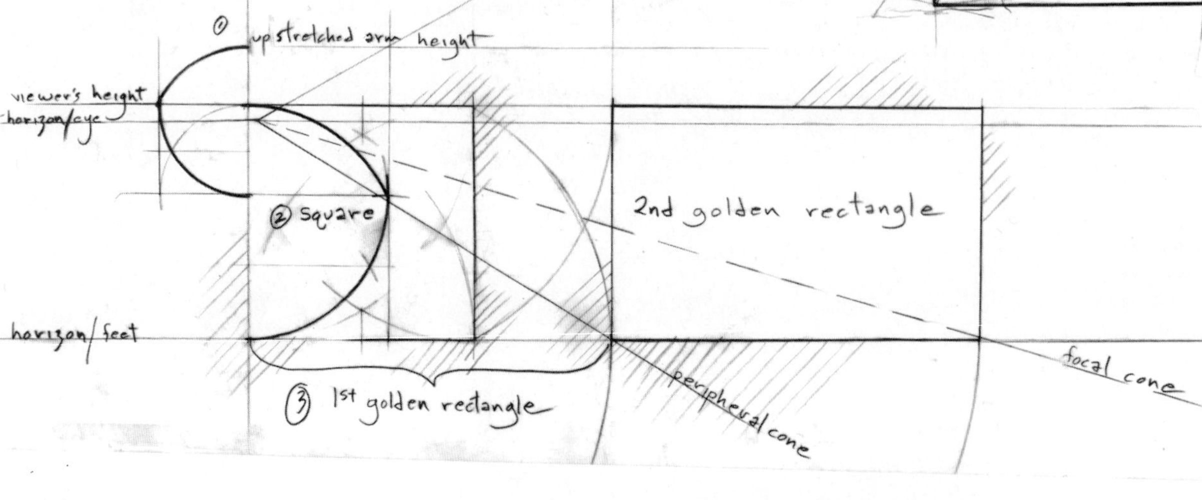

① upstretched arm height
viewer's height
horizon/eye
② square
2nd golden rectangle
horizon/feet
③ 1st golden rectangle
peripheral cone
focal cone

167. Everything in the world fits. Design it all together at once, and living order goes; relate a little here and a little there, a little now and a little then, and living order comes.

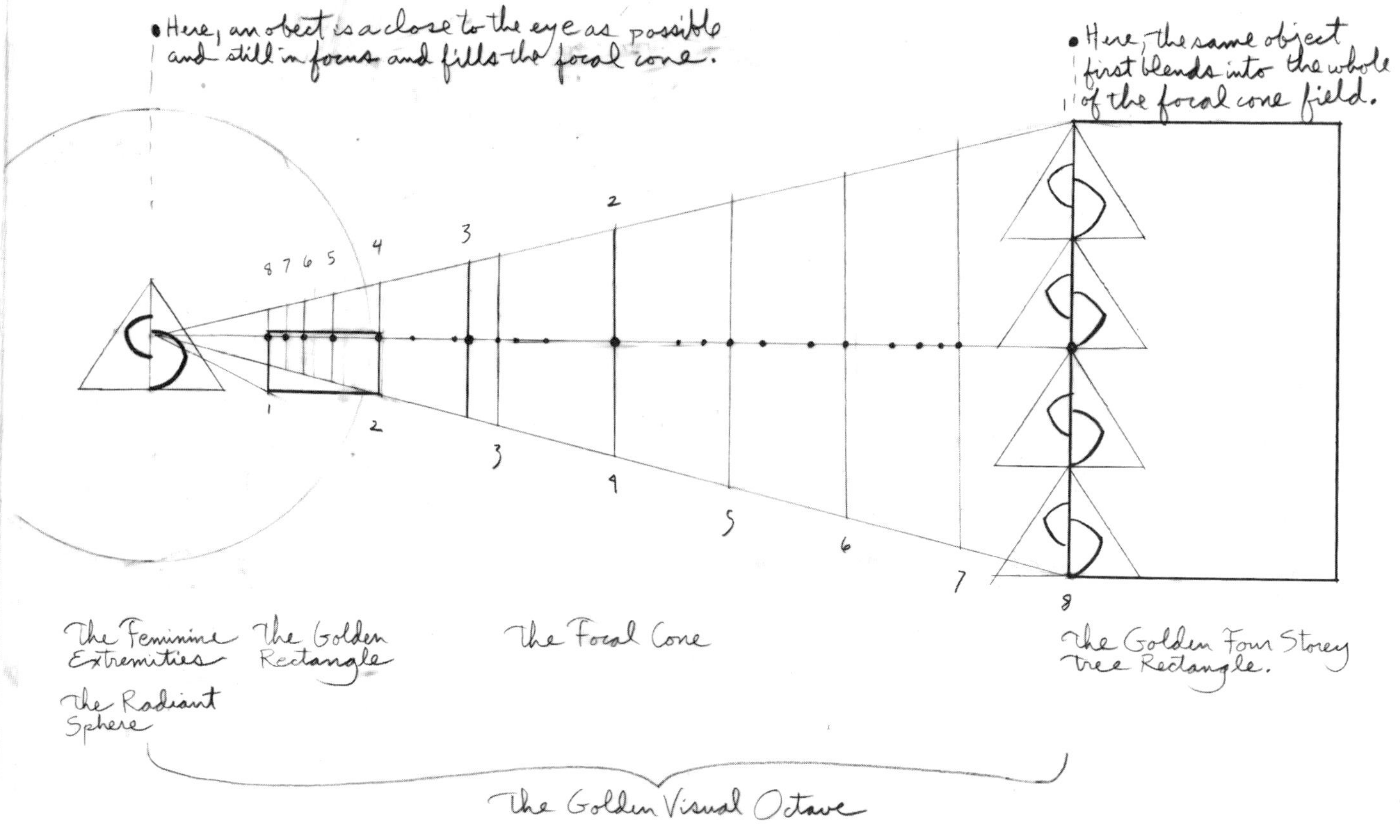

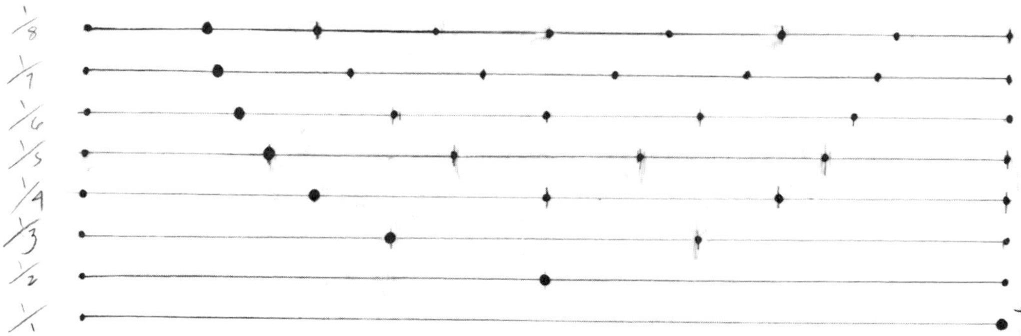

Just as Poetry can approximate music, we can continually approach a correct result, as in mathematics by successive approximates we solve an equation, this symbol is approximate.

The visual octave is analagous to the musical octave which is a musical interval embracing eight diatonic degrees.

An octave is the harmonic conbination of two tones eight intervals apart.

The Architecturally useful value of this Golden Visual Octave is that it is nearly, but not exactly correct, not dogmatic, yet bringing together.

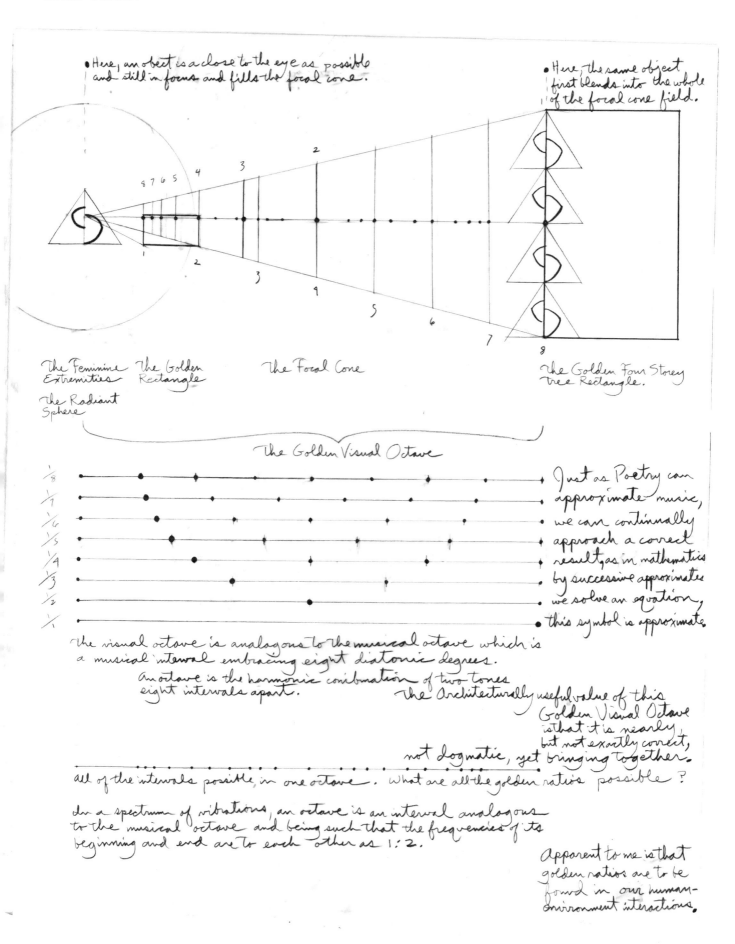

all of the intervals possible in one octave. What are all the golden ratios possible?

In a spectrum of vibrations, an octave is an interval analagous to the musical octave and being such that the frequencies of its beginning and end are to each other as 1:2.

Apparent to me is that golden ratios are to be found in our human-environment interactions.

168. We are shadowcasters. Our focal and peripheral cones intersect our shadowlabe and fix sixteen shadow lengths in divine proportion to our height, eight to our left and eight to our right.

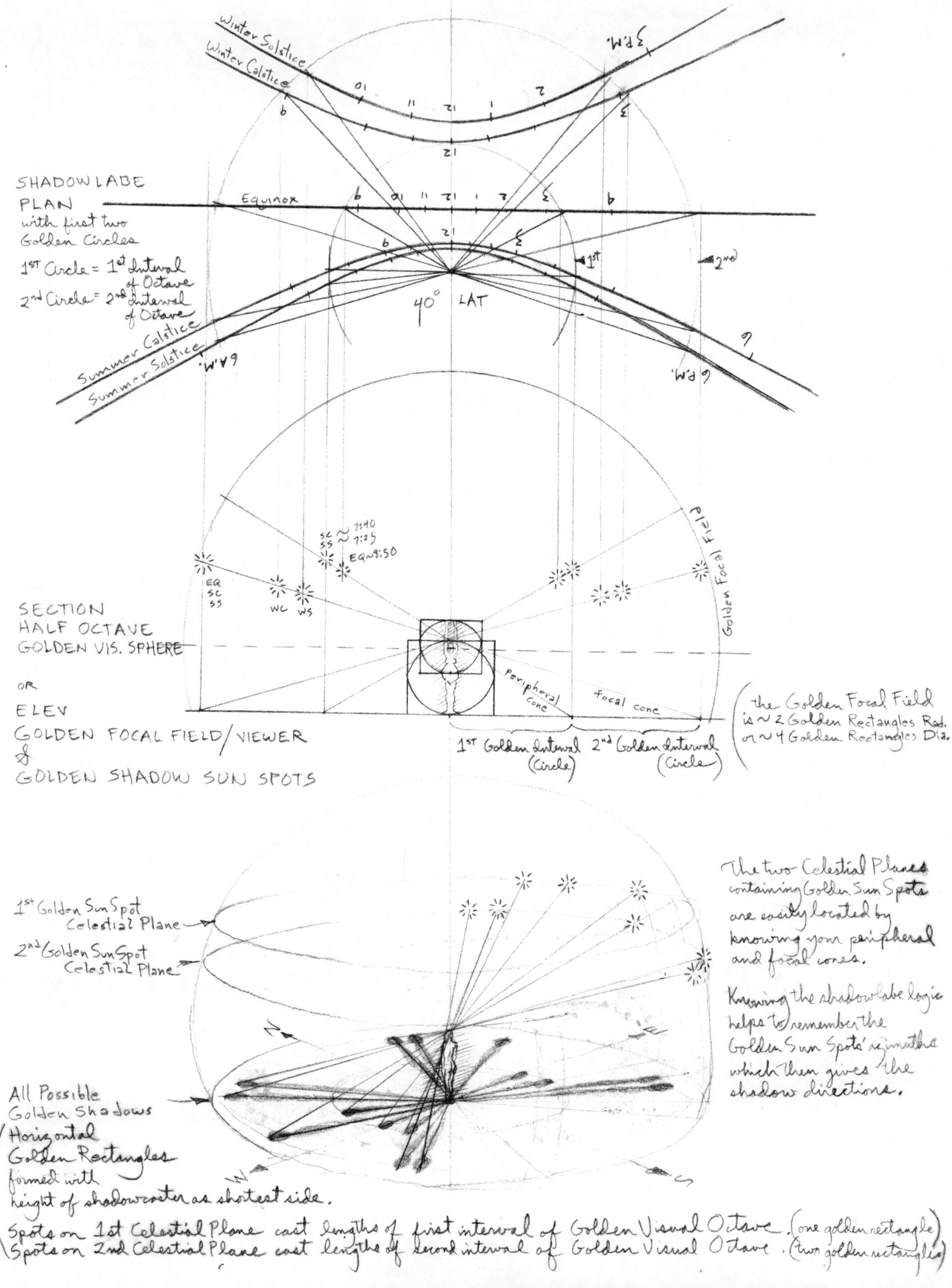

169. These divine shadow fans are formed of different azimuth combinations for each latitude. We need only remember our latitude to know the sun on a site by our self.

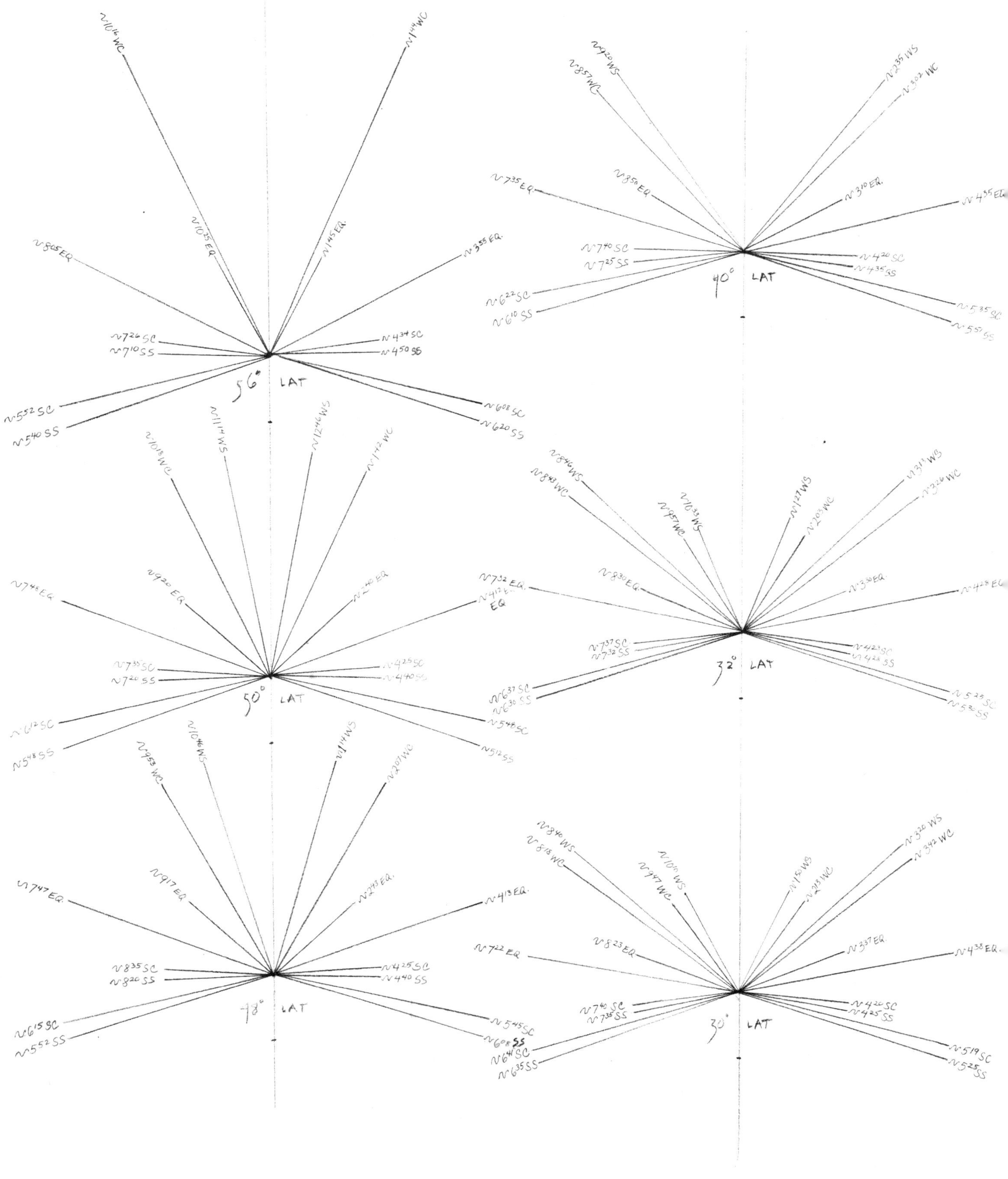

170. The side view of the Golden Visual Octave teaches the relation of the focal and peripheral cones to the intervals of the octave and the ranges of branches and roots. The plan view teaches relations of our golden shadows to our radiant comfort to trees' canopy range.

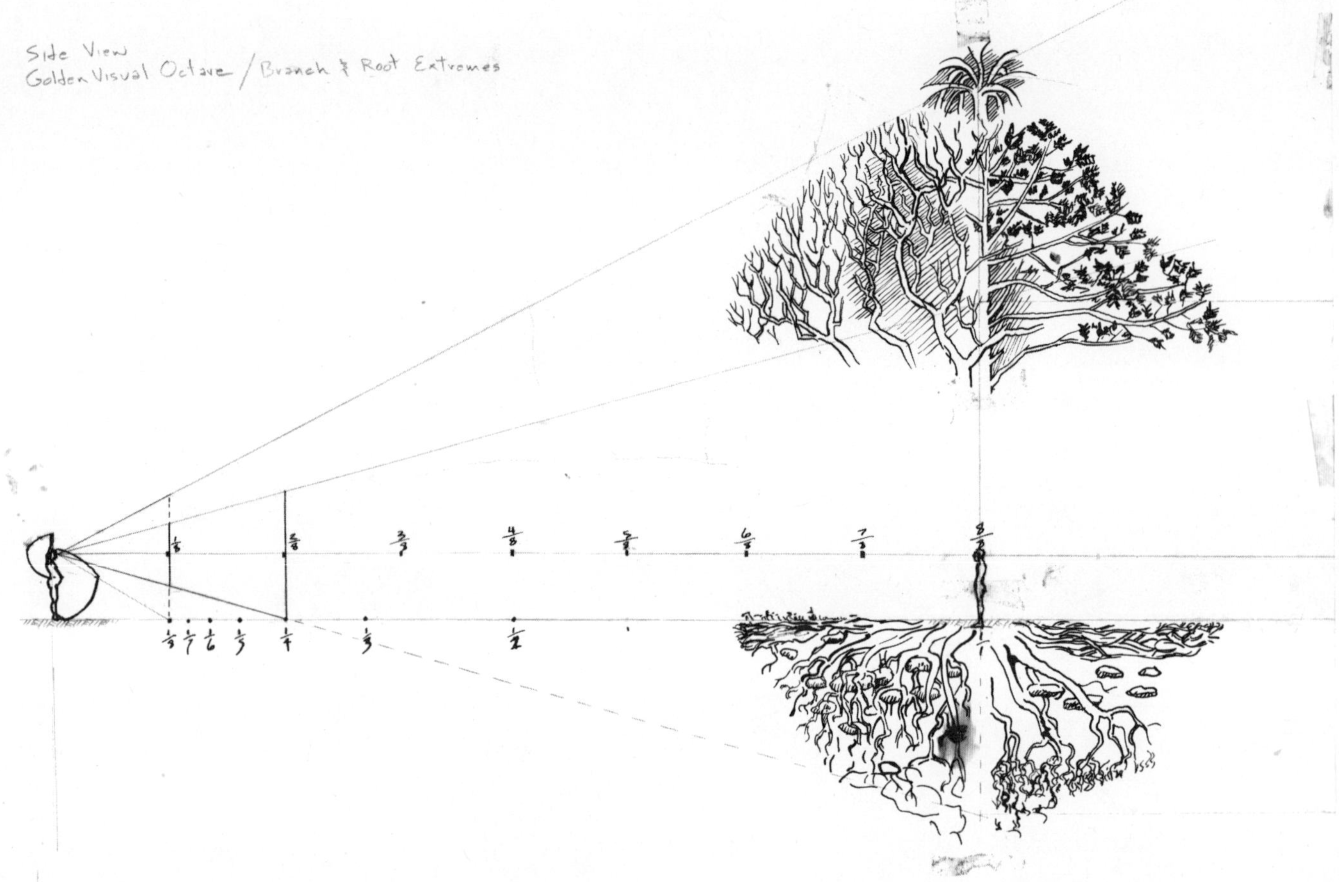

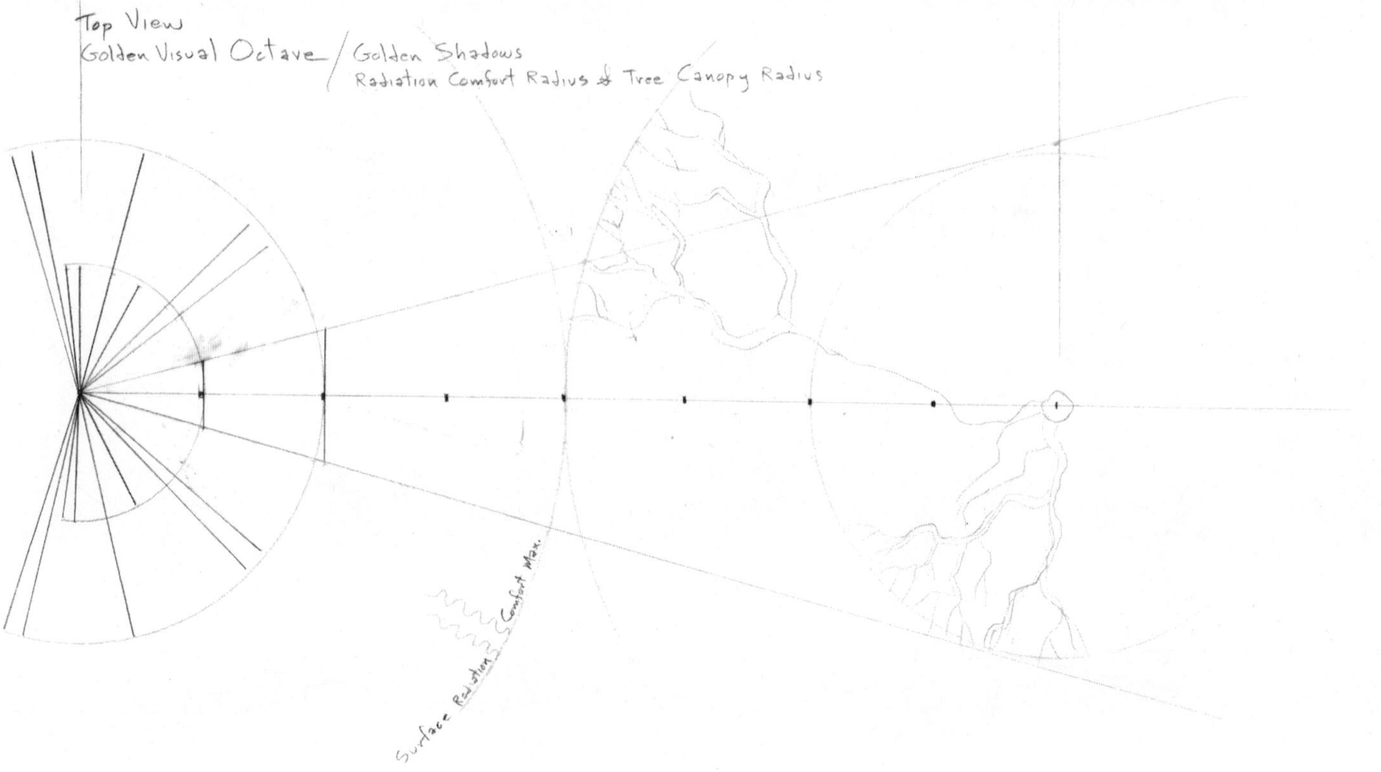

171. The end view relates man and woman geometric harmonies to their Golden Shadow Sun Spots fixed in their Octave's focal area. Three-grace-like positions fix the Golden Visual Octave: The beloved's head fills the focal cone, stands at the first and second interval and blends with the whole visual field at the eighth interval.

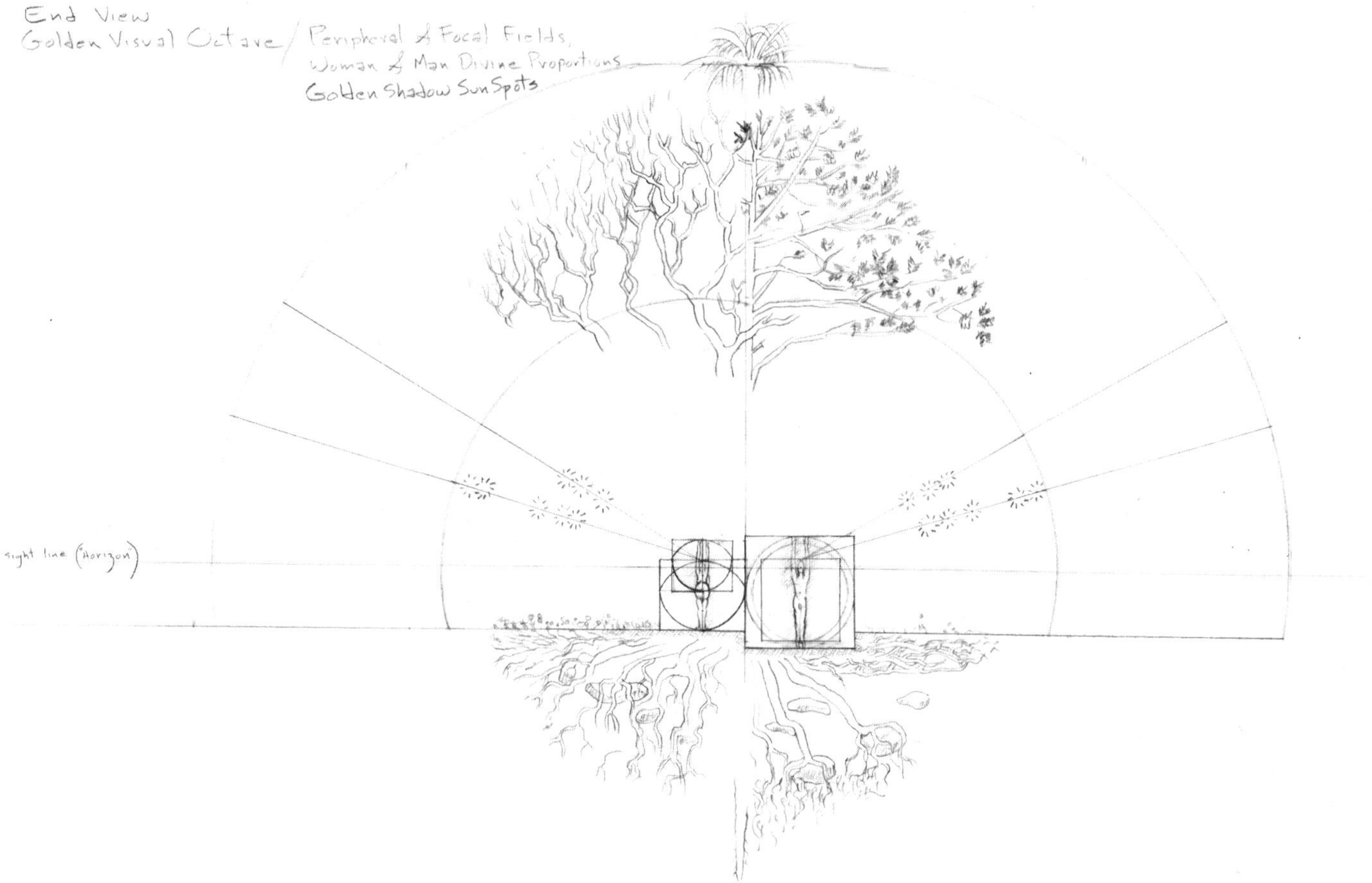

172. Another collection of notes will be published which will demonstrate design results from the interactions of principles in this collection. To me, though these principles are old, they are symbols of something to come.

173. Symbols make visible what we've learned to be true and useful. Though the symbols in this book are not avant garde, I believe they should be the cutting edge of art & architecture.

BIBLIOGRAPHY

1. Alexander, Christopher, Ishikawa, Sara, and Silverstein, Murray. A Pattern Language. New York: Oxford, 1977.
2. Allen, Edward. The Responsible House. Cambridge, MA: MIT Press, 1974.
3. American National Red Cross, First Aid. New York: Doubleday & Co., October, 1974.
4. Anderson, Bruce. Solar Energy: Fundamentals in Building Design. New York: McGraw-Hill Book Company, 1977.
5. Anderson, Bruce. The Solar Home Book, Andover, MA: Brick House, 1976.
6. F. Arnold, Henry. Trees and Urban Design. Van Nostrand Reinhold Company, 1980.
7. Aronin, J. Climate and Architecture. New York: Reinhold Publishing, Inc., 1953. (Out of print.)
8. Ashrae. Handbook of Fundamentals. New York: American Society of Heating, Refrigerating, and Air-conditioning Engineers, 1972.
9. Baer, S. Sunspots. Albuquerque: Zomeworks Corporation, 1975.
10. Balcomb, D., et al. ERDA's Pacific Regional Solar Heating Handbook. San Francisco: Government Printing Office, 1975.
11. College of Architecture, Arizona State University, Solar-Oriented Architecture. Washington, D.C.: American Institute of Architects Research Corporation, 1975.
12. College of Architecture, Arizona State University. Earth Integrated Architecture. Tempe: College of Architecture, Arizona State University, 1975.
13. Conklin, G. The Weather Conditioned House.
14. Crowther, R. Sun Earth. Denver: Crowther/Solar Group, 1976.
15. Daniels, F. Direct Use of the Sun's Energy. New Haven: Yale University Press, 1964.
16. Davis, A. J. and Schubert, R.P. Alternative Natural Energy Sources in Building Design. New York: Van Nostrand Reinhold Company, 1981.
17. Da Vinci, Leonardo. The Notebooks of Leonardo Da Vinci, Vols. 1 and 2, tr. by Jean P. Richter. New York: Dover Pubns., 1970.
18. Dekorne, J. The Survival Greenhouse. El Rito: The Walden Foundation, 1975.
19. Detwyler, Thomas R. Man's Impact on Environment. New York: McGraw-Hill, 1971.
20. Dreyfuss, Henry. The Measure of Man, New York: Whitney, 1966.
21. Drysdale, J. W. Climate and House Design, Australia: 1945.
22. Duffie, J. A., and Beckman, W. A. Solar Energy Thermal Processes. New York: John Wiley & Sons, Inc., 1974.
23. Eccli, J., Ed. Low Cost, Energy-Efficient Shelter. Emmaus, PA: Rodale Press, 1976.
24. Egan, D. Concepts in Thermal Comfort. Englewood Cliffs, NJ: Prentice Hall, Inc., 1975.
25. ERDA. Passive Solar Heating and Cooling Conference and Workshop Proceedings. Springfield: NTIS, 1976.
26. Fathy, H. Architecture for the Poor. Chicago: University of Chicago Press, 1973.

27. Fisher, R., and Yanda, B. Solar Greenhouse. Santa Fe: John Muir Publications, 1976.

28. Fitch, J. M. American Building, The Environmental Forces that Shape It. New York: Schocken Books, Inc., 1975.

29. Givoni, B. Man, Climate, and Architecture. Amsterdam: Applied Science Publishers, Elsevier Publishing, 1969.

30. Hackleman, Michael A., and House, David W. Wind & Windspinners. Culver City, Ca.: Peace Press, 1975.

31. Harington, Donald. The Architecture of the Arkansas Ozarks. Boston: Little, Brown & Company, 1975.

32. "House Beautiful; Climate Control Project." Bulletin of the American Institute of Architects, March 1950.

33. Jordan, R. C., and Liu, B. Y., Ed. Applications of Solar Energy for Heating and Cooling of Buildings. New York: ASHRAE, 1977.

34. Kern, K. The Owner Built Home. New York: Charles Scribners Sons, 1972.

35. Kreider, J. F., and Kreith, F. Solar Heating and Cooling. Washington, D.C.: Hemisphere Publishing Co., McGraw-Hill Book Company, 1975.

36. Le Corbusier. Oeuvre Complète, 8 volumes. Zurich: Les Editions d'Architecture, 1964.

37. Le Corbusier. Problems de L'ensoleillement. Techniqueset Architecture, 1943.

38. Le Corbusier. Towards a New Architecture. London: The Architectural Press, 1976.

39. Leckie, J., et al. Other Homes and Garbage. San Francisco: Sierra Book Clubs, 1975.

40. Lee, Douglas. Physiological Objectives in Hot Weather Housing. Housing & Home Finance Agency.

41. Life Techniques Conference (3rd Annual) New Mexico; October, 1974.

42. Lynch, Kelvin. Site Planning, Cambridge, MA: MIT Press, 1962.

43. Margolin, Malcolm. The Earth Manual. Boston: San Francisco Book Company/ Houghton Mifflin Book, 1975.

44. Mazria, Edward. The Passive Solar Energy Book. Emmaus, PA: Rodale Press, 1979.

45. McCulagh, James C. The Solar Greenhouse Book. Emmaus, PA: Rodale Press, 1978.

46. Mother Earth News. Handbook of Homemade Power. New York: Bantam Books, 1974.

47. Olgyay, A., and Olgyay, V. Solar Control and Shading Devices. Princeton: Princeton University Press, 1957.

48. Olgyay, V. Design with Climate. Princeton: Princeton University Press, 1963.

49. Portola Institute. Energy Primer. Menlo Park: Portola Institute, 1974.

50. Rapoport, A. House Form and Culture. Englewood Cliffs, NJ: Prentice-Hall, Inc., 1969.

51. Reynolds, John. Windmills and Watermills. London: Davenport Askew & Co. Ltd., 1970.

52. Robinette, Gary O. Landscape Planning for Energy Conservation. Reston, VA: Environmental Design Press, 1977.

53. Robinette, G. O. Plants/People/and Environmental Quality. Washington, D.C.: U.S. Department of the Interior, National Park Service, 1972.

54. Skurka, W., and Naar, J. Design for a Limited Planet. New York: Ballentine Books, 1976.

55. Steadman, P. Energy, Environment and Building. Cambridge: Cambridge University Press, 1975.

56. Stoner, Carol Hupping, Ed. Producing Your Own Power. Emmaus, PA: Rodale Press Inc., 1974.

57. Studies in Illumination: IV, Daylight in Buildings, Public Health Bulletin #218, U.S.P.H.S.

58. Total Environmental Action. Solar Energy Home Design in Four Climates. Harrisville: Total Environmental Action, 1975.

59. United States Weather Atlas.

60. Van Dresser, Peter. Homegrown Sundwellings. Santa Fe: The Lightning Tree-Jene Lyon, 1977.

61. Villecco, M., Ed. Energy Conservation in Building Design. Washington D.C.: American Institute of Architects Research Corporation, 1974.

62. Vitruvius. The Ten Books on Architecture, tr. by Morris H. Morgan. New York: Dover Pubns., 1960.

63. Wade, Gary. Homegrown Energy, Willits, CA: Oliver Press, 1974.

64. Watson, D. Designing and Building a Solar House. Charlotte, VT: Garden Way Publishing Co., 1977.

65. Williams, J. Richard. Solar Energy. Ann Arbor: Ann Arbor Science Publishers, 1974.

66. Yanda, Bill, and Fisher, Rick. Solar Greenhouse. Santa Fe: John Muir Publications, 1976.

67. Zimmerman, Martin H., and Brown, Claud L. Trees, Structure & Function. New York: Springer-Verlag, 1977.

Index

Accents, 123
Acid-alkaline, 64
Adobe, 112
Adam-Purple, 64
Agnosticism, 97, 99
Air
  engines, 19
  film, 105
  heating of, 19, 103
  mass, 52
  moisture, 58
  movement, 95, 103
  pumps, 19
  structure, 94
  temperature fluctuation in space, 116, 118
  temperature and sky conditions, 106, 107
  weather prediction, 52
Albedo, 53
Alert, 136
Alexander, Christopher, 122, 128
Altitudes, 20, 21, 106, 107
Altruism, 97, 99
Ambient Air Temperature, 118, 119
American Society of Heating, Refrigerating, & Air-conditioning Engineers, 32, 115
Anatomy, of pregnant woman, 6-11
Angle of incidence, 30-46, 108, 109
Angle of sight, 126
Angle of Sun's rays,
Anthropometrics, 140, 141, 144, 145, 150, 154, 155, 156, 157
  and temperature, 117
Anthropomorphic proportions, 148
Anti-dessicant sprays, 70
Approach, 122
Aquifer, 59
Aquinas, 99
Aridity, 52, 95
Aristotle, 97, 98, 99
Arnold, Henry F., 160
Arrowfoil, 57
Artesian, 59
Asceticism, 97, 99
Atheism, 97, 99
Atmosphere, 106, 107, 111, 138
Atomism, 97, 99
Attention, analysis of,
Augustine, 99
Ausculation, 6
Azimuth, 20, 30-46

Backache, countering one, 157, 158
Backfill, 93
Baer, Steve, 114
Balance, as economy of motion, 124, 125
Balmy, 52
Bandaging, 143
Basal metabolism, 104
Bearing, 87
Behavior, 148, 149
Benchmark, 91
Bentham, 97, 99
Berkeley, 99
Birth, 6-11
  birthing position, and divine proportion/pentagon, 151
  delivery, 10
  examination, 6
  foetal positions, 8
  labor, 9
  separation, 11
  structure, 7
Body, human
  and heat exchange with its surroundings, 118, 119
  and temperature, 114
  basics (skeleton per organ systems), 135
  dimensions of, 140, 141 (anthropometrics, 144-149, 152, 153)
  first aid, 143
  metabolism, 104
  movement of, 122
  structure of, 132-149
  surfaces, 118
Bones, human, 132-137, 139, 140
Bonzai, 85
Borders, 43
Branches, 70, 72-75, 82, 83, 84, 85
  coexistence, 73
Branching systems, 66, 67, 70, 72, 73
  decurrent, 67
  excurrent, 67
  shrubbiness vs. treeness, 72
Breathing, 105, 135
Breeze, 53, 94, 95
Brick, 112, 71
  common, 116
  conductivity of, 116
  magnesium, 116
British Thermal Units, 30, 31, 103, 104, 110, 111, 112, 113, 114, 115
  absorbed by earth, 32-34

Calculations
  ambient air temperature, 118, 119
  anthropometrics, 140-159
  average outdoor air temperature, 103
  areas, 93
  BTU's, calories, Langelys, 31
  Divine Proportions, 151-169
  effective storage mass, 114, 117
  focal cones, 124-7, 164
  greenhouse slopes, 108, 9
  heat capacity, 112
  heat conductivity, 116
  heat gain, 19-49, 111
  heat loss, 110, 111
  heat radiation, 117
  MRT, mean radiant temperature, 118, 119
  peripheral vision/hearing, 120
  relative humidity, 58
  seating, 136-8
  shadow-lengths, 21
  shadowlabe, 21
  skylight, 106, 7
  soil, 62-5
  space, 122-31
  stairs, 130-1
  view axis, 85
  vision cones, 120-1
  volumes, 93
  water flow, 60
  water head, 60
  wind speed, 55
  windmill power, 56
Calorie, 30
Calperi, 49
Calstice, 26, 39, 44, 48, 49
  and sunny spots, 45
  and temperature changes (climatic), 51
Canopy, 82
Ceilings, 80, 81, 82, 118, 119, 121, 128, 129
  shadows, 16
Celestial Sphere, 20, 35
Chill, 103, 142
  prediction of, 52
Chinaberry, 83
Chinese Cedar, 83
Chinese Juniper, 79
Circle, 4, 88, 114, 147, 149, 150, 152
  in proving divine proportion, 152
  man's basic proportions in relation to, 4, 144
  of sun, 88
Cladrastis Lutea, 81
Clay, 62, 63, 65
  non-plastic,
  organic,
  plastic,
Climate
  and resulting structures, 95
  architectural, 19
  as design factor for architecture, 102
  definition of, 18
  form and space basics, 95
  macro, 52
  micro, 53
  standard variables in, 94
  structure of, 94, 95
  U.S.A. temperature/sunshine charts, 50, 51
Climatic interpretation, 52, 53
Clock compass, 88, 89
Closure, visual, 85, 122, 125, 128
Clothing, 142
Clouds, 52, 55, 59, 103, 106, 107, 111
Cloudy climate
  design allowances for, 103
  greenhouse wall angles for, 109
  reflection, 107
Collarbone, 135
Color, 74-85, 94, 123
Colorado Bluespruce, 80
Comfort
  birthing,
  hearing, 120, 123
  heat sensation, 118, 119
  lying, 137, 148
  sensation and, 123
  sitting, 136, 138
  social gathering, 122, 129
  stair climbing, 130, 131
  temperature and, 113, 117, 118, 119
  viewing, 126

Compass, 86, 87, 88, 89
determining N-S axis without a compass
features and functions,
use of,
Compost, 64
Compression, 139
Conceptualism, 97, 99
Concrete, 71, 111, 112, 113
conductivity of, 116
Conduction, 104, 105, 113, 115, 116
Construction
of divine proportions of woman's body, 153
of golden rectangle of tree, 160-161, 163
of golden visual octave, 166, 167
Contour maps, 91
Contrapuntal, 123
Contrasts, 43
Convection, 104, 105
Coolers, 19
Critical idealism, 97, 99
Critical realism, 97, 99
Criticism, 97, 99
Crown, 10
Cultivation, 53
Cut and fill calculations, 93
Cycles, 48, 51, 52, 53, 54, 84, 85
seasonal, 12
soil, 64, 93
Cyclonic climate, 52
Cypress, 71

Dams, how to locate and build, 60
Day length, 40
DaVinci, 144, 150, 152
Deciduous trees, 76, 77
Decision-making factors, 102, entire book
Declination, 86, 87
Decurrent, 66, 67
Degree days, 116
Dehydration, 19
Delight, 138
Density, 94, 112
affecting heat capacity,
affecting thermal radiation,
of construction materials, 95
Determinism, 97, 99
Dialectical materialism, 97, 99
Dickenson, Scott, 137
Diffusion, 43
solar gain, 115
Digestion, 135
Direction, 86, 87
Distillers, 19
Divine extremity, 159
Divine proportion
of fully pregnant woman, 151, 152, 153, 154, 155, 156, 157, 158
of tree, 160, 161
of visual octave, 166, 167
Divisions, 145
Dogmatism, 97, 99
Doorway, 7, 8, 94
Douglas fir, 71, 79, 80
Drainage,
slopes and, 59
underground, 70

Dualism, 97, 99
Dumpy level, 92

Earth, 13, 103
as building material, 112
orbit, 12, 35
packed/conductivity of, 116
seasons, 12
stabilized,
structure, 94
thermal lag, 39
views, 13, 35-42
Earthwork claculations, 93
Eastern Red Cedar, 80
Eastern White Cedar, 80
Eclipse, 15
Ecosphere, of sun, 12
Egoism, 97, 99
Einstein, ii intro
Elaegnus Augustifolia, 76
Electromagnetic radiation, 118
Electromagnetic specturm, 18
Elmus Americana, 80
Emissivity, 117, 118
Empiricism, 97, 99
Energy, 93
heat, body as source of, 104
solar, 19
water, 59-61
wind, 54, 55-57
Entrance, 112
Epicurus, 97, 99
Equical, 26, 39, 48, 49, 50, 51
Equinox, 30, 31, 41, 48, 49, 50, 51, 88, 89
shadow lengths, 21, 41
Espalier, 85
Euonymus Alatus, 78
European Larch, 80
European Mountain Ash, 81
Evaporation, 104, 105
Evergreens, 78, 79
Evolution, of mature humans, 155
Evolutionism, 97, 99
Excavation/Backfill, 93
Excurrent, 66, 67
Existentialism, 98, 99
Extension, 8
Extremities, 144, 145, 146, 147, 150, 153, 166
divine, 159
of men, 154
of men and women not fitting divine proportions, 155
of pregnant woman, 154, 155, 156
of woman, 153, 154
reasonable, 145

Feathering, 57
Feminine intensity, 159, 165
Fibonnaci numbers, 152
Fire, 94
First aid, 143
Flexibility, in bandaging, 143
Flexible columns, 134
Flexion, 10
Floors, 94, 118, 119, 121
Focal cone, 120, 121, 124, 164, 165, 166, 167,
focal area, 125

Focal field, 164
Foot-candles, 106
Footcare, 142
Fontanelles, 8
Forsythia, 74
Foundation, 142
Fraxinus Excelsior, 82
Fraxinus Pennsylvania Lanceolata, 80

Freeze, 70
Frost, 70
Frost bite, 104, 142
Fruit trees, 75
Fundamental sight, 164
Fundamental tone, 165, 166
Fundus, 6
Fungi, 67
Funnel effect, 95
Fusion, 118

Gale, 55
Geometry, 90, 91
of surveying,
Ginkgo, 81
Ginkgo Biloba, 81
Glass, 105, 110, 114, 115, 119
Glazing, 105, 110, 111, 114, 115
Gleditsia Triacanthos, 77, 81
Golden mean, 165
Golden octave, 165
Golden ratio, 151
Golden rectangle, 128, 153, 158, 159, 165, 166, 167
and visual octave, 164
of a tree, 160, 161, 162
of woman and tree, 162, 163
Golden rule, 165
Golden section, 152, 161
Golden symbol of woman, man, and child unity, 143-158
Golden visual octave, 145, 165, 166, 167
to construct, 166
Grading plan, 93
Granite, 53
Graphic enlargement, 93
Gravel, 62, 63, 65
Gravity
center of, 154
of a body, 140
Green Ash, 80
Greenhouse effect, 19, 105, 106, 108, 109, 110, 111
common thermal wall in,
sizing of,
thermal mass in,
Greenhouses, 105, 108, 109
Grey Birch, 81
Ground cover, 70
Gypsum board, 112

Harmonic combination, 124, 125
Harmonics, 164
Harmony, 124, 125
of trinity, man, woman, child, 159
of visual octave, 165-167
Head, 60
Health and climate, 95
Hearing, visual, 120, 121

Heat, 93, 105
and climate, 142
and clothing, 142
body as source of, 104
calculating, 110, 111
capacity, 103, 112, 113
fluctuations in a body, 142
fluctuations in a space, 114
from people, 104
loss of, 104, 110, 114
sensation, 118
storage of, 115
transfer of, 116
Heat capacity, 112
Heat loss calculations, 110, 111, 112, 115
Heating and cooling principles
of body, 142
of earth, 19
of space, 114
of water, 19
over and underheating of a space, 115
Heliothermic axis/planning, 52, 53
Henry Dreyfuss Assoc., 137, 141
Heraclitus, 98, 99
Hibiscus Syriacus, 74
Homopuntal, 123
Honey Locust, 77, 81
Horizon, 13, 20, 31, 89, 90, 120, 164, 166
Horizontal overhang, shading mask for, 16
Horsepower, 56, 60
Hot air engine, 19
Hot-arid zone, 95
Hot-humid zone, 95
Human behavior, and movement, 148, 149
Human body
and clothing, 142
and temperature, 114, 118
comfort of, 104, 118, 122, 124, 128, 130, 137, 143, 157, 158
extremities of, 144
more proportions of, 140-159
Human, evolution of mature, 155
Human interaction, 128, 129
in spaces, 122
Human proportions, 146, 147
Humanism, 97, 99
Humidity, 58, 94, 104, 142
Hurricane, 55
Hypothalamus, 104

Idealism, 98, 99
Images, 43
Indirect gain systems, 115
Instrumentalism, 98, 99
Insulation, 94, 103, 105, 111, 112, 119, 142
Interval, 123
Intuitionism, 98, 99

James, 97, 99
Japanese Honeysuckle, 75
Juniperis Chinensis, 79

Kant, 97, 99
Kierkegaard, 99

Ladder, 130
Landscaping and airflow, 54, 95
Langely, 30
Latitude, 109
Lavis Decidua, 80
Le Corbusier, 152
Leveling, 92
Leverage, 132
Light, 30, 43, 94, 123
conversion, 19
incidental, 32
Ligustrum amureusa, 78
Little leaf Linden, 77
Loam, 65
Lonicera Japonica, 75
Lush slope, 52
Lynch, Kevin, 52, 53

Machiavelli, 99
Maclura Pomifera, 81
Macroclimate, 52
Magnetic variations, 87
Manna Ash, 82
Mapmaking, 86, 91
Marsh/draining, 59
Masonry, 71, 111, 112, 113
containers, 71
Materialism, 98, 99
Mazria, Edward, 117
Mean radiant temperature, 117, 118, 119
Measurements, musical, 123
Medical science, 64
Meliorism, 98, 99
Melody, 123
Membrane, 8
Memory, 125
Metabolism, 104
Meter, 123
Microclimate, 53
Mill, 97, 99
Modular man, 152
Moisture, 58, 94, 103, 104, 142
Molecular vibration, 104
Monay, 94
Monism, 98, 99
Mountain Ash of Europe, 82, 83
Movement, 122, 123, 124
as space perception, 123
as structured by bones, 132, 133
ascending, 130, 131
body, 134, 136, 137, 138, 139
body/space, 122
defined per musical terms, 124
landscaping, 85
of earth, 93
walking and sight, 121
Mulch, 53
Muscles, 139
Music, 122, 123, 124
Musical modeling, 124
Musical octave, 124, 125, 167
Musical terms, 124
defined per movement,
Muybridge, 131
Mysticism, 98, 99

Natural cover and temperature moderation, 70
Naturalism, 98, 99
Navel, 11, 144, 145, 146, 153, 156, 157, 158, 161, 162
change in position during pregnancy, 156
in construction of golden rectangle of pregnant woman, 158
Neutral monism, 98, 99
New York, 34
Nietzsche, 99
Nominalism, 98, 99
North, 86, 87, 90
finding without a compass, 88
magnetic, 87
true, 34, 86
Norway Maple, 80
Nutrition, 64, 66, 67

Octave, 123, 145. *See also* Musical octave, Visual octave
musical, 124, 145
visual, 125, 164, 165, 166, 167
Olive trees, 83
Olgyay, Victor, 55
Optimism, 98, 99
Order, living, 167
Organs, receptor, 106
Organ systems, 135
Organic, 63
Organisms, 100
Organization, living, preface
Oriental Cherry, 75
Orientation, 31, 32, 33
light contrasts, 43
of building, 95

Pagoda tree, 77
Palpation, 6
Pantheism, 98, 99
Pascal, 99
Peach tree, 75
Pelvis, 6, 7, 135, 140, 141, 154
Pendulifolia, 83
Pentagon, 151
Penumbre, 14, 43
Peripheral vision cone, 120, 121, 126, 127, 164, 165, 166
Perpendicularity, 98, 99
Perspiration, 142
Pessimism, 98, 99
Phenomenalism, 98, 99
Philosophy survey, 96-99
Photochemical, 19
Photosynthesis, 19, 43, 106
Photovoltaic, 19
Picea Pungens, 80
Pin Oak, 81
Pitch, 123
Planimetric, 93
Plants, 105
light and growth, 43, 106, 107
structure, 94
Plastic, 63
Plato, 97, 99
Platonic solids, 151, 166
Pluralism, 98, 99
Polaris, 86
Polypuntal, 123
Populus Alba, 77
Porous materials, 95
Positivism, 98, 99

Posture(s), 7, 134, 136, 137, 138, 139
good and bad, 157
healthy in fully pregnant woman, 158
of pregnant woman, 136, 137, 138, 141, 155, 156, 157, 158, 159
Practical idealism, 97, 99
Pragmatism, 97, 98, 99
Pregnancy, 151, 155, 156, 157, 158, 159
Prevailing winter winds, 54, 95
Profile, 91
Proportions, 4, 43, 117, 128, 130, 141, 144, 145, 146, 147, 148, 149, 152
anthropomorphic, 148, 149
climatic architecture, 95
divine, 151, 152
human, 148, 149
of glass to material mass, 113
of human body, 120, 121
of light, 43
soil testing, 63
trees, 73-79
visual octave, 164
Protagoras, 98, 99
Prunus Serrulata, 75
Pseudotsuga Mengiesii, 79
Pseudotsuga Taxifolia, 80
Prunus Persica, 75
Psyche, 9, 101
Purple, Adam, 64
Psychology, 101
Pyrus Calleriana, 75

Quercus Alba, 80, 81
Quercus Palustis, 81

Radiation. *See also* Solar radiation.
electromagnetic, 118
hitting a greenhouse, 108, 109
long wave (heat), 105
of body temperature, 104
of vision cone, 120
Rain, 52
Rationalism, 98, 99
Real estate, 94
Reasonable extremities, 144, 145
Reflection, 107, 108, 109, 112
Reflectivity of various surfaces, 53
Relativism, 98, 99
Relaxed, 137
Religion, 96, 97, 98, 99, 101
Reproductive organs/system, 7
Respiration, 135
Revolution, 142
Rhododendron, 78
Rhus aromatica, 74
Rhus Typaina, 81
Rhythm, 123
Robinia Pseudoacacia, 81, 82, 83
Roofs, 94, 95
greenhouse, 108, 109
Roots, 66
containers, 71
Rosamultiflora, 74
Rose, 74
Rose of Sharon, 74
Rotation
of birthing head, 10
of earth, 12
Rowan, 82, 83
Russian Olive, 76

Sand, 62, 63, 65, 112
Saphora Japonica, 77
Sartre, 99
Savonius, 57
Scale, 123, 126, 127
relationships, 164
Scholar tree, 77
Seasonal heat lag, 109
Seasons, 12
Seating, 128, 129, 136, 137, 138
Seeds, 67
Semi-privacy, 128, 129
Sensation, 138
Separation-individuation, 101
Sequence, 85
Shadow
and vision cone, 121
eclipse, 15
equinox, 41
length tables, 20, 21
measurement of, see shadowlabe
overhangs, 16
paths, 41
penumbra, 14
Skeleton, 135, 139, 140, 141
Skepticism, 98, 99
Skiagraphy, 139
Skiascope, 139
Skull, 135
foetal, 8
Sky, 20, 106, 107
Sky dome, 20, 106
Slope, 32-46, 52, 53, 59, 90
of greenhouse walls, 108, 109
Snow, 47, 52
Sociology, 97, 99
Socrates, 97, 99
Solar altitude, 20
Solar azimuth, 20
Solar collector, 19, 31
Solar gain, direct vs. diffused, 115
Solar radiation
absorption of, 30-32
affected by seasons, 50,51
earth's atmosphere and, 118
intensity of on a surface, 32-46
on earth's surface, in BTU's, 32-34, 36, 38, 40, 42, 44, 46
reflection of, 30-32
Solar radiation calculator, 32-46
Solar sextant, 17. *See also* Shadowlabe.
Solstice, 26, 30, 31, 48, 49, 50, 88, 89
Sorbus Aucuparia, 81
Sorbus Fastigata, 83
Sound, 120, 132
South, 88, 89
Space, 94, 95, 122, 123
air temperature, 102
and relation to man, 119
ascending and descending, 130, 131
human interaction in, 128, 129
laws of structuring, 132
prediction and measurement of. *See* Shadowlabe or curves and vision cone edges.
solstice, 41
tree, 82
umbra, 14
walls, 16
Shadowlabe, 15-30, 89
and vision cone, 121
how to construct, 21
Shape of building, 95
Shrubs, 74-75. *See also* specific latin and common names of trees.
Sight, 43, 85, 92, 106, 107, 120, 124, 125, 126, 127, 164, 165, 166, 167
focal, peripheral, 164, 165
Sight proximities, 126, 127
Silhouette, 84
Silt, 62, 63, 65
Site analysis, 52, 53, 93
leaving and arriving, 122
movement requirements, 120, 128, 130, 138
of best site to build dam, 60
of pregnancy, 6
of wind, 55
open and closed feeling, 95
temperature with respect to man, 119
transitional dimensions, 122
Site structure, 94
Specific heat, 112
Spheres, 151
Spine, 134, 136, 137
Square, 144, 146, 147, 149, 150, 151
in proving divine proportion, 152
man's basic proportions in relation to, 4
Staghorn sumac, 81
Stairs, 130, 131
Steel, 112
Stems, 72
Stone, 71, 112
Storms, 52
Stride, 120, 131
Structure
climate, 95
fontanelles, 8
life variables, 94
membrane, 8
of good posture, 139
organ protection, 7
pelvic canal, 7
site, 93
Sumac, 74
Summer, 121
shadow curves and vision,
Sun, 12, 15
angle of incidence, 108, 109
calculating position of. *See* Shadowlabe
celestial sphere, 13, 20
climate, 18
clock compass, 88-89
conversion, 19
dehydration, 19
eclipse, 15
energy output of,
movement of, across sky dome (celestial sphere), 13
path diagrams, or monthly paths of, 32-46

Sun (continued)
position of in relation to earth's, 12, 13
photosynthesis, 19
simulation, 17
source of greenhouse effect, 105
umbre and penumbre, 14
Sun-arc angles, 46, 47
Surfaces
area calculations for, 93
clothing, 142
flesh, 138
heated, 112, 115
of bones, 133
skin, 104
wall, 118, 119
Surveying, site, 90, 91, 92
equipment, 92
philosophy, 96-99
Suture, 11
Symbiosis, 64
Symbolism, 168, 169
Symbol of man, woman, child unity, 159, 160
Symmetry, 128

Tangent tables, 20, 21, 91
shadowlabe use of, 20, 21
Taxus Baccata, 78
Technology, 100
Temperate zone,
Temperature, 103, 104, 119
ambient air, 118
average outdoor, 103
comfort and, 113
diurnal, 53
effective, 103
fluctuations in, 116
mean radiant, 119
of a point in space, 118
ranges of, 48, 49
sinks, 53
U.S., 50, 51
Tempo/musical modelling, 124
Tension, 139
Texture, 80-85, 94, 123
Theism, 98, 99
Thermal interchange between body and environment, 117, 118
Thermal lag, 39
Thermal mass, 115, 116, 117
Thermal storage, 110, 111, 112, 113, 114, 115
Thermodynamics, 103
Thermoelectric, 19
Thermoheliodon. *See* Shadowlabe and Btu charts
Thermoionic, 19
Threshold, 122
Tilia Cordata, 77
Tile, 112
Tilt
of earth, affecting solar radiation, 12
of reflectors, 31
of south wall of greenhouse, 108
Tilth, 64
Timbre (tone color), 123
Time, 86, 88, 122, 123
Tone, 124
Topiary, 85
Topsoil, 67
Transcendentalism, 98, 99
Transit level, 91, 92
Transitional spaces, 122
Transmittance, 109
Tree of Heaven, 83
Trees. *See also* specific latin and common names.
and divine proportion, 160, 161, 162, 163, 166
as windbreak, 53
canopy textures of, 82, 83
ceilings, 80
divine proportions of, 160, 161
protection factors, 70, 71
shade from, 73-84
soil requirements, 160
structure of, 66, 67
triangle, pregnant woman, 162
walls, 80
weave and density of, 84
Triangle, 144, 150, 162, 163, 166
Trinity, man, woman, child, 159
sun, site, self, 2, 3, 171
True north, 34, 86
True south, 34, 86

Umbilical, 11
Umbra, 14, 43
Unity, of man, woman, child, 159
Utilitarianism, 97, 99

Van Dresser, Peter, 114
Vaporization, 105
Ventilation, 95
Venturi effect, 95
Vertebrae, 134, 135
View, 123
Vision, 120, 126, 127
Vision cone, 120, 121, 124
Visual field, 125
Visual "hearing," 120
Visual octave, 125, 164, 165
divine, 167
golden, 166
Vitruvius, 117, 144, 152
man's chest area, 117
Volume calculations, 93
Voluntarism, 98, 99

Waitplace, 122
Walk, 120
Walking, 131
Walls, 81, 94, 95, 111, 118, 119, 121
conductivity of, 116
masonry, 117
radiation, 118
shadows, 16
thickness, 116, 122
water, 117
Warm slope, 52
Water, 53, 58, 59, 66, 104, 105
capacity as heat storage, 103, 111, 112, 113, 117
collection, 59
conductivity of common water, 113
convection in, 113
water wheel types, 61
well location, 59
Water flow meter, 60
Water table, 59, 66
Weather
bureau, 55
prediction, 52
slope types, 53
Weir, 60
Wells, 59
how to locate and make,
White Colorado Fir, 79
White Oak, 80, 81
White Poplar, 77
Wind, 52-57, 94, 95, 103, 105
converter/wheeltypes, 57
effect of topography, 54
force of, 55
movements of, 54
protection from, 53
storm track, 52
thermodynamics, 103
trees and, 70
Windbreaks, 53
Wind effect and landscaping, prevent wind damage for transplanted trees, 70
Windmill, 56
Windows, 85, 94, 111, 119
Windroses, 55
Winged Euonymous, 78
Winter, 121
Woman, 150-153
body dimensions, 140
divine proportions of, 154, 158
extremities of, 153
figure and movement, 148, 149
posture of pregnant, 156, 157
Wood, 112
Worms, 64
Wright, David, 14

Yew, 78

Zeno of Citrium, 98, 99
Zeno of Elea, 98, 99